A Level
Advancing
Biology
for OCR
Year 2

B

Series Editor
Fran Fuller

Authors
Michael Fisher
Dawn Parker
Jennifer Wakefield Warren

OXFORD
UNIVERSITY PRESS

Great Clarendon Street, Oxford, OX2 6DP, United Kingdom

Oxford University Press is a department of the University of Oxford. It furthers the University's objective of excellence in research, scholarship, and education by publishing worldwide. Oxford is a registered trade mark of Oxford University Press in the UK and in certain other countries

British Library Cataloguing in Publication Data
Data available

978-0-19-835767-4

10 9 8 7 6 5 4 3 2 1

Paper used in the production of this book is a natural, recyclable product made from wood grown in sustainable forests. The manufacturing process conforms to the environmental regulations of the country of origin.

Printed in China

This resource is endorsed by OCR for use with specification H422 A Level GCE Biology B (Advancing Biology). In order to gain endorsement this resource has undergone an independent quality check. OCR has not paid for the production of this resource, nor does OCR receive any royalties from its sale. For more information about the endorsement process please visit the OCR website www.ocr.org.uk

AS/A Level course structure

This book has been written to support students studying for OCR A level Biology B. It covers the A Level Year 2 only modules from the specification. There is an index at the back to help you find what you are looking for.

AS exam

A level exam

Year 1 content

1 Development of practical skills in biology
2 Cells, chemicals for life, transport, and gas exchange (Chapters 1–7)
3 Cell division, development, and disease control (Chapters 8–15)

Year 2 content

4 Energy, reproduction, and populations (Chapters 16–22)
5 Genetics, control, and homeostasis (Chapters 23–31)

A Level exams will cover content from Year 1 and Year 2 and will be at a higher demand. You will also carry out practical activities throughout your course.

Contents

How to use this book

This book contains many different features. Each feature is designed to support and develop the skills you will need for your examinations, as well as foster and stimulate your interest in biology.

Terms that you will need to be able to define and understand are highlighted by **bold text**.

Application features

These features contain important and interesting applications of biology in order to emphasise how scientists and engineers have used their scientific knowledge and understanding to develop new applications and technologies.

1   All application features have a question to link to material covered with the concept from the specification.

Study Tips

Study tips contain prompts to help you with your understanding and revision.

Extension features

These features contain material that is beyond the specification. They are designed to stretch and provide you with a broader knowledge and understanding and lead the way into the types of thinking and areas you might study in further education. As such, neither the detail nor the depth of questioning will be required for the examinations. But this book is about more than getting through the examinations.

1   Extension features also contain questions that link the off-specification material back to your course.

Synoptic link

These highlight the key areas where topics relate to each other. As you go through your course, knowing how to link different areas of biology together becomes increasingly important. Many exam questions, particularly at A Level, will require you to bring together your knowledge from different areas.

Summary Questions

1   These are short questions at the end of each topic.

2   They test your understanding of the topic and allow you to apply the knowledge and skills you have acquired.

3   The questions are ramped in order of difficulty.

Worked example:

Worked examples take you through a calculation step-by-step, you can then test your skills using the summary questions and chapter practice questions.

Practice questions at the end of each chapter, including questions that cover practical and math skills.

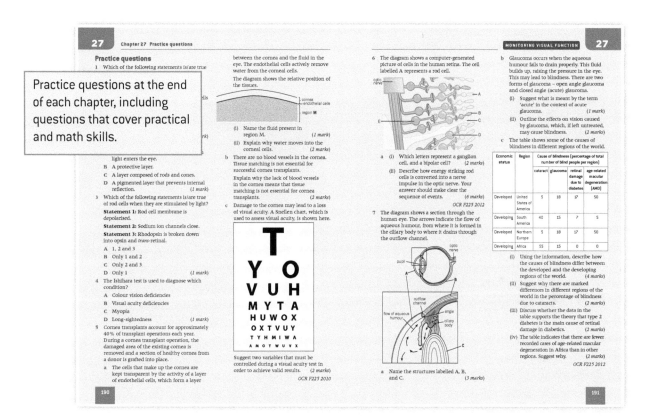

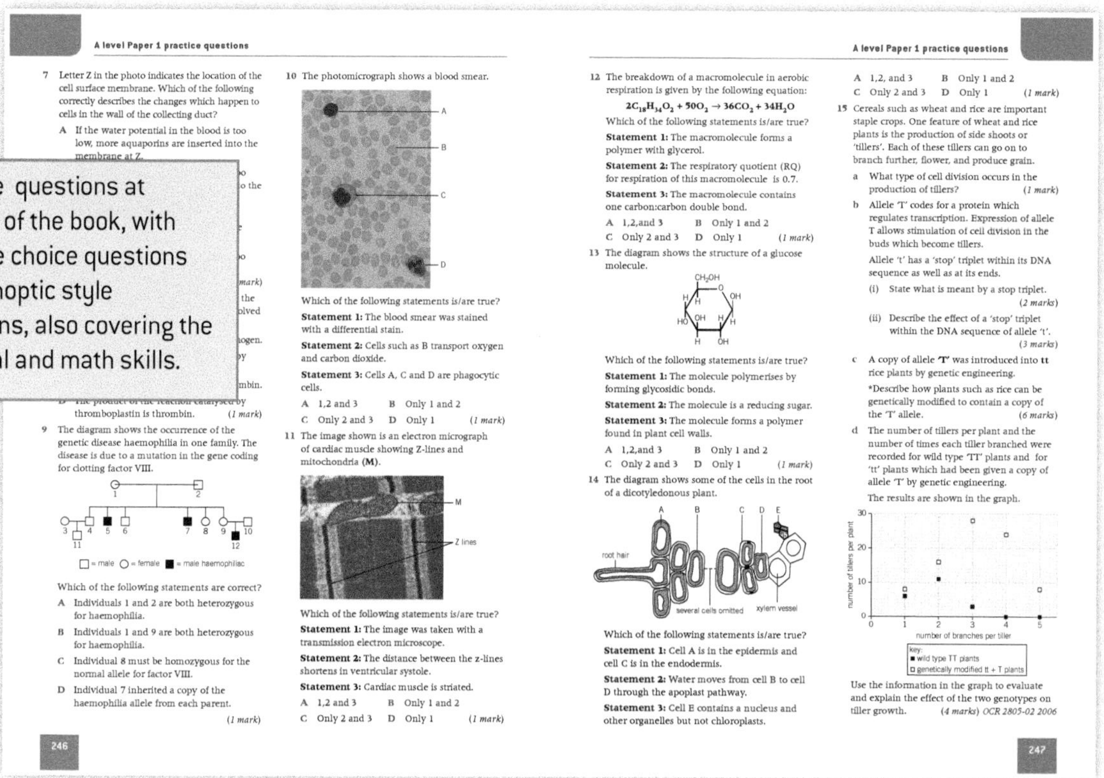

A level Paper 2 practice questions

Pre-released Case Study

CRISPR: a tasty new DNA editing technique?

DNA editing techniques, such as gene therapy and gene knockouts, have existed for many years. However, a new procedure called CRISPR has been developed in recent years. Some scientists believe CRISPR will revolutionise the manipulation of genomes.

CRISPR is a natural defence mechanism used by bacteria to protect themselves from viruses. The basic CRISPR machinery comprises a DNA-cutting enzyme and a guide molecule (a short piece of RNA). When viral DNA is detected in its cytoplasm, a bacterium produces a short RNA fragment that matches the viral DNA. The RNA binds to the viral DNA and forms a complex with an enzyme called Cas-9, which cuts and disables the viral DNA.

Scientists have adapted the bacterial mechanism to enable DNA to be edited either *in vitro* or *in vivo*. The guide RNA can be designed to match a particular target sequence and therefore be specific to a target gene.

CRISPR causes genes to be edited when a cell attempts to repair the cuts in DNA caused by Cas-9. Often the repair is inexact, which means the base sequence is altered (i.e., a mutation is introduced) and the gene no longer works. This is equivalent to a gene knockout and enables researchers to study the role of the disabled gene. By introducing a replacement DNA sequence once Cas-9 has cut the cell's DNA, faulty alleles can be replaced by functional versions. The introduction of new DNA has a low rate of success, but could represent a new method of gene therapy.

The CRISPR method could be used as somatic cell gene therapy to combat diseases in the future. Scientists are researching the possibility of correcting disease-causing alleles in monogenic conditions such as cystic fibrosis and haemophilia. One advantage of CRISPR is that it has the potential to target many genes at once, which may be useful when tackling complex polygenic diseases. Other diseases that could be treated with CRISPR include cancer (by targeting genes such as BRCA1) and blood diseases such as beta thalassemia. HIV therapy may be possible.

The gene coding for the CCR5 protein could be switched off in T lymphocytes. CCR5 is a protein used by HIV to enter T lymphocytes.

CRISPR has also been used in stem cells and in gametes of various species for research purposes. The CRISPR components can be injected into egg cells with sperm using Intracytoplasmic Sperm Injection (ICSI).

1 The following questions are based on the pre-released case study *CRISPR: a tasty new DNA editing technique?*

   a Describe how the enzyme Cas-9 is able to catalyse the breakdown of DNA. *(3 marks)*

   b Suggest and explain an advantage of CRISPR over traditional gene therapy techniques. *(2 marks)*

   c Copy and complete the table using ticks (✓) to indicate the features of the three genetic techniques listed. The first row has been completed for you. Rows may contain more than one tick.

Technique	CRISPR DNA editing	Insertion of a gene into a plasmid	RNA interference using siRNA
Requires DNA ligase		✓	
Can produce transgenic DNA			
Involves the binding of RNA to DNA			
Can stop or reduce the expression of a gene			

   (3 marks)

   d The CRISPR method could be used in the future to target genes such as mutant BRCA1 alleles.

   Explain why using CRISPR to target mutant BRCA1 alleles reduces the risk of breast cancer but is not guaranteed to prevent breast cancer. *(2 marks)*

   e Suggest how switching off the CCR5 gene would reduce the spread of HIV infection within a patient's body. *(2 marks)*

   f Why would blood diseases such as beta thalassemia be easier to treat with CRISPR than diseases of other tissues? *(1 mark)*

   g ICSI can be used to introduce CRISPR components into an egg cell. State one other procedure for which ICSI can be used. *(1 mark)*

2 The process of vernalisation in some plant species ensures that they flower at the correct time of year.

   In thale cress (*Arabidopsis thaliana*) the process of vernalisation is controlled by epigenetic gene silencing.

   A period of cold temperature switches off the expression of the Flowering Locus C (FLC) gene. The FLC gene codes for a transcription repressor that stops the expression of the genes that control flowering in plants. Flowering can therefore begin once the FLC gene is switched off.

   a The FLC gene is switched off through epigenetic chromatin remodelling.

   Suggest what chromatin remodelling involves and how this stops the expression of FLC. *(2 marks)*

   b A team of scientists investigated the effect of vernalisation on the timing of flowering in four varieties of daikon (*Raphanus sativus*). The four varieties are called Jumbo Scarlet, E40, Minowase, and Everest.

   The scientists recorded the number of days it took each plant to flower after the end of treatment. The results of their experiments are shown in the table.

Treatment	Number of days to flowering (following the end of treatment)			
	Jumbo Scarlet	E40	Minowase	Everest
no vernalisation (18.5 °C)	no flowering	39	47	42
vernalisation (6.5 °C) for 15 days	34	29	29	27
vernalisation (6.5 °C) for 20 days	35	26	32	29

   (i) State and explain which variety of daikon should be chosen when early flowering is a desirable trait. *(2 marks)*

   (ii) State and explain which variety should be chosen as a model organism to study the effects of the environment on vernalisation and flowering. *(2 marks)*

   (iii) State two additional factors that need to be considered to confirm the validity of this data. *(2 marks)*

   c Pea plants (*Pisum sativum*) are able to flower without vernalisation. Their flowers can be purple or white.

   Flower colour in *P. sativum* is controlled by a gene locus that has two gene variants.

   P = a dominant allele coding for purple flowers

   p = a recessive allele coding for white flowers.

   (i) A purple-flowered pea plant is crossed with a white-flowered pea plant.

   The genotype of the purple-flowered plant is unknown. Calculate the range of probabilities that an offspring of these two plants will have white flowers. *(2 marks)*

   (ii) In one population of pea plants, 25% of the plants have white flowers.

   Calculate the percentage of plants that are heterozygous for the flower colour genotype. *(3 marks)*

   (iii) The photograph shows a flowering pea plant.

   Suggest whether the flower is adapted for insect or wind pollination. Explain your answer using evidence from the photograph. *(3 marks)*

250

251

A level Paper 1 practice questions

7 Letter Z in the photo indicates the location of the cell surface membrane. Which of the following correctly describes the changes which happen to cells in the wall of the collecting duct?

   A If the water potential in the blood is too low, more aquaporins are inserted into the membrane at Z.

Practice questions at the end of the book, with multiple choice questions and synoptic style questions, also covering the practical and math skills.

   D The product of the reaction catalysed by thromboplastin is thrombin. *(1 mark)*

9 The diagram shows the occurrence of the genetic disease haemophilia in one family. The disease is due to a mutation in the gene coding for clotting factor VIII.

= male   = female   ■ = male haemophiliac

   Which of the following statements are correct?

   A Individuals 1 and 2 are both heterozygous for haemophilia.

   B Individuals 1 and 9 are both heterozygous for haemophilia.

   C Individual 8 must be homozygous for the normal allele for factor VIII.

   D Individual 7 inherited a copy of the haemophilia allele from each parent.

   (1 mark)

10 The photomicrograph shows a blood smear.

   Which of the following statements is/are true?

   Statement 1: The blood smear was stained with a differential stain.

   Statement 2: Cells such as B transport oxygen and carbon dioxide.

   Statement 3: Cells A, C and D are phagocytic cells.

   A 1,2 and 3   B Only 1 and 2
   C Only 2 and 3   D Only 1   *(1 mark)*

11 The image shown is an electron micrograph of cardiac muscle showing Z-lines and mitochondria (M).

   Which of the following statements is/are true?

   Statement 1: The image was taken with a transmission electron microscope.

   Statement 2: The distance between the z-lines shortens in ventricular systole.

   Statement 3: Cardiac muscle is striated.

   A 1,2 and 3   B Only 1 and 2
   C Only 2 and 3   D Only 1   *(1 mark)*

12 The breakdown of a macromolecule in aerobic respiration is given by the following equation:

$$2C_{18}H_{34}O_2 + 50O_2 \rightarrow 36CO_2 + 34H_2O$$

   Which of the following statements is/are true?

   Statement 1: The macromolecule forms a polymer with glycerol.

   Statement 2: The respiratory quotient (RQ) for respiration of this macromolecule is 0.7.

   Statement 3: The macromolecule contains one carbon:carbon double bond.

   A 1,2,and 3   B Only 1 and 2
   C Only 2 and 3   D Only 1   *(1 mark)*

13 The diagram shows the structure of a glucose molecule.

   Which of the following statements is/are true?

   Statement 1: The molecule polymerises by forming glycosidic bonds.

   Statement 2: The molecule is a reducing sugar.

   Statement 3: The molecule forms a polymer found in plant cell walls.

   A 1,2,and 3   B Only 1 and 2
   C Only 2 and 3   D Only 1   *(1 mark)*

14 The diagram shows some of the cells in the root of a dicotyledonous plant.

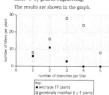

   Which of the following statements is/are true?

   Statement 1: Cell A is in the epidermis and cell C is in the endodermis.

   Statement 2: Water moves from cell B to cell D through the apoplast pathway.

   Statement 3: Cell E contains a nucleus and other organelles but not chloroplasts.

   A 1,2, and 3   B Only 1 and 2
   C Only 2 and 3   D Only 1   *(1 mark)*

15 Cereals such as wheat and rice are important staple crops. One feature of wheat and rice plants is the production of side shoots or 'tillers'. Each of these tillers can go on to branch further, flower, and produce grain.

   a What type of cell division occurs in the production of tillers? *(1 mark)*

   b Allele 'T' codes for a protein which regulates transcription. Expression of allele T allows stimulation of cell division in the buds which become tillers.

   Allele 't' has a 'stop' triplet within its DNA sequence as well as at its ends.

   (i) State what is meant by a stop triplet. *(2 marks)*

   (ii) Describe the effect of a 'stop' triplet within the DNA sequence of allele 't'. *(2 marks)*

   c A copy of allele '**T**' was introduced into **tt** rice plants by genetic engineering.

   *Describe how plants such as rice can be genetically modified to contain a copy of the 'T' allele. *(6 marks)*

   d The number of tillers per plant and the number of times each tiller branched were recorded for wild type 'TT' plants and for 'tt' plants which had been given a copy of allele 'T' by genetic engineering.

   The results are shown in the graph.

   Use the information in the graph to evaluate and explain the effect of the two genotypes on tiller growth. *(4 marks) OCR 2805-02 2006*

246

247

viii

Kerboodle

This book is supported by next generation Kerboodle, offering unrivalled digital support for independent study, differentiation, assessment, and the new practical endorsement.

If your school subscribes to Kerboodle, you will also find a wealth of additional resources to help you with your studies and with revision.

- Study guides
- Maths skills boosters and calculation worksheets
- On your marks activities to help you achieve your best
- Practicals and follow up activities to support the practical endorsement
- Interactive objective tests that give question-by-question feedback
- Animations and revision podcasts
- Self-assessment checklists

Revise with ease using the study guides to guide you through each chapter and direct you towards the resources you need.

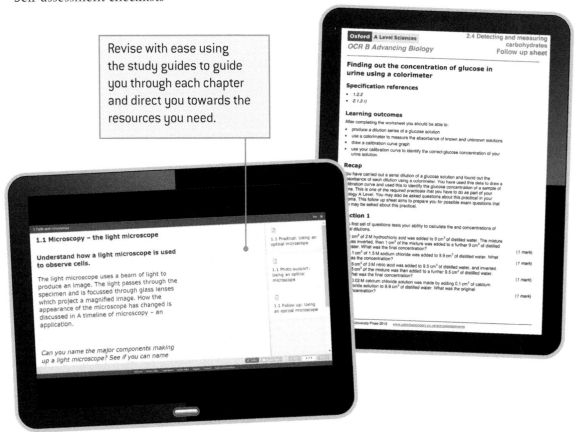

For teachers, Kerboodle also has plenty of further assessment resources, answers to the questions in the book, and a digital markbook along with full teacher support for practicals and the worksheets, which include suggestions on how to support and stretch students. All of the resources are pulled together into teacher guides that suggest a route through each chapter.

The need for energy

All living organisms require a supply of energy to drive metabolic reactions and other processes that occur inside their cells. Examples include:

- Movement – this can be achieved in a variety of ways, for example, movement of cilia, movement of flagella, muscle contraction, movement of chromosomes during nuclear division.

- Homeostasis – endotherms need thermal energy to maintain a stable body temperature that will ensure optimum conditions for enzyme-controlled reactions.

- Anabolic reactions – the synthesis of large, complex molecules from smaller, simpler molecules – for example, the synthesis of polypeptides, complex carbohydrates and nucleic acids.

- Active transport – energy is required to move molecules against their concentration gradient.

- Chemical activation – the addition of phosphate group(s) makes the molecule more reactive, for example, the phosphorylation of glucose at the start of glycolysis.

- Bioluminescence – some organisms can convert chemical potential energy into light energy, for example, glow worms and fireflies.

- Secretion – the release of molecules from cells requires energy to form and move vesicles to the cell surface membrane during exocytosis.

Overview of respiration

Aerobic respiration forms ATP from the breakdown of glucose, using oxygen. The process involves four key stages, one of which occurs in the cytoplasm and three of which occur in mitochondria.

The four key stages of respiration are:

1 Glycolysis
2 Link reaction
3 Krebs cycle
4 Oxidative phosphorylation

Respiration involves many different enzymes and coenzymes. Decarboxylase and dehydrogenase enzymes are involved in many stages of respiration.

- **Decarboxylase enzymes** – hydrolyse the carboxyl group of a molecule, which usually produces carbon dioxide.

- **Dehydrogenase enzymes** – remove hydrogen atoms from certain molecules and pass them to other molecules, for example, coenzymes such as **NAD** and **FAD**.

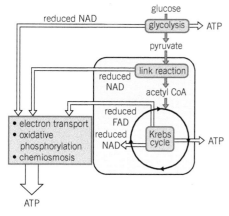

▲ Figure 1 An overview of respiration

The coenzymes NAD and FAD transfer hydrogen atoms from one molecule to the next. In doing so the coenzymes become reduced and then reoxidised, transferring chemical potential energy in the process.

Releasing energy

Rather than rapidly releasing all of the energy in glucose in one go, the chemical bond energy in glucose is gradually released in a series of small steps. During these small steps, energy that is released is used to make **adenosine triphosphate** (ATP).

ATP is referred to as a phosphorylated nucleotide and can be hydrolysed to ADP (adenosine diphosphate) and inorganic phosphate (Pi), which releases $30.6 \, kJ \, mol^{-1}$ of energy.

The hydrolysis of ATP is catalysed by the enzyme ATPase. The energy released from the hydrolysis of ATP is used by the cell to drive metabolic reactions in the cell. AMP (adenosine monophosphate) and ADP can be converted back to ATP by the addition of inorganic phosphate groups. ATP is broken down and resynthesised 1000 times each day.

Glycolysis

Glycolysis means the 'splitting of sugar'. It consists of a series of linked enzyme-controlled reactions that take place in the cytosol of the cell. The first reaction in glycolysis involves the phosphorylation of glucose to form glucose-6-phosphate. The conversion of glucose-6-phosphate is endergonic (a reaction that absorbs energy from the surroundings) and the hydrolysis of ATP is exergonic (a reaction that loses energy to the surroundings during the process). The reactions are coupled and the combined reaction proceeds spontaneously. Phosphorylation of glucose both prevents it from being able to diffuse out of the cell and also reduces the activation energy that will be needed for the next reaction in the pathway.

In summary, glycolysis splits the hexose molecule, glucose, into two triose pyruvate molecules. ATP is also formed from ADP by the gain of inorganic phosphates from a substrate molecule that is not AMP or ADP – this method of synthesising ATP is called substrate level phosphorylation. During glycolysis two ATP molecules are hydrolysed but four ATP molecules are formed – hence there is a net gain of two ATP for each glucose molecule that enters glycolysis. Hydrogen atoms that are released are picked up by the coenzyme NAD to form reduced NAD.

Glycolysis is common to both aerobic and anaerobic respiration. The presence or absence of oxygen determines the pathway pyruvate follows at the end of glycolysis.

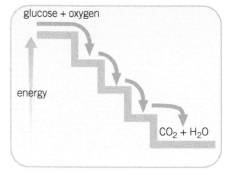

▲ Figure 2 *Respiration gradually releases energy in glucose by a series of small steps*

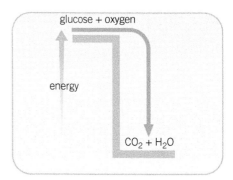

▲ Figure 3 *In combustion (burning) the energy is released all in one go*

Study tip

Glycolysis is an example of a metabolic pathway where the product of one reaction becomes the substrate for the next.

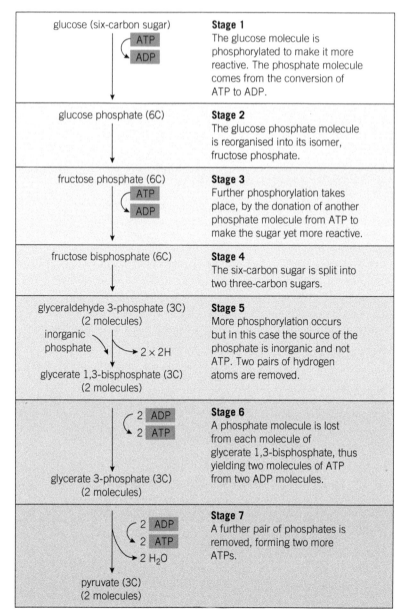

Stage 1
The glucose molecule is phosphorylated to make it more reactive. The phosphate molecule comes from the conversion of ATP to ADP.

Stage 2
The glucose phosphate molecule is reorganised into its isomer, fructose phosphate.

Stage 3
Further phosphorylation takes place, by the donation of another phosphate molecule from ATP to make the sugar yet more reactive.

Stage 4
The six-carbon sugar is split into two three-carbon sugars.

Stage 5
More phosphorylation occurs but in this case the source of the phosphate is inorganic and not ATP. Two pairs of hydrogen atoms are removed.

Stage 6
A phosphate molecule is lost from each molecule of glycerate 1,3-bisphosphate, thus yielding two molecules of ATP from two ADP molecules.

Stage 7
A further pair of phosphates is removed, forming two more ATPs.

▲ Figure 4 *Stages of glycolysis*

Summary questions

1   a   State the important properties of ATP that make it ideal as a
        short-term energy molecule.   (*3 marks*)
    b   Suggest why all the energy released from the hydrolysis of one
        molecule of ATP is not available for use in other cellular reactions.
                                                                       (*1 mark*)

2   Explain why the hydrolysis of ATP is more efficient than the direct
    hydrolysis of glucose to obtain the energy required by the cell.   (*3 marks*)

3   Suggest why increasing the concentration of glucose in a cell does not
    necessarily increase the amount of pyruvate produced.   (*2 marks*)

16.2 The link reaction and the Krebs cycle

Specification reference: 4.1.1

Link reaction

In the presence of oxygen, the two pyruvate molecules formed at the end of glycolysis are passed across the outer and inner mitochondrial membrane, by active transport, into the mitochondrial matrix. Each pyruvate contains chemical potential energy, which can be used to synthesise ATP indirectly. Three things happen to the pyruvate molecule in the mitochondrial matrix:

- The pyruvate is decarboxylated – a molecule of carbon dioxide is removed from pyruvate, and this diffuses out of the mitochondrion and cell.
- The pyruvate is also dehydrogenated using dehydrogenase enzymes. Each pyruvate loses a pair of hydrogen atoms. These are picked up by the coenzyme NAD to form reduced NAD.
- The two carbon remainder of the pyruvate molecule combines with coenzyme A to form **acetyl coenzyme A** (acetyl CoA).

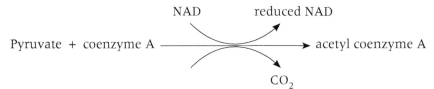

The biochemistry of the Krebs cycle

Acetyl coenzyme A (acetyl CoA) enters the Krebs cycle, which also takes place in the mitochondrial matrix. Acetyl CoA combines with a four-carbon compound, oxaloacetate, to produce 6C citrate and reform coenzyme A. A series of enzyme-controlled reactions decarboxylate and dehydrogenate the citrate to reform oxaloacetate. During these reactions each turn of the Krebs cycle produces three molecules of reduced NAD, one molecule of reduced FAD, as well as one ATP molecule (by substrate level phosphorylation) for each acetyl CoA molecule.

The regeneration of oxaloacetate is essential to enable the continuation of the Krebs cycle.

> **Learning outcomes**
>
> Demonstrate knowledge, understanding, and application of:
> → an outline of the link reaction
> → an outline of the Krebs cycle.

> **Study tip**
>
> The decarboxylation of pyruvate in the link reaction is sometimes referred to as oxidative decarboxylation – as hydrogen atoms are lost at the same time. This also distinguishes it from the decarboxylation of pyruvate, which happens in anaerobic respiration in yeast cells.

> **Study tip**
>
> Remember, the reactions involved in the link reaction can be described in four ways:
>
> decarboxylation – a molecule of carbon dioxide is removed (from pyruvate)
>
> dehydrogenation – a pair of hydrogen atoms is removed (from pyruvate)
>
> oxidation –pyruvate is oxidised
>
> reduction – hydrogen is gained by NAD (to form reduced NAD)
>
> Remember, reduction and oxidation reactions always occur together: when one molecule is oxidised another must be reduced – this is called a **redox reaction**.

Synoptic link

You will need to remember the structure and function of the mitochondria, which is covered in Topic 1.6, Cell ultrastructure.

Synoptic link

You will need to remember how enzymes work and the factors that affect the rate of enzyme-controlled reactions, which you learnt about in Topic 3.4, The structure of enzymes.

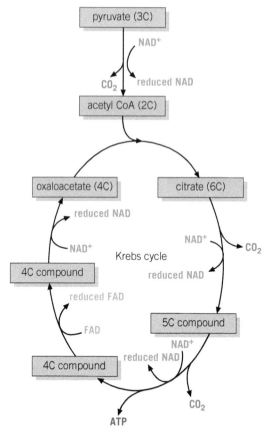

▲ Figure 1 *The Krebs cycle*

The importance of the Krebs cycle

The Krebs cycle converts the product of the link reaction, acetyl CoA, into carbon dioxide, reduced FAD, reduced NAD and ATP. The process of dehydrogenation to form reduced FAD and reduced NAD is essential in the later formation of ATP. Some amino acids and fatty acids can also be used to form acetyl CoA, which can also enter the Krebs cycle.

The most important role of the Krebs cycle is to provide hydrogen atoms that can be used in the electron transport chain to provide energy for the formation of ATP.

Summary questions

1 State what happens to each of the products of the Krebs cycle. (*5 marks*)

2 a Explain the importance of the dehydrogenase enzymes and coenzymes in the link reaction and Krebs cycle. (*3 marks*)

   b State precisely where the link reaction and Krebs cycle occur and explain why they do not occur in the cytoplasm. (*2 marks*)

3 a Suggest how the enzymes of the Krebs cycle are regulated to control the production of reduced coenzymes. (*2 marks*)

   b Using examples from glycolysis and the link reaction, give two examples of molecules that are produced by oxidation reactions. State the type of enzyme responsible for the reaction and its substrate. (*2 marks*)

16.3 Oxidative phosphorylation and the electron transport chain

Specification reference: 4.1.1

Oxidative phosphorylation

In this last stage of aerobic respiration, the reduced coenzymes that were in the earlier stages, and oxygen are used to produce ATP via oxidative phosphorylation. This occurs on the inner membranes of the mitochondria and involves a series of electron carriers known as the electron transport chain.

Learning outcomes

Demonstrate knowledge, understanding, and application of:

→ an outline of the process of oxidative phosphorylation.

The electron transport chain

The inner membrane of a mitochondrion is folded to form **cristae**. Fixed to the inner membrane is a series of electron carriers (specialised proteins) that make up the **electron transport chain**. Each reduced NAD molecule releases its hydrogens. Each hydrogen atom splits into a proton (hydrogen ion, H^+) and electron (e^-). The electrons are picked up by the first electron carrier protein in the chain. This carrier has now been reduced and the reduced NAD has been oxidised. The NAD can be re-used in the Krebs cycle or in other reactions to pick up hydrogen again.

The first electron carrier passes electrons to the next carrier in the chain. This results in the first carrier becoming oxidised and the second carrier becoming reduced. The electrons continue to be passed down the electron transport chain from carrier to carrier, releasing energy, which is used to make ATP. At the end of the electron transport chain each electron combines with two protons and an oxygen atom to form water, in a reaction that is catalysed by cytochrome oxidase. The oxygen atom that acts as the final electron acceptor is reduced and forms a covalent bond with two hydrogen atoms giving a molecule of water.

The transfer of electrons between carriers in the electron transport chain is coupled to the movement of protons, in a process referred to as 'proton pumping'. As the electrons are passed down the series of electron carriers, the energy that is released is used to transfer protons across the inner membrane of the mitochondrion to the intermembrane space.

Chemiosmosis

Proton pumping builds up a high concentration of protons in the space between the inner and outer mitochondrial membranes, which generates a concentration gradient across the inner membrane. It also produces an **electrical gradient**, as the protons have a positive charge. Consequently, the protons flow down an electrochemical gradient through ATP synthase molecules, which are embedded in the inner membrane of the mitochondrion. This flow releases energy that is used by ATP synthase to phosphorylate ADP. As the protons (hydrogen ions) are used in the formation of water, the proton gradient across the

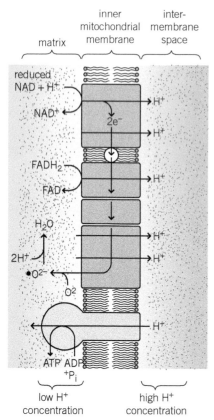

▲ **Figure 2** *Synthesis of ATP according to the chemiosmosis theory*

inner mitochondrial membrane is maintained so that chemiosmosis can continue. Apart from the specialised protein channels in the form of ATP synthase the inner membrane is impermeable to protons.

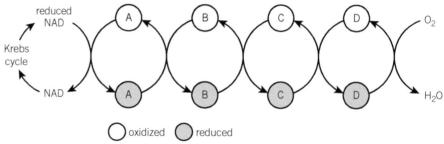

▲ **Figure 1** *Summary of the electron transport chain. The arrows represent oxidation and reduction reactions taking place and the letters represent electron carriers in the chain*

For every two hydrogens that are donated to the electron transfer chain by reduced NAD, three ATP molecules can potentially be made. The hydrogens donated by reduced FAD can potentially result in two ATP molecules. However, some energy from ATP is used to:

- transport ADP from the cytoplasm to the matrix
- transport pyruvate from the cytoplasm to the matrix
- transport ATP from the matrix to the cytoplasm for use by the cell.

This means that there is a net ATP production of 2.5 ATP per reduced NAD and 1.5 ATP per reduced FAD.

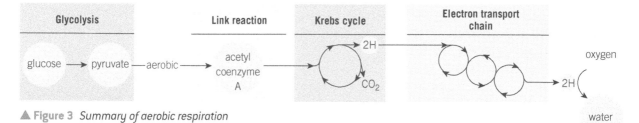

▲ **Figure 3** *Summary of aerobic respiration*

Synoptic link

You will need to remember the structure of cell membranes, which you learnt about in Topic 1.7, Interrelationship of organelles.

You will also need to remember how enzymes work, which was covered in Topic 3.4, The structure of enzymes.

Summary questions

1. Explain why the movement of protons through ATP synthase is called chemiosmosis. *(2 marks)*

2. Describe why the following have numerous mitochondria packed with cristae:
   a. hepatocytes (liver cells)
   b. striated muscle cells
   c. spermatozoan tails
   d. neurones. *(4 marks)*

3. Suggest how prokaryotic cells can complete oxidative phosphorylation. *(2 marks)*

16.4 Anaerobic respiration

Specification reference: 4.1.1

Anaerobic respiration

In anaerobic respiration, molecules other than oxygen are used as electron acceptors. In mammalian cells this type of respiration occurs when there is a shortage of oxygen, for example, during vigorous exercise. Anaerobic conditions can occur in any type of cell that is deprived of oxygen. In plants, waterlogged soils can result in anaerobic conditions in root cells.

Anaerobic respiration in muscle cells

In the absence of oxygen, the electron transport chain cannot occur as there is no final electron acceptor available. This causes a build-up of reduced NAD. NAD is not recycled and therefore is not available for the Krebs cycle to continue. This in turn means that there is no CoA recycled and available for the link reaction. As a result, there is a build-up of pyruvate and a lack of NAD and FAD in the muscle cells.

In the absence of oxygen, the cells use pyruvate as an alternative hydrogen acceptor, which enables reduced NAD to be reoxidised to NAD, and allows glycolysis to continue. Consequently, a small amount of ATP can still be produced by substrate level phosphorylation. In this pathway the pyruvate is converted to lactate as shown in Figure 1.

> ## Learning outcomes
>
> Demonstrate knowledge, understanding, and application of:
>
> → anaerobic respiration in muscle cells
>
> → anaerobic respiration in yeast cells.

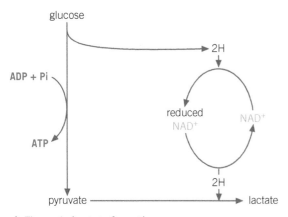

▲ Figure 1 *Lactate formation*

The build-up of lactate in cells can lead to the inhibition of glycolysis and eventually even this small supply of ATP may stop. The lactate build-up can also cause cramp and fatigue, called lactic acidosis. The lactate will be transported in the plasma to the liver. Once oxygen becomes available again, lactate is oxidised back to pyruvate, using the enzyme lactate dehydrogenase, and forming reduced NAD. Approximately one fifth of the pyruvate is respired aerobically to release energy. The remaining pyruvate is converted into glucose-6-phosphate and then glycogen.

9

Anaerobic respiration in yeast cells

Anaerobic respiration in yeast is called **fermentation**. Pyruvate is decarboxylated in the cytoplasm and converted to ethanal and carbon dioxide. The ethanal is then used as the hydrogen acceptor to form ethanol, allowing the reduced NAD to be converted back to NAD. If the ethanol levels build up they can become toxic to the yeast. Ethanol cannot subsequently be broken down by yeast cells to yield any additional energy.

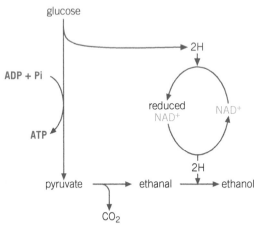

▲ Figure 2   *Alcoholic fermentation*

Synoptic link

You will need to remember cell ultrastructure and organelle structures, which you learnt about in Chapter 1, Cells and microscopy.

Study tip

Construct a spider diagram to show the stages of aerobic respiration and then add to your diagram the two different forms of anaerobic respiration that occur in different types of cell.

Comparison of aerobic and anaerobic efficiency

The hydrolysis of one mole of glucose to carbon dioxide and water releases approximately 2880 kJ of energy. The hydrolysis of one mole of ATP releases 30.6 kJ of energy. It is possible to assess the relative efficiencies of aerobic respiration and anaerobic respiration by comparing the net gain of ATP. In aerobic respiration, 32 ATP molecules are produced, whereas in anaerobic respiration there is a net gain of two ATP molecules.

$$\text{Efficiency of aerobic respiration} = 32 \times 30.6 \times \frac{100}{2880} = 34.0\%$$

$$\text{Efficiency of anaerobic respiration} = 2 \times 30.6 \times \frac{100}{2880} = 2.1\%$$

Comparing the two: $\frac{34}{2.1}$, aerobic respiration is 16.2 times more efficient than anaerobic respiration.

Summary questions

1   Explain why alcoholic fermentation is of considerable importance to humans.   *(2 marks)*

2   Yeast cell cultures grow much more slowly in anaerobic conditions – explain why.   *(2 marks)*

3   Describe and explain the different locations of pyruvate decarboxylation in yeast cells.   *(2 marks)*

Respiratory substrates

In the previous topics the breakdown of glucose has been concentrated on, but glucose is not the only respiratory substrate. A respiratory substrate is any molecule that can be oxidised via the metabolic pathways of glycolysis, the link reaction and the Krebs cycle, to result in the formation of ATP. Cells can oxidise molecules other than sugars to release energy. Most fatty acids and glycerol from lipids, and amino acids from proteins can also, in certain circumstances, be used as respiratory substrates without first being converted to carbohydrates.

Respiration of lipids

Lipids are hydrolysed to glycerol and fatty acids. The glycerol molecules can then be phosphorylated into glyceraldehyde (triose) phosphate and broken down through glycolysis. The three fatty acids that are released from the hydrolysis of each triglyceride (lipid molecule) are broken down in the matrix of the mitochondria into two carbon fragments by a process called beta oxidation, during

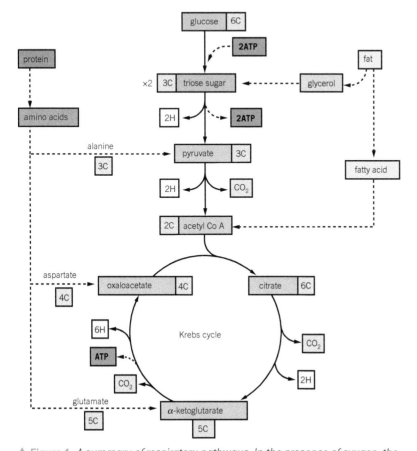

▲ **Figure 1** *A summary of respiratory pathways. In the presence of oxygen, the hydrogen will be passed on resulting in ATP production by oxidative phosphorylation*

which NAD and FAD are reduced. The two carbon fragments then combine with coenzyme A to form acetyl coenzyme A. Acetyl coenzyme A can then enter the Krebs cycle. One gram of fat releases more than twice as much energy as one gram of carbohydrate.

▼ Table 1 *Mean energy values of respiratory substrates*

Respiratory substrate	Mean energy value $[kJ\ g^{-1}]$
carbohydrate	15.8
lipid	39.4
protein	17.0

Respiration of proteins

Proteins, or rather, amino acids, are only used as a source of energy in extreme conditions, for example, starvation. Proteins are hydrolysed into their constituent amino acids. The amino acids are then **deaminated** in the liver. Deamination removes the amine group (NH_2) from the rest of the amino acid molecule. The remaining part of the amino acid then enters the respiratory pathway according to the number of carbons present, for example, 5C and 4C residues enter the Krebs cycle, 3C residues are converted into pyruvate.

Amino acids that will convert into pyruvate or into components of the Krebs cycle are known as glucogenic amino acids and can be converted into glucose. This is called gluconeogenesis. Most fatty acids cannot be converted into glucose because acetyl CoA cannot be converted back into pyruvate in animal cells. However, by using fatty acids as a respiratory substrate, glucose can be conserved.

Respiratory quotient

The respiratory quotient (or **RQ** or **respiratory coefficient**) is the ratio of carbon dioxide given off (eliminated) to oxygen used up (consumed) during respiration in a given time period.

$$RQ = \frac{CO_2}{O_2}$$

It is important that the CO_2 and O_2 are given in the same units and that the same period of time is used. Acceptable units would be either moles, or volume of gas at standard temperature and pressure.

An RQ value of one or less indicates that aerobic respiration is occurring. Any RQ value above one generally suggests that anaerobic respiration has occurred or that carbohydrates are being converted into lipids. If only anaerobic respiration is taking place the RQ = ∞.

▼ Table 2

Respiratory substrate	Respiratory quotient
carbohydrates	1
fat	0.7–0.72
proteins	0.8–0.9

For example, if a tissue is respiring lipids aerobically, the RQ should be approximately 0.7. Stearic acid is a fatty acid that can be respired. Stearic acid has the chemical formula $C_{17}H_{35}COOH$. The RQ can be calculated in the following way:

Step 1: Write a basic equation for the reaction:
$$C_{17}H_{35}COOH + ?O_2 \rightarrow ?CO_2 + ?H_2O$$

Step 2: Balance the number of carbon atoms in the products and reactants: 18C atoms in reactants, therefore 18 needed in the products, so
$$C_{17}H_{35}COOH + ?O_2 \rightarrow 18CO_2 + ?H_2O$$

Step 3: Now balance the number of hydrogen atoms: 36H atoms in reactants, therefore 36 needed in the products, so
$$C_{17}H_{35}COOH + ?O_2 \rightarrow 18CO_2 + 18H_2O$$

Step 4: Now balance the number of oxygen atoms: products contain 54 oxygen atoms, therefore 54 needed in the reactants. Two oxygen atoms are present in the stearic acid, therefore 52 remain to be accounted for, hence $26O_2$.
$$C_{17}H_{35}COOH + 26O_2 \rightarrow 18\ CO_2 + 18H_2O$$

Step 5: Now calculate the RQ:
$$RQ = \frac{18}{26} = 0.69$$

Respirometers

A respirometer can be used to measure the rate of respiration of a living organism by measuring the rate of exchange of oxygen and/or carbon dioxide (Figure 2). Respirometers allow the investigation of how factors such as age, chemicals, or temperature affect the rate of respiration. They are designed to measure respiration either at the level of a whole organism, or at the cellular level. In the simple respirometers shown in Figure 2, a graduated pipette is attached via a bung to the respirometer chamber. At the start of the investigation a small amount of coloured fluid is inserted (using a fine syringe) into the tip of the graduated pipette. The starting position of the meniscus is recorded. At predetermined time intervals (e.g., every minute) the new position of the meniscus is recorded. Over a specified time period (e.g., 10 minutes) the total volume of oxygen taken up can be measured and used to measure the rate of respiration.

Simple respirometers can be setup to determine the RQ of organisms, for example, germinating peas. Three respirometers, such as those in Figure 2, are placed in a water bath to maintain a constant temperature. In Figure 2, for example:

- The left hand respirometer contains germinating peas.
- The middle respirometer contains dried peas (dormant).
- The right hand respirometer contains glass beads, which acts as a control. Any changes in the position of the meniscus in this chamber will be due to absorption of the carbon dioxide by the

▲ Figure 2 *An experiment using simple, single-chambered respirometers*

Summary questions

1 Explain why the aerobic respiration of lipids produces a higher release of energy per gram compared with the oxidation of glucose. *(4 marks)*

2 a Show why the RQ value for the aerobic respiration of glucose is 1. *(2 marks)*

  b Calculate the RQ for the complete oxidation of a polyunsaturated fatty acid with the formula $C_{55}H_{104}O_6$. *(2 marks)*

3 a Suggest why the values for RQs in living organisms usually vary. *(1 mark)*

  b Suggest why the RQ value is usually between 0.8 and 0.9. *(2 marks)*

  c Explain what would happen to this value if the individual continued to exercise for a prolonged period of time. *(2 marks)*

soda lime, and temperature changes causing the volume of gas within the chamber to alter. This can be subtracted from the measurements taken in the other chambers to gain a more accurate value for the rate of respiration.

The apparatus in Figure 2 can also be used to investigate the effect of temperature on the respiration rate for both dormant and germinating peas.

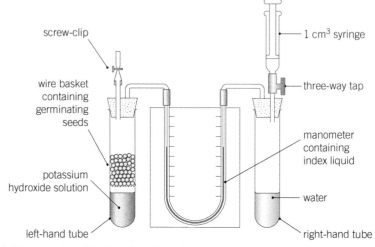

▲ **Figure 3** *A double-chambered respirometer*

Double-chambered respirometer

In a double-chambered respirometer one side of the apparatus contains the respiring organisms (e.g., woodlice). The other side contains the same volume of glass beads. This helps to compensate for any changes in atmospheric pressure that would move the manometer fluid (Figure 3). The syringe is used to adjust the level of manometer fluid. The tap is open initially as the fluid is adjusted, but is closed during the experiment. With the tap open, the apparatus is placed in a water bath and allowed to equilibrate. The fluid is adjusted, the tap is then closed, and a timer is started. At the end of, for

example, three minutes, the distance moved by the fluid in the manometer is noted. The tap is then opened and the fluid is re-set so that repeat readings can be carried out at the same temperature and at different temperatures to measure the effect of temperature on oxygen consumption. Table 3 shows a set of experimental results.

> Copy and complete the table to calculate the mean rate of oxygen consumption at the two temperatures investigated.

▼ Table 3

Temperature (°C)	Start point of manometer fluid (cm³)	Finish point of manometer fluid (cm³)	Change in volume (cm³)	Rate of oxygen consumption (cm³ O₂ min⁻¹)	Mean rate of oxygen consumption (cm³ O₂ min⁻¹)
5	15.00	16.70	2.7	0.9	
5	17.00	19.00			
5	14.00	16.30			
10	12.00	15.00			
10	16.00	19.20			
10	17.00	20.00			

Practice questions

1 Which of the following statements is/are true of the conversion of pyruvate to acetyl coenzyme A in the link reaction?

Statement 1: Pyruvate is reduced.

Statement 2: Pyruvate is decarboxylated.

Statement 3: Pyruvate is dehydrogenated.

A 1, 2 and 3

B Only 1 and 2

C Only 2 and 3

D Only 1 (*1 mark*)

2 Which of the following statements is/ are true of the electron transport chain in mitochondria?

Statement 1: The first electron carrier is reduced by reduced NAD.

Statement 2: Oxygen acts as the terminal electron acceptor.

Statement 3: Energy is used to pump protons out of the intermembrane space.

A 1, 2 and 3

B Only 1 and 2

C Only 2 and 3

D Only 1 (*1 mark*)

3 Which molecule acts as the hydrogen acceptor during alcoholic fermentation?

A Glucose

B Pyruvate

C Ethanol

D Ethanal (*1 mark*)

4 What is the respiratory quotient of palmitic acid ($CH_3(CH_2)_{14}COOH$) to 2 significant figures?

A 0.69

B 0.70

C 0.71

D 0.72 (*1 mark*)

5 Describe how the respiratory quotient of a plant can be determined using a respirometer. (*6 marks*)

6 The figure shown represents the first stage of respiration.

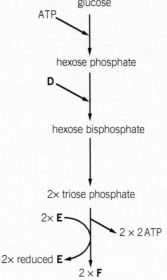

a (i) Name the stage represented by the figure. (*1 mark*)

   (ii) State precisely where in the cell this stage takes place. (*1 mark*)

   (iii) Identify the components D, E, and F. (*3 marks*)

b In anaerobic conditions, compound F does not proceed to the link reaction.

Describe the fate of compound F during anaerobic respiration in an animal cell and explain the importance of this reaction. (*5 marks*)

OCR F214 2010

7 One way of calculating the rate of respiration is to measure the volume of oxygen taken up over a period of time.

A student carried out an experiment to investigate the effect of temperature on the rate of respiration in soaked (germinating) pea seeds and dry (dormant) pea seeds.

A simple piece of apparatus called a respirometer was used.

The potassium hydroxide solution in this apparatus absorbs carbon dioxide. If the apparatus is kept at a constant temperature, any changes in the volume of air in the respirometer will be due to oxygen uptake.

a State the stage or stages of aerobic respiration during which:

   (i) carbon dioxide is produced *(1 mark)*

   (ii) oxygen is used *(1 mark)*

b The student set up three respirometers, A, B, and C, in water baths at two different temperatures. The respirometers were left for 10 minutes in order to equilibrate. The contents of each respirometer are shown in the table.

Temperature (°C)	Respirometer	Contents
15	A	30 soaked pea seeds
	B	Glass beads + 30 dry pea seeds
	C	Glass beads
25	A	30 soaked pea seeds
	B	Glass beads + 30 dry pea seeds
	C	Glass beads

At each temperature, respirometer C, which contained only glass beads, was a control. Respirometer B, at each temperature, also contained some glass beads.

   (i) Suggest why, at each temperature, respirometer B contained **some** glass beads. *(2 marks)*

   (ii) Suggest how the student determined the quantity of glass beads to place in respirometer B at each temperature. *(2 marks)*

c After the student had left each respirometer to equilibrate, a small volume of coloured fluid was introduced into each graduated tube.

The respirometers were then left in the appropriate water baths for 20 minutes and maintained at the correct temperature. During this time, the coloured fluid in the graduated tube moved.

The level of the coloured fluid in each respirometer was recorded at the start of the experiment and after 20 minutes.

The results are summarised in the table.

Temperature °C	Respirometer	Reading at start (cm³)	Reading after 20 minutes (cm³)	Difference (cm³)	Corrected difference (cm³)	Rate of oxygen uptake (cm³ min⁻¹)
15	A	0.93	0.74	0.19	0.16	0.08
	B	0.93	0.86	0.07	0.04	0.002
	C	0.91	0.88	0.03		
25	A	0.94	0.63	0.31	0.27	
	B	0.93	0.84	0.09	0.05	0.003
	C	0.95	0.91	0.04		

   (i) The table is incomplete. Calculate the missing value for the rate of oxygen uptake for soaked pea seeds (A) at 25°C. Show your working and give your answer in $cm^3 min^{-1}$ *(2 marks)*

   (ii) Explain why there is an increased rate of respiration in soaked seeds at 25°C compared with the soaked seeds at 15°C. *(2 marks)*

   (iii) Suggest a reason for the difference in the rate of respiration between soaked and dry pea seeds. *(2 marks)*

OCR F214 2012

8 The compound 2,3,5-triphenyl-tetrazolium chloride (TTC) is an electron acceptor. TTC will diffuse into actively respiring cells and accept electrons from the electron transport chain.

When TTC accepts electrons and becomes reduced it changes from colourless to pink. The tissues in which this reaction takes place will be stained a pink colour.

a State the precise location of the electron transport chain in the cell. *(1 mark)*

b A student carried out an investigation into the respiratory activity of plant tissue. She used three groups of germinating broad bean seeds. These were first treated as shown in the table.

Seed	Treatment
Group A	kept at 22°C for 24 hours before the investigation
Group B	kept at 6°C for 24 hours before the investigation
Group C	kept at 22°C for 24 hours and then placed in water at 90°C for 5 minutes before the investigation

The groups of seeds were then sliced longitudinally and placed, cut surface down, in a shallow dish containing a small volume of TTC solution. The cut surfaces remained in contact with the solution for 10 minutes.

The seeds were then removed from the dish. The excess TTC solution was wiped off and the cut surfaces of the seeds in each group were observed.

The appearance of the seeds in each group is shown here. The shaded areas are the regions where the tissues have stained a pink colour.

seeds in group A   seeds in group B   seeds in group C

(i) Describe the differences observed in the seeds in Groups A, B and C.

(1 mark)

(ii) Suggest reasons for the results observed in the seeds in Group A.

(2 marks)

(iii) Suggest reasons for the differences in the amount of staining observed in the seeds in Groups B and C when compare to those in Group A.

(2 marks)

c If oxygen is not present or is in short supply, respiration can take an anaerobic pathway after glycolysis. In plant cells, this pathway is the same as the one used in yeast cells.

(i) Name the hydrogen acceptor in this pathway. *(1 mark)*

(ii) Name the intermediate compound in this pathway. *(1 mark)*

(iii) Name the products of this pathway.

(1 mark)

(iv) Explain why this pathway is important for the plant cell. *(2 marks)*

OCR F214 2012

17 METABOLISM AND EXERCISE
17.1 The consequences of exercise on the body
Specification reference: 4.1.2

Having learnt in previous topics how energy is released in cells, in this chapter you will learn how the body adapts to the energy requirements of exercise.

Exercise

The UK government recommends that adults between the ages of 19 and 64 years should aim to be active daily. Over a week, an individual should aim to be active for least 150 minutes (2½ hours) of moderate intensity activity in bouts of 10 minutes or more. One way to do this is to exercise for 30 minutes on at least five days a week. Similar benefits can also be achieved through 75 minutes of vigorous intensity activity spread across the week, or through combinations of moderate and vigorous intensity activity. Individuals should also undertake physical activity to improve muscle strength on at least two days a week and they should minimise the amount of time that is spent being sedentary (sitting) for extended periods.

Aerobic exercise has many immediate and long-term benefits, and long-term exercise ('training') results in adaptations by the body.

Immediate effects of exercise

Short-term aerobic activity results in changes to both the circulatory and respiratory systems (Table 1).

Study tip

Remember – long-term exercise means that the person has undergone a physical training programme over a period of time, for example, weeks or months.

▼ Table 1

Change	Explanation
Heart rate increases	Caused by an increase in adrenaline secretion and the stimulation of the sympathetic nervous system. Even before exercise begins the heart rate increases in anticipation. This is known as the anticipatory response. It is caused by the release of neurotransmitters called adrenaline and noradrenaline. After the initial anticipatory response, heart rate increases in direct proportion to exercise intensity until either a plateau or the maximum heart rate is reached. In theory, maximum heart rate is estimated using the formula 220 minus age. But this is only an estimation – it is not particularly accurate and is rarely attained.

Cardiac output (heart rate × stroke volume) increases proportionally with exercise intensity. |
Vasodilation of arterioles in skeletal muscles	Signalled by the secretion of nitric oxide by the arteriolar endothelium in response to the fall in oxygen levels.
Increase in blood flow to active muscles	Caused by the dilation of arterioles supplying oxygenated blood to active muscles.
Increase in stroke volume	Caused by more blood returning to the left atrium of the heart (the larger the volume of blood filling the ventricle in diastole the greater the volume of blood pumped out during systole).

▼ Table 1 *continued*

Change	Explanation
Reduced blood flow to the digestive system	Occurs because there is a finite volume of blood within the body. More blood is diverted to active skeletal muscles and less flows to areas such as the digestive system. At rest, 15–20% of circulating blood supplies skeletal muscle. During vigorous exercise this increases to 80–85% of cardiac output.
Vasodilation of arterioles supplying the surface of the skin	Caused by the secretion of adrenaline, resulting in arterioles under the skin receiving more blood to enable the body to lose heat by radiation. Evaporation of sweat aids the cooling of the body.
Increased breathing rate and depth of breathing (tidal volume)	This increase in ventilation brings more air into the alveoli. This increases the concentration gradient, allowing more oxygen to diffuse into the blood, and more carbon dioxide to diffuse out. Increase in the acidity of the blood is detected by chemoreceptors, which send impulses to the respiratory centre in the medulla of the brain. The response is an increase in the rate and extent of the contractions of the diaphragm and intercostal muscles.

Long-term effects of exercise

The human body adapts to long-term, regular exercise (training) in a variety of ways, as shown in Table 2.

▼ Table 2

Circulatory system	Respiratory system	Skeletal system
Increased VO$_2$max: the maximum volume of oxygen that can be taken in, transported, and utilised increases. This is due to the changes described in this table (you will learn more about VO$_2$ max in Topic 17.2, Aerobic fitness).	Increased maximum breathing rate: after training, athletes can achieve a faster breathing rate than untrained individuals.	Increase in cross-sectional area of slow-twitch muscle fibres: these muscles rely on aerobic respiration to produce ATP. They do not increase in number but the size of these muscle fibres increases.
Increased heart size: the mass and volume increase, cardiac muscle undergoes hypertrophy, particularly in the left ventricle. As well as the chamber size increasing, studies show that the myocardial wall thickness also increases. This increases the force of ventricular contraction.	Increased tidal volume: this is to maintain a large concentration gradient to ensure an increase in the rate of oxygen supply. This means trained athletes can breathe faster and deeper if they need to. However, due to changes in the effective gas exchange surface, the tidal volume of athletes at rest is frequently lower than that of an untrained person.	Increase in number and size of mitochondria in muscle fibres: this increases the number of respiratory enzymes in the muscle fibres, which increases the rate of the Krebs cycle and oxidative phosphorylation.
Decreased resting heart rate: resting heart rate can decrease significantly following training in a previously sedentary individual. During a 10 week exercise program, an individual with an initial resting heart rate of 80 beats/min can reasonably expect to see a reduction of about 10 beats/min in their resting heart rate. Highly conditioned athletes can have resting heart rates in the low 30s.	Increased vital capacity: development of the intercostal muscles and diaphragm results in a larger achievable vital capacity.	Increased capillary network surrounding muscle fibres: this increases the number of capillaries and the ratio of capillaries : muscle fibres, which increases the volume of blood in the muscles and so improves oxygen delivery.

▼ **Table 2** *continued*

Circulatory system	Respiratory system	Skeletal system
Increased stroke volume: at rest, stroke volume averages 50–70 cm^3/beat in untrained individuals, 70–90 cm^3/beat in trained individuals, and 90–110 cm^3/beat in world-class endurance athletes.	Increased density of capillaries in the lungs: this gives an increased effective gas exchange surface. Hence, more oxygen is able to diffuse into the blood in a given period of time. This means ventilation rate (breathing rate and tidal volume) is reduced at rest.	Increased efficiency in lipid metabolism in muscle fibres: This conserves carbohydrates in muscle cells.
Decreased heart rate recovery time: After a period of training, the time it takes for the heart rate to recover to its resting value is reduced.		Increased myoglobin and glycogen stores: increased myoglobin (respiratory pigment found in skeletal muscles) increases the amount of oxygen stored in muscles. Increased glycogen stores improve energy release to active muscles.
Increased number of red blood cells: this increases the ability of the blood to transport oxygen.		Increased vascularisation of muscles: increased size and number of blood vessels serving the skeletal muscles. Increased number of capillaries, resulting in more effective blood redistribution and gas exchange in the muscles.

Synoptic link

You will need to remember the structure and functions of blood vessels and the heart, which you learnt about in Chapter 5, The heart and monitoring heart function, and Topic 6.1, The transport system in mammals.

You will learn more about homeostasis (the regulation of temperature) in Topic 29.2, Regulation of body temperature and metabolic rate.

You will learn more about the structure of muscles and how they contract, in Topic 17.6, Skeletal muscle structure, and Topic 17.7, Sliding filament theory – muscle contraction.

Study tip

Remember, arterioles contain smooth muscle in their walls that relax in response to nitric oxide or stimulation from the hypothalamus.

Summary questions

1 At rest the cardiac output is about 5 dm^3 min^{-1}. During intense exercise this can increase to 40 dm^3 min^{-1}. Calculate the percentage increase in cardiac output. Give your answer to 1 significant figure. *(2 marks)*

2 Suggest why it is essential to remove excess carbon dioxide from the blood. *(4 marks)*

3 a Evaluate the use of recovery rates as a tool for tracking the effects of a training program. *(4 marks)*
  b Explain how the altered metabolism of lipids by muscle cells can result in a reduced BMI. *(2 marks)*

17.2 Aerobic fitness

Specification reference: 4.1.2

Aerobic fitness

Having learnt about the consequences of exercise on the body in the previous topic, you will now consider aerobic fitness. Whilst different individuals show different responses to training, it is generally accepted that 20 minutes of aerobic exercise per day will result in an improvement in aerobic fitness for the average person. This level of activity will then need to be sustained to ensure that the person maintains their improved fitness level. Activities can include walking, running, cycling, running up and down stairs, swimming, etc.

Factors affecting aerobic fitness

Aerobic fitness is how efficiently oxygen is used by the body – for example the amount taken up, transported in the blood, pumped by the heart, and how well muscles use the oxygen to provide energy. Various factors will affect an individual's aerobic fitness, such as:

- Age – age matters because it affects strength, speed, and endurance. Age defines aspects of fitness as peak potentials are reached at different times in our lives. As you age, physical efficiency declines as well.
- Gender – males and females have different physical builds and as a consequence have different levels of aerobic fitness.
- Participation in an exercise programme.

Other factors that can affect aerobic fitness include – smoking, quality of nutrition, use of stimulants (e.g., caffeine), alcohol consumption, depression, and motivation.

Other health benefits of improved aerobic fitness

Training can improve aerobic fitness by causing the body to make adaptations that are beneficial to health. These were discussed in the previous topic. There are many other recognised benefits of regular aerobic exercise, such as:

- strengthening the skeletal muscles
- improving the efficiency of the circulatory system
- reducing blood pressure
- improving mental health (e.g., reducing stress, lowering the incidence of depression, increasing cognitive capacity)
- reducing the risk of diabetes
- high-impact aerobic activities (such as jogging or using a skipping rope) stimulating bone growth, reducing the risk of osteoporosis.

To achieve an initial improvement in aerobic fitness it is important that the following factors are considered:

- frequency of exercise
- time (duration)
- intensity of exercise
- type of exercise.

▲ Figure 1 *Water aerobics class – low impact aerobic exercise*

These factors can be abbreviated to F.I.T.T. factors, which are considered when designing a fitness programme. Because aerobic exercise is a 'stressor' to which the body adapts, improvement in aerobic fitness requires progressive overload – increasing the frequency, intensity and/or duration of exercise over time. The success of a training programme can be measured by monitoring changes in the resting heart rate, and/or recovery times for an individual. These correlate well with an individual's VO_2max and can be used to estimate VO_2max. Direct measurement of VO_2max requires specialist equipment.

Measuring VO_2max as an indicator of aerobic fitness

VO_2max is the maximum rate at which oxygen can be taken in, transported, and utilised, as measured during incremental exercise, most usually when the individual is on a motorised treadmill. The maximum oxygen consumption reflects the aerobic physical fitness of the individual.

VO_2max can be expressed either as:

- an absolute rate, for example, litre of oxygen per minute ($dm^3 \, min^{-1}$)

- a relative rate, for example, millilitres of oxygen per kilogram of body mass per minute ($ml \, kg^{-1} \, min^{-1}$).

VO_2max does not generally vary linearly with body mass, so comparisons of the performance capacities of individuals that differ in body mass must be carried out with appropriate statistical procedures.

To measure the VO_2max, an individual is put through a graded exercise test. Prior to any exercise test it is essential that an appropriate risk assessment is undertaken of the subject to determine if there are any pre-existing medical conditions, for example, asthma. During the assessment the exercise intensity increases progressively. At the same time, the ventilation, and oxygen and carbon dioxide concentration of the inhaled and exhaled air is measured. The individual's VO_2max is reached when their oxygen consumption remains at steady state despite an increase in workload. It is important that the assessment involves physical effort sufficient in both duration and intensity to fully test the aerobic energy system.

In schools, VO_2max is estimated rather than measured directly by using factors such as changes in heart rate during exercise.

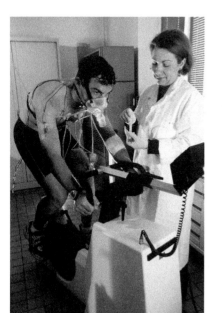

▲ **Figure 2** *Researcher preparing to test a subject (on exercise bike) for research into the effects that extreme conditions have on human physiology. Monitoring equipment is used to record data such as breathing rate, heart rate, blood oxygen level, respiratory gas exchange, body temperature, and heart function*

Summary questions

1 a Using Figure 3 describe the changes in VO$_2$max shown in
      the graph.   (*3 marks*)

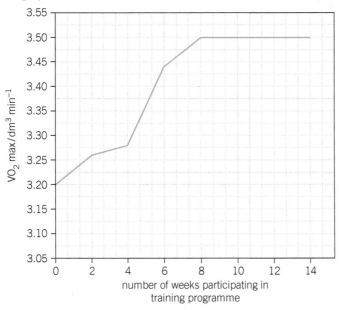

▲ Figure 3

   b Calculate the percentage increase in VO$_2$max between week 2
      and week 8. Give your answer to one decimal place.   (*2 marks*)

2 Outline a method that could be used in a school to investigate the
   effect of a training programme on resting heart rate in a group of
   female students.   (*9 marks*)

3 Explain why levels of lactate in blood plasma would be much lower
   during intensive exercise in subjects at the end of the training
   programme.   (*3 marks*)

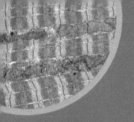

17.3 Oxygen deficit and oxygen debt

Specification reference: 4.1.2

In Chapter 16, Cellular respiration, you learnt that glucose could also be broken down anaerobically to release energy. In this chapter you look at some of the consequences of that.

Oxygen deficit

Anaerobic respiration will result in the production of lactate. This could result in the lactate pathway and glycolysis stopping due to a fall in pH. Lactate is removed from muscles and broken down in liver cells. The breakdown of lactate in the liver requires a supply of oxygen. The difference between the oxygen demand of the active muscles and the oxygen that they actually receive is called the oxygen deficit. When the oxygen supply does not meet the oxygen demand, the muscle cells resort to anaerobic respiration to ensure that ATP is still produced, albeit at a lower level.

Excessive post-exercise oxygen consumption (EPOC)

EPOC is the increased volume of oxygen consumed following vigorous exercise. In the past, the term 'oxygen debt' was used to describe the oxygen required for lactic acid/lactate metabolism. However, calorimeter experiments have disproved a direct association between lactate metabolism and elevated oxygen uptake.

After exercise, EPOC is used to restore the body to its resting state and enable it to adapt to the exercise just performed. The oxygen is required for the following:

- re-oxygenating haemoglobin
- re-oxygenating myoglobin
- balancing hormones
- replenishing glycogen stores in the muscles
- carrying out any necessary cell repair
- regenerating ATP
- converting lactate to glucose or glycogen
- meeting the demands of the increased metabolic rate as a result of thermogenesis in brown adipose tissue (from the increase in body temperature that occurs during exercise) and the increased heart rate that remains immediately after exercise has finished.

EPOC can be found by calculating the difference between the total volume of oxygen consumed during the recovery period and the total volume of oxygen that would be consumed over the same period with the body at rest.

The EPOC effect is highest just after the exercise is completed and decreases over time. EPOC increases with the intensity of the exercise, and with the duration of the aerobic exercise.

Synoptic link

You learnt about lactate in Topic 16.4, Anaerobic respiration.

Synoptic link

You will need to recall the immediate effects of exercise. These were covered in Topic 17.1, The consequences of exercise on the body.

Some studies have suggested that resting metabolic rate (RMR) remains elevated for 24 hours after high-intensity interval training, due to excess post-exercise oxygen consumption. HIIT may improve maximal oxygen consumption (VO_2max) more effectively than doing only traditional, long aerobic workouts.

EPOC will be less in someone who is aerobically fit, so measuring the time it takes for the heart rate to decline has been used to estimate aerobic fitness. This is the basis of the Harvard Step Test, which is carried out as follows:

Using a step between 30–50 cm high, step up, placing both feet on the step and then down, placing both feet together on the floor. This is one cycle, and the stepping rate should be one cycle every 2 seconds. This is carried out for 5 minutes.

At the end of 5 minutes, wait 1 minute and then take a pulse for 15 seconds. This is repeated at 2 minutes, and then 3 minutes after the exercise. Pulse rates are converted to beats per minute and added together and a score is calculated:

$$\frac{30\,000}{total\ beats}$$

30 000 is used as this is the duration of the stepping (300 seconds) multiplied by 100 – which avoids results appearing as fractions. The score can be compared to standards, such as those in Table 1, to estimate fitness. Individuals with a higher score have a higher aerobic fitness.

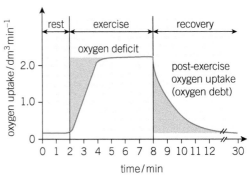

▲ Figure 1 *Graph to show oxygen uptake before, during, and after exercise*

▼ Table 1

Gender	High	Above average	Average	Below average	Low
male	above 90	90–80	79–65	64–55	below 55
female	above 86	86–76	75–61	60–50	below 50

The test could be used to assess the effect of the F.I.T.T factors on aerobic fitness. As fitness improves, so does recovery time, and the 'total beats' figure will be resulting in a higher score if fitness has improved.

Summary questions

1   Suggest how training could reduce an individual's EPOC.   (*2 marks*)

2   Explain why an EPOC occurs even if the person is not exercising vigorously.   (*3 marks*)

3   Explain why an increased ventilation rate occurs in response to changes in the amount of carbon dioxide in the blood.   (*3 marks*)

Synoptic link

You will need to remember the effects of anaerobic respiration. These were covered in Topic 16.4, Anaerobic respiration.

You will learn how different areas of the brain are responsible for changes in heart rate and ventilation rate, in Topic 29.3, Regulation of heart rate.

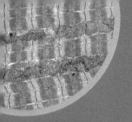

17.4 Enhancing athletic performance

Specification reference: 4.1.2

Learning outcomes

Demonstrate knowledge, understanding, and application of:

supplementary methods of enhancing athletic performance, including:

→ carbohydrate loading

→ blood doping

→ recombinant erythropoietin

→ steroids.

Enhancing athletic performance

In previous topics you learnt about the factors affecting fitness which in turn affects athletic performance. The right nutrition can help individuals to excel in sports, recover faster, decrease their risk of injury, and even reduce muscle soreness. While some methods, such as carbohydrate loading, are acceptable, others, such as recombinant erythropoietin, blood doping, and steroid treatment, are illegal in competitive activities. These work by changing the body's normal function.

Carbohydrate loading

A carbohydrate loading diet, also called a carb-loading diet, is a strategy for increasing the amount of glycogen stored in the muscles and therefore to improve athletic performance. This will mean that the glycogen can be hydrolysed rapidly in the muscle cells at the precise site where the ATP is needed. This ensures faster release of energy, when compared to relying on blood glucose.

Prior to an event, athletes will consume foods that are high in carbohydrates. The carbohydrate is broken down into glucose and then absorbed into the blood plasma. Some of this glucose is then transported to the muscles fibres where it is converted to glycogen for storage. Athletes tend to follow a strict regime:

- Carbodepletion – for a period of days in advance of the event the athlete reduces carbohydrate intake in favour of protein and fat.

- Carbohydrate loading – in the two or three days immediately before competing the athlete switches to a high carbohydrate diet. Successful carbohydrate loading requires the athlete to consume 6 to 10 g of carbohydrate per kilogram of body muscle mass, which can amount to 10% of their total body mass. This will also mean, however, that the body's mass may also increase by as much as 2 kg.

- After the event the athlete recovers by consuming carbohydrates and proteins.

Carbohydrate loading will not enable muscles to work harder or faster – but it will enable them to work for longer, and as such is of benefit to endurance athletes, but not to sprinters or jumpers. Although both lipids and carbohydrates can be broken down aerobically, glucose (from stored glycogen) can be more rapidly mobilised and broken down than lipids. This means that the rate of energy released from glucose is much faster. If the glucose supplies fall, the rate of exercise will have to fall to a level that the mobilisation and breakdown of lipids can support.

▲ Figure 1 *Sources of carbohydrates shown here include – pasta, potatoes, bread, pulses (beans), rice, and cereals such as wheat and oats*

RhEPO

The body naturally responds to low oxygen levels due to, for example, high altitude or reduced blood volume, by secreting the hormone erythropoietin (EPO) from cells surrounding capillaries in the renal cortex in the kidney. EPO controls erythropoiesis (red blood cell production). A rise in EPO in the blood causes red blood cell production to increase in the bone marrow.

An injection of recombinant human EPO (RhEPO) into the blood enables athletes to artificially increase their red blood cell levels. RhEPO is one of several erythropoiesis-stimulating agents that target the bone marrow to stimulate an increase in the red blood cell level within a few days. However, use of RhEPO can lead to severe cardiovascular problems as well as kidney failure.

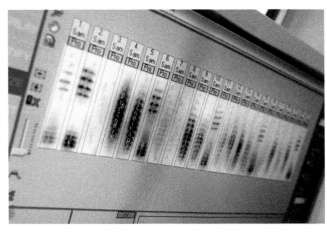

▲ **Figure 2** *Close-up of a screen displaying the results of a test for the presence of the performance-enhancing hormone erythropoietin (EPO). This test looks for raised levels of red blood cells in a sample. If the red blood cells constitute more than 50% of the volume of blood it is presumed that the athlete had taken EPO. Flow cytometry can be used to obtain these results*

Blood doping

Blood doping is a method that some athletes use to raise their red blood cell levels in order to enhance athletic performance. A higher concentration of red blood cells in the blood can improve an athlete's aerobic capacity (VO_2max) and hence endurance. Many methods of blood doping are illegal, particularly in professional sport.

Approximately $1\,dm^3$ of blood is removed several months before the event (this allows the athlete's body to naturally replenish blood by secreting extra erythropoietin). The removed red blood cells are separated from the plasma and stored at cold temperatures. Prior to the event the cells are warmed to body temperature and introduced back into the athlete's blood. As well as being unethical and illegal in many sports, increasing the red cell concentrations above natural levels can increase blood viscosity, which increases the chances of thrombosis, pulmonary embolism, and stroke. This is referred to as hyperviscosity syndrome, and can lead to a reduced cardiac output, and actually decrease the oxygen content of the blood.

Steroid enhancement

Steroids are complex molecules made naturally from cholesterol. Steroids are non polar and lipid soluble, so they can diffuse through cell membranes. Steroid hormones, oestrogen, testosterone, and progesterone stimulate anabolic reactions, for example protein synthesis, and promote tissue growth and repair.

Anabolic steroids, such as nandrolone and stanozolol, are artificially produced and are similar in structure to natural steroids. These artificial steroids can be used to increase muscle mass, and also make athletes more aggressive, competitive, and able to train for longer periods of time.

Synoptic link

You learnt about flow cytometry and how it is used to count cells in Topic 1.5, Counting cells.

Synoptic link

You will need to remember the importance of blood types from Topic 3.9, Blood donation.

Study tip

Remember, 'recombinant' means that the EPO glycoprotein has been produced using genetically modified cells (Topic 25.3, Genetic engineering – bacterial cells).

Synoptic link

You will need to remember how molecules cross cell membranes. You learnt about this in Chapter 1, Cells and microscopy.

Long term use of steroid hormones can lead to a decrease in the body's own production of testosterone as well as a decrease in the ability of the body's immune system to respond to pathogens. Liver damage can also occur, as well as erectile dysfunction and cardiovascular damage.

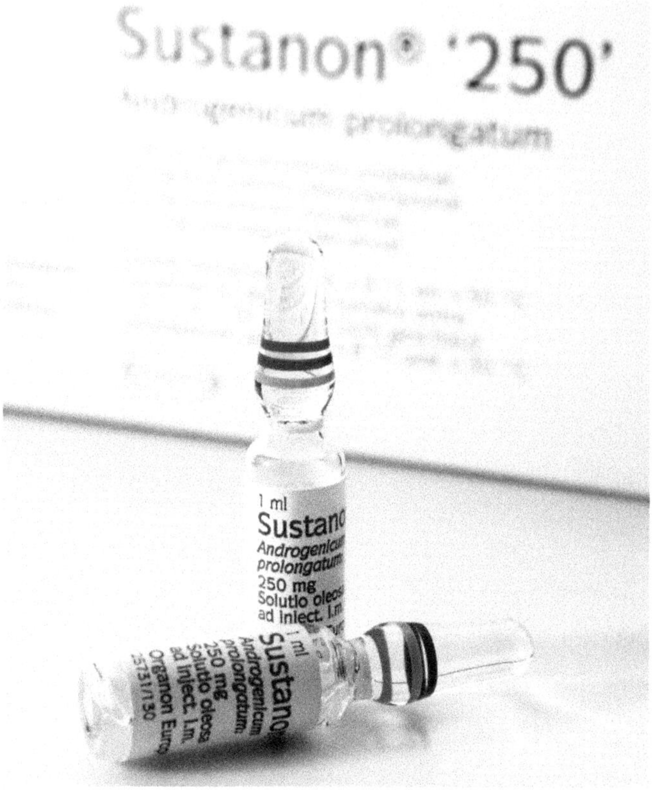

▲ **Figure 3** *Bottles containing Sustanon 250, a popular anabolic steroid used for muscle building*

Synoptic link

You will learn how proteins can be produced using genetic engineering. This is covered in Topic 25.3, Genetic engineering – bacterial cells.

Study tip

Make sure that you can discuss the ethical issues associated with using enhancement techniques. Remember that there are legal enhancement substances such as creatine phosphate. Creatine has not been banned in any professional sport because many everyday foods contain it and it would be hard to ban those foods. In addition, the body produces about 2 g of creatine of its own every day.

Study tip

Remember, glycogenolysis is the hydrolysis of the glycosidic bonds between adjacent glucose molecules within glycogen. This releases glucose molecules, which can then be respired to produce ATP.

Study tip

Remember, you cannot produce, make or create energy. Cells produce ATP, which is then used to release energy. The energy that is released is then used by muscle cells to contract.

Summary questions

1   a   Suggest why carbohydrate loading is not suitable for football players.
    (1 mark)

    b   State four properties of glycogen that make it ideal as an energy storage product for athletes.
    (4 marks)

2   Explain why it is important that athletes who choose to use blood doping as a mechanism to improve their performance use their own red blood cells.
    (3 marks)

3   Suggest how an anabolic steroid can enter muscle cells and cause an increase in protein synthesis.
    (4 marks)

17.5 Respiratory pigments and oxygen transport

Specification reference: 4.1.2

Respiratory pigments

In Chapter 16, Cellular respiration, you learnt about the role of oxygen in the release of energy from substrates such as glucose. A respiratory pigment is a molecule that is capable of binding reversibly with oxygen. There are two main respiratory pigments in humans:

- haemoglobin – used to transport oxygen, located inside erythrocytes, and capable of binding up to four oxygen molecules
- myoglobin – used to store oxygen, located inside skeletal muscle cells, and capable of binding one oxygen molecule.

Haemoglobin is responsible for transporting 98% of the oxygen around the body, with the remaining 2% being carried in solution.

Carriage of oxygen

Each prosthetic group (**haem group**) in a molecule of haemoglobin can bind to one oxygen molecule. When all four haem groups are occupied with oxygen the haemoglobin molecule is said to be **saturated**.

$$Hb \quad + \quad 4O_2 \quad \rightleftharpoons \quad HbO_8$$
haemoglobin   oxygen   oxyhaemoglobin

Haemoglobin has an oxygen-binding capacity of $1.34\,cm^3$ O_2 per gram of haemoglobin. The presence of haemoglobin increases the oxygen-carrying capacity of blood by a factor of 70 as the solubility of oxygen in plasma is very low. Oxyhaemoglobin is formed in the pulmonary capillaries adjacent to the alveoli of the lungs. The oxygen combined with haemoglobin in the red blood cells then travels through the blood stream to the cells that require it. The oxygen is released (**dissociates**) in respiring tissues and is used as the final electron acceptor in the production of ATP by the process of oxidative phosphorylation.

Myoglobin only binds one molecule of oxygen as there is only one haem group associated with the single polypeptide chain. Oxymyoglobin acts as an oxygen reserve, only releasing its oxygen when oxygen levels in the skeletal muscles become very low. Myoglobin has a higher affinity for oxygen than haemoglobin.

Oxygen dissociation curves

A dissociation curve describes the relationship between the partial pressure of oxygen and the saturation of the respiratory pigment. Partial pressures can be calculated using the following equation:

$$\frac{(V)_{gas}}{(V)_{total}} = \frac{(P)_{gas}}{(P)_{total}}$$

Where V is the volume (of the gas and of the mixture) and P is the pressure.

Synoptic link

You learnt about the structure of haemoglobin in Topic 3.3, Protein structure – haemoglobin.

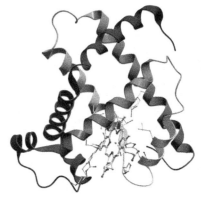

▲ **Figure 1** *Molecular model of myoglobin. Myoglobin consists of eight alpha helices connected by loops (shown in different colours) and a porphyrin ring with an iron ion at its centre*

This means in air, if atmospheric pressure is 101 kPa, and the oxygen concentration is 21%, the *partial* pressure of oxygen will be $(21/100) \times 101$ kPa.

Adult haemoglobin

The dissociation curve for adult haemoglobin is described as S-shaped or sigmoidal (Figure 2). This is because when the first oxygen molecule attaches to its binding site, it causes a conformational change in the haem group in one of the four sub-units of the haemoglobin. This change is transmitted though the molecule to the remaining three subunits producing a change in their oxygen binding sites. Consequently it is easier for further oxygen molecules to bind to the haem sites in the remaining three subunits. This is described as **co-operative binding** and results in the sigmoid curve.

At very low partial pressures of oxygen (pO_2), the affinity of haemoglobin for oxygen is low, that is, it is difficult for the oxygen to combine with the first haem group. This is shown by the shallow part of the dissociation curve at low partial pressures of oxygen. When the first oxygen molecule has bound to the haem group the increased affinity is shown by the steeper gradient of the dissociation curve – the second, third, and fourth oxygen molecules will bind more easily. At very high partial pressures of oxygen the dissociation curve plateaus as, even though the sites have a high affinity for oxygen, the haemoglobin is now almost fully saturated.

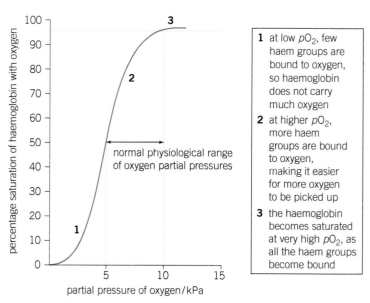

▲ Figure 2 *Oxygen dissociation curve for adult haemoglobin*

Fetal haemoglobin

The developing fetus must obtain sufficient oxygen from the mother's blood in the placenta. Fetal and maternal blood do not mix but are separated by a layer of modified epithelial cells in the placenta that allow metabolite exchange. When fetal oxygen demand is highest in the later stages of pregnancy, there is evidence that maternal and fetal blood-flow is organised in a counter-current system in some mammals to maximise the efficiency of exchange. The fetal dissociation curve is positioned to the left of the adult dissociation curve because fetal haemoglobin has a higher affinity for oxygen than adult haemoglobin (Figure 3). Oxygen binds more readily with fetal haemoglobin than with adult haemoglobin because two of the subunits in fetal haemoglobin differ to those in adult haemoglobin.

Myoglobin

As myoglobin has a higher affinity for oxygen than haemoglobin, it will only release its oxygen at very low partial pressures of oxygen. This means that at any partial pressure of oxygen there will be a higher oxygen saturation level for myoglobin than that for adult haemoglobin. This feature is essential for myoglobin to act as an oxygen store. When skeletal muscles are working hard, the oxygen demand may be greater than the oxygen supply from the blood. Oxymyoglobin will then dissociate to release oxygen and enable aerobic respiration to continue for a longer period of time.

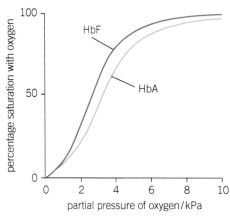

▲ **Figure 3** *Oxygen dissociation curves for fetal (HbF) and adult haemoglobin (HbA)*

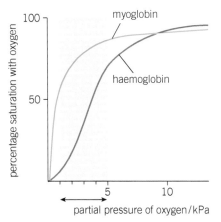

▲ **Figure 4** *Oxygen dissociation curves for myoglobin and adult haemoglobin*

> ## Study tip
>
> The further the dissociation curve moves to the right, the more readily the haemoglobin dissociates with (gives up) oxygen.
>
> The further the dissociation curve moves to the left, the more readily the haemoglobin associates with (picks up) oxygen.

Factors that affect the dissociation of oxygen from haemoglobin

Carbon dioxide

Carbon dioxide is transported in three ways in the blood:

- in aqueous solution in plasma (~5%)
- combined with the amine groups of the four polypeptide chains in haemoglobin as carbaminohaemoglobin (~10%)
- as hydrogen carbonate (bicarbonate, HCO_3^-) ions dissolved in the plasma (~85%).

Most of the carbon dioxide that is produced by cellular respiration diffuses into red blood cells. The zinc-containing enzyme carbonic anhydrase catalyses the reaction between carbon dioxide and water – carbonic acid is formed, which dissociates into hydrogen carbonate and hydrogen ions:

$$CO_2 + H_2O \rightleftharpoons H_2CO_3 \rightleftharpoons H^+ + HCO_3^-$$

In conditions where there is a high concentration of carbon dioxide, such as actively respiring tissues, the reaction proceeds to the right. This causes a high number of hydrogen ions (H^+) to be produced, which could decrease the pH of the blood plasma.

The H^+ ions in the cytosol of the red blood cells combine with the haemoglobin to form haemoglobinic acid. The haemoglobin acts as a buffer – it mops up the additional hydrogen ions. The formation of haemoglobinic acid causes the oxyhaemoglobin to release its bound oxygen. The effect of carbon dioxide on the oxygen saturation levels of haemoglobin is known as the **Bohr effect**. The increased acidity shifts the dissociation curve to the right, causing a greater release of oxygen from haemoglobin at the same partial pressure of oxygen. This means that as the concentration of carbon dioxide increases, for example, during vigorous exercise, the dissociation of oxygen from oxyhaemoglobin in that tissue also increases to ensure that sufficient oxygen is delivered.

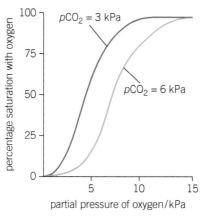

▲ **Figure 5** *The Bohr effect*

Synoptic link

You will need to remember how molecules can move by diffusion, which you learnt about in Chapter 1, Cells and microscopy.

The build-up of hydrogen carbonate ions inside the red blood cell generates a diffusion gradient causing them to diffuse into the plasma.

pH

A decrease in pH (increase in H^+ ion concentration) shifts the dissociation curve to the right, due to the formation of haemoglobinic acid, while an increase shifts it to the left. As muscles respire anaerobically, the production of lactate raises the H^+ ion concentration, and pH in the muscle tissue can fall quickly.

Temperature

Hyperthermia causes the curve to shift to the right, while hypothermia causes a shift to the left. Increasing temperature will weaken the association between the oxygen molecules and the haem groups. The higher temperature will disrupt the hydrogen and ionic bonds in the tertiary and quaternary structure of the haemoglobin. The higher the temperature, the less saturated the haemoglobin is with O_2. During exercise heat is produced due to increased metabolic activity. This promotes the release of oxygen.

Hiroshima haemoglobinopathy

Haemoglobinopathies are single-gene disorders that cause an abnormal structure of one of the globin chains in haemoglobin. The genes are carried on autosomes and show co-dominant patterns of inheritance. A common haemoglobinopathy is sickle-cell anaemia – it is estimated that 7% of the human population are carriers for this disorder.

An uncommon variant of haemoglobin was first observed in Hiroshima in Japan. It is caused by a genetic substitution mutation in the triplet for residue 143 in the beta chain, resulting in the amino acid histidine being replaced by aspartate. This defect results in an altered structure in the beta chain of the globin molecules within haemoglobin causing it to have an increased affinity for oxygen. This reduces the Bohr effect as the 'Hiroshima haemoglobin' releases less oxygen than that of the normal variant of haemoglobin when carbon dioxide increases. This results in tissues receiving less oxygen and the individual with Hiroshima haemoglobin experiencing the same side effects as those experienced by a person with mild to moderate anaemia. As a consequence the body produces more red blood cells in an attempt to compensate for the lower oxygen supply.

Would the oxygen dissociation curve for Hiroshima haemoglobin be to the right or left of the dissociation curve for normal adult haemoglobin? Explain your answer.

Synoptic link

You will need to remember how molecules cross cell membranes, which you learnt about in Chapter 1, Cells and microscopy.

You learnt about the role of oxygen in aerobic respiration, in Topic 16.1, Glycolysis.

Summary questions

1   a   State three features that both myoglobin and haemoglobin possess, which show that they are globular proteins.   (4 marks)
    b   State two features that differ in the quaternary structure of haemoglobin and myoglobin.   (2 marks)
    c   Explain why carbonic anhydrase, haemoglobin, and myoglobin can all be said to have a quaternary protein structure.   (2 marks)

2   The partial pressure of oxygen in fetal blood leaving the placenta is approximately 3 kPa. Using Figure 3 estimate the difference in oxygen saturation between fetal and adult haemoglobin at this partial pressure.

3   Explain why the S-shaped dissociation curve is physiologically important for humans.   (4 marks)

17.6 Skeletal muscle structure

Specification reference: 4.1.2

Skeletal muscle

In previous topics you learnt about athletic performance. Athletic performance relies on the action of muscles. Humans have three types of muscle:

- Cardiac muscle – this is only found in the wall of the heart and is myogenic.

- Smooth muscle – this is found in the walls of blood vessels and in the walls of the digestive tract, ureter, urethra, bladder, and uterus. It is described as involuntary muscle and neurogenic as it is controlled by the autonomic nervous system.

- Skeletal muscle – this is found attached to the skeleton by tendons and is used to generate movement. It is described as voluntary and neurogenic as it is controlled by the motor cortex in the brain.

Skeletal muscle is the most abundant tissue in humans. It makes up ~23% of the body mass in females and ~40% in males.

Skeletal muscle structure

Individual muscles are made up of hundreds of cylindrical muscle fibres. These cells have a diameter of about 50 µm but can vary in length from a few millimetres to many centimetres. Each muscle fibre is *multinucleate* and is made up from numerous parallel myofibrils, which appear striated (stripy) when viewed under an electron microscope. The striations occur due to the presence of overlapping strands of a contractile protein called **myosin** and strands of a smaller protein called **actin**. Each repeating unit of cross striations is called a sarcomere and is usually 2.5–3.0 µm in length. Each sarcomere is a

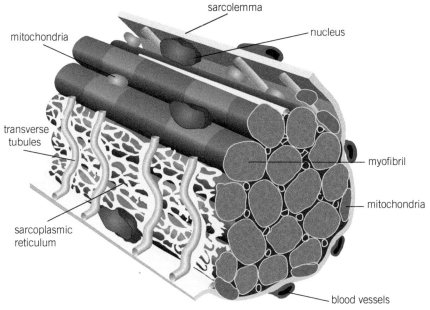

▲ Figure 1 *Histology of a muscle fibre, approx ×10 000 magnification*

contractile unit. Between the myofibrils is the sarcoplasm containing a system of membranes – the sarcoplasmic reticulum.

The myofibril has alternating anisotropic and isotropic bands:

- Anisotropic band (A band): made up of actin and myosin filaments. It is a dark region in the sarcomere where the actin and myosin overlap. In the centre of the A band in a relaxed muscle is a lighter region called the H zone, which only consists of myosin filaments.
- Isotropic band (I band): made up only of actin filaments. It possesses a central line called the Z line. The Z lines mark the end of one sarcomere and the start of the next sarcomere.

Each myofibril is surrounded by a network of small tubes containing calcium, which branch from the sarcoplasmic reticulum, and which are vital for muscle contraction.

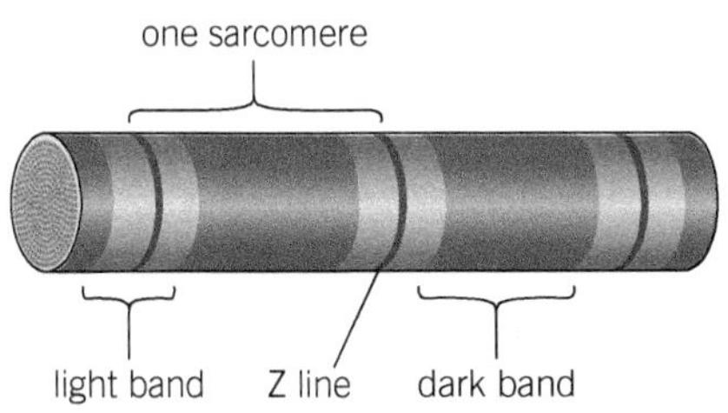

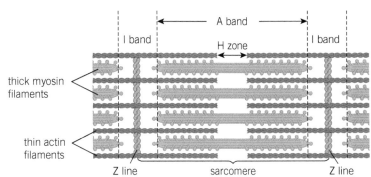

▲ **Figure 2** *Ultrastructure of a muscle fibre*

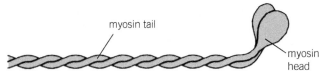

▲ **Figure 3** *Ultrastructure of a sarcomere*

Proteins within muscle fibres

Actin

Actin is a globular protein that forms thin filaments, typically 5 nm in diameter and 2.0 µm in length. Actin molecules can link together to form long chains. Two chains lie side by side and are twisted around each other, anchored by the Z lines. Actin is also associated with two accessory proteins:

- **tropomyosin** – a long, thin molecule, which lies in the groove between the two actin chains
- **troponin** – a globular protein that binds to actin at regular intervals.

▲ **Figure 4** *A single myosin molecule*

Myosin

Myosin is a fibrous protein, which forms thick filaments that lie side by side to each other. Each myosin molecule is ~10 nm in diameter and ~2.5 µm long. The molecule comprises a long rod-shaped fibre with a head at the end. Half of the molecules of myosin have their heads at one end, the other half have the heads at the other end.

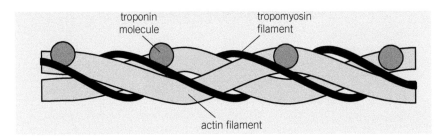

▲ **Figure 5** *Arrangement of actin, troponin, and tropomyosin*

Study tip

Remember: d**A**rk = anisotropic band, **I**ight – isotropic band.

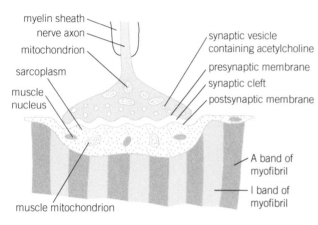

myelin sheath
nerve axon
mitochondrion
sarcoplasm
muscle nucleus
muscle mitochondrion

synaptic vesicle containing acetylcholine
presynaptic membrane
synaptic cleft
postsynaptic membrane
A band of myofibril
I band of myofibril

▲ **Figure 6** *Neuromuscular junction*

Neuromuscular junction

Skeletal muscle fibres will not contract unless they receive an impulse from a somatic motor neurone. The place where the motor neurone meets the muscle cell is called a neuromuscular junction. When a nerve impulse is received at the neuromuscular junction it triggers the release of a neurotransmitter (acetylcholine), which diffuses across the gap and binds with receptors on the sarcolemma at the motor end-plate. This causes depolarisation of the sarcolemma and results in skeletal muscle cells contracting.

Synoptic link

You will learn about the roles and actions of the somatic and autonomic nervous systems and about nerve impulses in Chapter 26, The nervous system.

Summary questions

1   Using the information in Figure 7, calculate the width of
    a) an A band, and b) a sarcomere, to the nearest whole µm.   (*4 marks*)

▲ **Figure 7** *Transmission electron micrograph of skeletal muscle, ×5000 magnification*

2   Using different size circles to represent actin and myosin, and a key, sketch a diagram to show the transverse (cross section) arrangement of these proteins in a sarcomere in:
    a   the H zone
    b   the overlap region of the A zone.   (*4 marks*)

3   Suggest and explain what differences in cell ultrastructure occur in a myofibril compared to an epithelial cell.   (*3 marks*)

17.7 Sliding filament theory – muscle contraction

Specification reference: 4.1.2

Skeletal muscle contraction

In the previous topic you learnt about the structure of skeletal muscles. In this topic you will learn how muscle contraction is brought about. The contraction of skeletal muscle fibres occurs as a result of actin and myosin filaments sliding over each other – the sliding filament theory. As the muscle fibre contracts, the myosin filaments pull the actin filaments (and the Z lines to which they are attached) towards the centre of the sarcomere, causing the sarcomere to shorten. As each sarcomere shortens, the overall muscle length of the muscle fibre reduces.

Sequence of events in muscle contraction

1 A nerve impulse reaches the neuromuscular junction.

2 Synaptic vesicles in the motor neurone fuse with the end-plate membrane and release the neurotransmitter acetylcholine.

3 The neurotransmitter binds to the sarcolemma (muscle membrane) and depolarises it.

4 The neurotransmitter is hydrolysed by enzymes.

5 An action potential is generated that passes down the T-tubules (extensions of the sarcolemma).

6 The action potential passes from the T-tubules to the sarcoplasmic reticulum, causing the release of calcium ions into the sarcoplasm.

7 Calcium ions bind to troponin, causing it to change shape.

8 The troponin moves the tropomyosin so that it no longer blocks the myosin binding site on the actin filament.

9 Myosin heads are normally attached to ADP. The calcium ions activate an enzyme called myosin kinase, which releases the ADP so the myosin head is free to bind. The myosin heads attach to the actin filament, forming **cross-bridges**.

10 The myosin head tilts, pulling the actin filaments over the stationary myosin filaments towards the centre of the sarcomere. This is called the **power stroke**.

11 ATP attaches to the myosin head causing it to detach from the actin filament.

12 Hydrolysis of ATP (to ADP and Pi) releases energy to move the myosin head outwards from the centre of the sarcomere, re-setting the myosin head. The myosin head acts as ATPase.

13 The myosin heads reattach, forming new cross-bridges to the actin at binding sites adjacent to the ones previously occupied. Over 100 cycles of attachment can occur per second.

14 Once the nerve impulse stops, calcium ions are reabsorbed.

15 Troponin reverts back to its original shape.

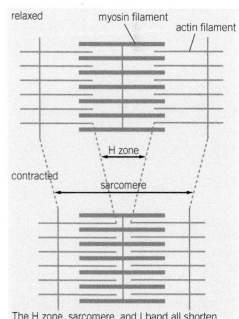

The H zone, sarcomere, and I band all shorten. The A band is unaltered

▲ **Figure 1** *Changes in a sarcomere during muscle contraction*

16 Tropomyosin blocks the attachment of the myosin head to the actin filament to prevent further muscle contraction.

17 Calcium ions are actively pumped into the cisternae of the sarcoplasmic reticulum and T-tubule system for use in further muscle contraction.

To reset the sarcomere, once the fibre is relaxed, the filaments will be pulled back to their original position by the action of an antagonistic muscle.

(a) The head of the myosin molecule is 'cocked', ready to attach to the actin filament

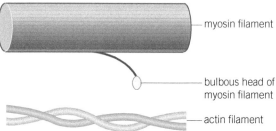

myosin filament

bulbous head of myosin filament

actin filament

(d) The myosin head detaches from the actin filament as a result of an ATP molecule fixing to the myosin head

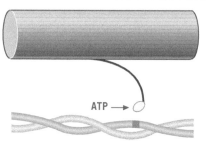

ATP →

(b) Myosin head attaches to a monomer unit on the actin molecule

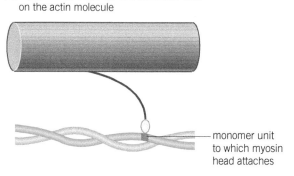

monomer unit to which myosin head attaches

(e) The ATP provides the energy to cause the myosin head to be cocked again. Hydrolysis of the ATP gives rise to ADP + Pi

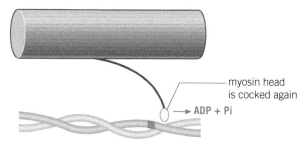

myosin head is cocked again

→ ADP + Pi

(c) The myosin changes position in order to attain a lower energy state. In doing so it slides the actin filament past the stationary myosin filament

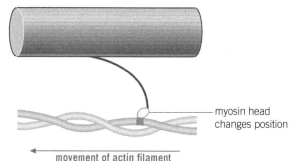

myosin head changes position

← movement of actin filament

(f) The cocked head of the myosin filament reattaches further along the actin filament and the cycle of events is repeated

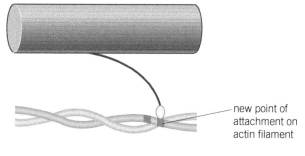

new point of attachment on actin filament

▲ **Figure 2** *Sliding filament theory*

➕ Studying muscle contraction using fluorescence

Many different research experiments have been carried out using fluorescence. These have led to a greater understanding of muscle histology and the mechanism of muscle contraction.

Fluorescence is the emission of light by a substance that has absorbed light or other electromagnetic radiation. It is a form of luminescence and, in most cases, the emitted light has a longer wavelength, and therefore lower energy, than the absorbed radiation. A fluorescence microscope can be used to detect fluorescence and generate an image.

In research experiments looking at muscle contraction, a fluorescently labelled derivative of the calcium binding subunit of troponin, TnC, was injected into striated muscle fibres taken from barnacles (*Balanus nubilus*). Studies have injected the large single muscle fibres with aequorin and shown that when the muscles were stimulated to contract there was a high bioluminescence when Ca^{2+} ions were released from the sarcoplasmic reticulum. The fluorescence signals increased in magnitude with increasing stimulus intensities. This research suggests that Ca^{2+} attaches rapidly to the troponin subunit but is released relatively slowly.

In a different experiment researchers used fluorescent dye to label myosin cross-bridges in rabbit muscle fibres, without impairing their function. Fluorescence polarisation was used to study cross-bridge orientation in rigor, relaxation, and contraction by monitoring the length of the sarcomeres. This study showed that torque (rotation) was generated at the actin–myosin boundary.

Fluorescent dyes have also been attached to myosin molecules to demonstrate how the myosin moves along the actin filaments. Some studies have researched this

mechanism using cells of the alga *Nitella axillaris*, and proved that the myosin–actin interaction was dependent on the availability of ATP.

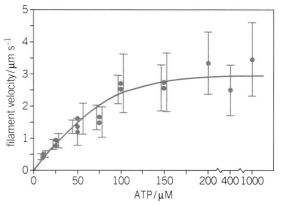

▲ **Figure 3** *Graph showing the velocity of myosin molecules with increasing ATP availability*

1 What do the vertical lines on each data point represent and how may they have been constructed? *(2 marks)*
2 What conclusion can be drawn about the reliability of the data as the ATP concentration increases? *(1 mark)*
3 Suggest why the velocity of the myosin molecules reaches a plateau. *(1 mark)*

Summary questions

1 a Describe the changes that occur to the length of
 i) actin filaments   ii) myosin filaments   iii) I band   iv) A band
 v) H zone as a result of muscle contraction. *(4 marks)*
 b Describe what happens to the Z lines as a result of muscle contraction. *(1 mark)*

2 Muscles can contract to exert a force but they are not able to pull themselves back to their original, relaxed position.
 a Explain why. *(3 marks)*
 b Suggest how the muscles return to their relaxed position. *(2 marks)*

3 After death, chemical changes can occur in the muscles that cause the corpse to become stiff and difficult to move or manipulate. This is called rigor mortis. Suggest why a body stiffens in rigor mortis. *(4 marks)*

Study tip

Remember, no actual shortening of any protein filament occurs – they slide over the top of each other to cause the reduction in sarcomere length.

In a contracting muscle, many muscle fibres will be contracting simultaneously, even when a whole muscle isn't contracting there will be some fibres contracting at any one time creating 'muscle tone'.

It is possible to investigate the effect of different concentrations of ATP on the change in length of muscle fibres by using excised strips of muscle tissue.

Practice questions

1 Which of the following statements is/are true of the role of calcium ions in skeletal muscle contraction?

Statement 1: Calcium ions are released into the sarcoplasm.

Statement 2: Calcium ions bind to tropomyosin.

Statement 3: Calcium ions inactivate an enzyme called myosin kinase.

  A   1, 2 and 3   B   Only 1 and 2

  C   Only 2 and 3   D   Only 1   *(1 mark)*

2 A person's resting cardiac output is $288\,000 \text{ cm}^3 \text{ h}^{-1}$. During intense exercise their cardiac output increases to $37 \text{ dm}^3 \text{ min}^{-1}$.

What is the percentage increase in cardiac output?

  A   471   B   571   C   671   D   771

                                       (1 mark)

3 Which of the following statements is/are true of the changes in a sarcomere during muscle contraction?

Statement 1: Myosin filaments shorten.

Statement 2: The H zone shortens.

Statement 3: I band shortens.

  A   1, 2 and 3   B   Only 1 and 2

  C   Only 2 and 3   D   Only 1   *(1 mark)*

4 The Bohr effect results in which of the following substances entering red blood cells?

  A   Chloride ions

  B   Oxygen

  C   Hydrogencarbonate ions

  D   Haemoglobin   *(1 mark)*

5 Which of the following statements is/are true of VO_2max?

Statement 1: It can be measured using a graded exercise test.

Statement 2: It is reached when oxygen consumption remains steady despite an increase in workload.

Statement 3: It can be measured in mol dm^{-3}.

  A   1, 2 and 3   B   Only 1 and 2

  C   Only 2 and 3   D   Only 1   *(1 mark)*

6 Athletes such as Chris Hoy, the 2008 Olympic cycling champion, increase their muscle mass and strength through specialised training regimes.

a The photomicrograph shown is of a sarcomere from a skeletal muscle fibre, approx ×5 000 magnification.

Four features have been labelled C, D, E, and F.

State one label letter that represents:

(i) an area containing only actin filaments   *(1 mark)*

(ii) an area containing both actin and myosin filaments   *(1 mark)*

b Outline the role played by calcium ions in the contraction of the sarcomere. *(5 marks)*

c Training increases the circumference of the biceps muscles.

A 10 week investigation was carried out to assess the effect of a **dietary supplement** on the increase in the circumference of the biceps muscles.

- Two groups of 10 men were chosen, groups G and H.
- Group G was given placebo tablets each day.
- Group H was given the dietary supplement in tablet form each day.
- Both groups had the circumferences of their biceps measured before and after the investigation.
- During the investigation a standard set of arm exercises was carried out by both groups each day.

The graph summarises the results of the investigation.

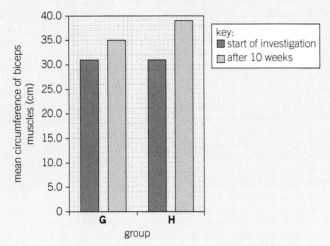

(i) Calculate the percentage increase in the mean circumference of the biceps muscles in group H. Show your working. (*2 marks*)

(ii) Suggest two factors, other than gender, that needed to be taken into account when designing this investigation. (*2 marks*)

(iii) Suggest one active ingredient of the dietary supplement and describe how it may have produced the results shown for group H. (*2 marks*)

OCR F224 2010

7 Myoglobin is a respiratory pigment that can combine with oxygen. Myoglobin is found in muscle cells

The diagram shows a molecule of myoglobin.

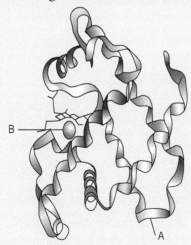

a Name structures A and B (*2 marks*)

b The graph shows oxygen dissociation curves for both myoglobin and haemoglobin.

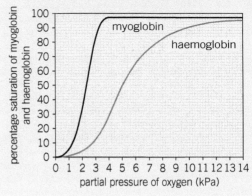

(i) Calculate the decrease in percentage saturation of both myoglobin and haemoglobin between 4 kPa and 2 kPa partial pressure of oxygen. (*1 mark*)

(ii) State the location in the human body where the percentage saturation of haemoglobin in the blood will be almost 100%. (*1 mark*)

(iii) Describe and explain the significance of the difference in affinity for oxygen between myoglobin and haemoglobin. (*3 marks*)

c Iron is an important element in the structure of both myoglobin and haemoglobin.

A rare disease called *haemochromatosis* (HC) results in there being an overload of iron in the blood. This build up of iron is toxic to the body and may cause liver, heart, or pancreatic disease.

- One form of HC is hereditary.

- Intake of dietary iron does not differ significantly between men and women.

- In the 16 to 45 age range, men show symptoms of HC earlier than women.

(i) Suggest why men show symptoms of hereditary HC at an earlier age than women. (*2 marks*)

(ii) A man who is known to have HC decided not to risk having children.

State two methods of contraception that he could use. (*2 marks*)

OCR F224 2010

18 FERTILITY AND ASSISTED REPRODUCTION

18.1 Human reproductive systems

Specification reference: 4.2.1

Synoptic link

This chapter builds on what you learnt about meiosis and fetal development in Chapter 9, Meiosis, growth, and development.

In this chapter you will learn about the formation of gametes, the menstrual cycle, and fertility. Here, you will examine the organs responsible for the production of gametes and for reproduction.

The human female reproductive system

Figure 1 shows the key parts of the female reproductive system, and Table 1 outlines their roles.

▼ **Table 1** Roles within the human female reproductive system

Organ	Function
cervix	the name is derived from the latin for neck – it provides a narrow opening into the uterus, protecting the fetus during pregnancy
ovary	produces the female gametes (oocytes, or eggs, or ova) in follicles
oviduct	this is where eggs are fertilised by sperm
uterus	this is where the embryo develops
vagina	stimulates the penis to ejaculate, provides a birth canal
vulva	protects the internal parts of the female reproductive system

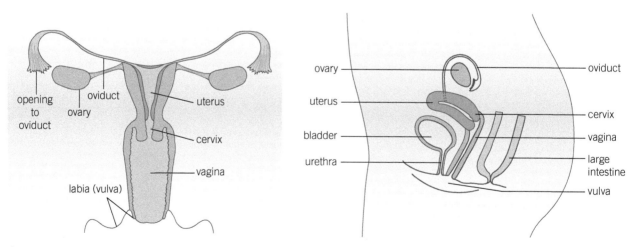

▲ **Figure 1** The human female reproductive system, front and side views

The human male reproductive system

The principal parts of the male reproductive system are illustrated in Figure 2. Their roles are outlined in Table 2.

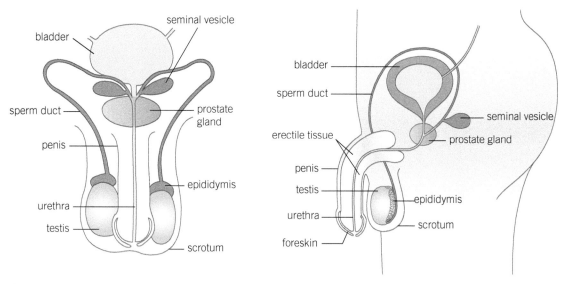

▲ **Figure 2** *The human male reproductive system, front and side views*

▼ **Table 2** *Roles within the human male reproductive system*

Organ	Function
epididymis	stores sperm prior to ejaculation
penis	penetrates the vagina and releases sperm
prostate gland	secretes an alkaline fluid to counteract the acidity of the vagina
scrotum	holds testes, maintaining a temperature 2 °C below normal body temperature
seminal vesicle	secretes a fluid containing proteins and fructose to nourish the sperm
sperm duct	transfers sperm during ejaculation
testis	produces sperm and testosterone
urethra	transfers semen during ejaculation and urine during urination

Examining the histology of reproductive systems

Histology is the study of the microscopic anatomy of cells and tissues. You should familiarise yourself with the histology of the ovary and the testis, and be able to identify the main features of each.

The histology of these organs can be presented in a few ways, such as in diagrams, photomicrographs, and electron micrographs. You may have the opportunity to view ovarian and testicular tissue under a light microscope, using the microscopy skills that you developed earlier in the A level course.

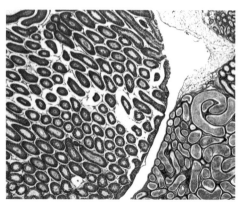

▲ **Figure 3** *Light micrograph of a testis (from a rat). The seminiferous tubules (dark pink) are the site of sperm production. The epididymis (light pink) is shown in the bottom right corner, ×2.5 magnification*

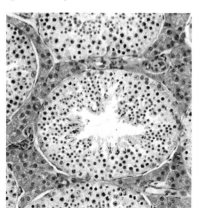

▲ **Figure 4** *Light micrograph of a seminiferous tubule in a testis, approximately ×100 magnification*

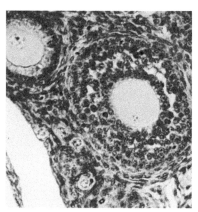

▲ **Figure 5** *Light micrograph of ovary tissue. An oocyte is present in the maturing follicle, ×50 magnification*

1   Produce a simplified, labelled diagram of Figure 3 to show the positions of the testis, one seminiferous tubule, and the epididymis.
2   Explain the appearance of the epididymis in Figure 3.

Synoptic link

You learnt about light microscopy in Topic 1.1, Microscopy – the light microscope.

Summary questions

1   In which part of the human reproductive systems:
    a   are ova produced?
    b   is sperm stored prior to ejaculation?
    c   does embryo implantation occur?
    d   are both urine and semen capable of being transferred?   (*4 marks*)

2   Cervical mucus normally has a pH of between 3 and 4. The pH rises just prior to ovulation and falls again shortly afterwards. Suggest
    a   an advantage of cervical mucus having a low pH (*2 marks*)
    b   an advantage of cervical mucus having a higher pH at ovulation.   (*2 marks*)

3   Ejaculated semen contains several different components. Describe the sources and explain the roles of each component.   (*9 marks*)

You looked at the sites of gamete production in the previous topic. Here you will examine in more detail how gametes form. The production of gametes is known as **gametogenesis** and involves meiotic cell divisions. You will also investigate the roles of hormones in gametogenesis.

Oogenesis

Oogenesis is the type of gametogenesis that occurs in females and results in the production of ova (eggs). It involves an initial multiplication phase, in which a germ cell divides by mitosis to form cells called oogonia. These cells grow in size to produce primary oocytes. At this point meiosis occurs – the primary oocytes (diploid chromosome number) divide in meiosis I to form secondary oocytes (haploid chromosome number). This meiotic division is initiated before the female is born, but pauses in prophase I for many years. Meiosis I is then reactivated and completed for a few primary oocytes in each menstrual cycle.

A second meiotic division (meiosis II) is initiated as soon as meiosis I is complete but pauses in metaphase II and is only completed if fertilisation takes place.

Structure and function of secondary oocytes

Secondary oocytes are large structures in the ovary that can be seen clearly under a light microscope. They arc released from follicles in the ovaries during ovulation.

Learning outcomes

Demonstrate knowledge, understanding, and application of:

→ the process of gametogenesis

→ the structure of secondary oocytes and sperm in relation to their functions

→ the role of hormones in gametogenesis.

Synoptic link

You learnt about meiotic cell divisions in Topic 9.1, Meiosis.

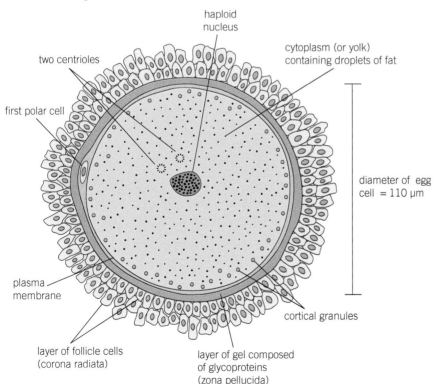

▲ **Figure 1** *The structure of a secondary oocyte*

Labels: haploid nucleus; two centrioles; first polar cell; cytoplasm (or yolk) containing droplets of fat; diameter of egg cell = 110 µm; plasma membrane; layer of follicle cells (corona radiata); layer of gel composed of glycoproteins (zona pellucida); cortical granules

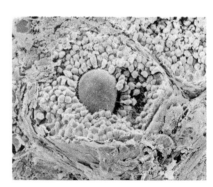

▲ **Figure 2** *An electron micrograph of ovary tissue. A secondary oocyte (pink) is present at the centre of a follicle (blue cells), approximately ×350 magnification*

Secondary oocytes are surrounded by cells from the ovarian follicle (the corona radiata), which protect the oocyte in the oviduct. The zona pellucida lies inside the corona radiata – it is composed of glycoproteins, which offer further protection and play a crucial role in fertilisation by sperm. You will learn about fertilisation in more detail in Topic 18.4, Fertility.

Spermatogenesis

Gametogenesis that produces sperm is known as *spermatogenesis*. Like oogenesis, this process begins with a multiplication phase in which a germ cell divides by mitosis. This produces spermatogonia cells, which grow into primary spermatocytes.

Primary spermatocytes divide by meiosis to produce mature spermatozoa (sperm cells). Unlike in oogenesis, the second meiotic division is initiated and completed directly after the first. No polar bodies are formed in spermatogenesis.

Structure and function of sperm

Synoptic link

A sperm cell moves using a flagellum. You learnt about the differences between prokaryotic and eukaryotic flagella in Topic 1.6, Cell ultrastructure.

Sperm cells are small in comparison with secondary oocytes and they are more difficult to observe under a light microscope. Sperm cells comprise three distinct sections: the head, midpiece, and tail.

Head – this contains a haploid nucleus, which fuses with the haploid nucleus of the secondary oocyte after fertilisation. The head is capped by an acrosome, which is a membrane-bound structure containing the hydrolytic enzymes that are required to digest the zona pellucida of the secondary oocyte.

Midpiece – this contains mitochondria. They provide the energy needed to move the tail.

Tail – this contains contractile filaments, made of microtubules in a 9 + 2 arrangement, which create whipping movements to enable sperm to swim.

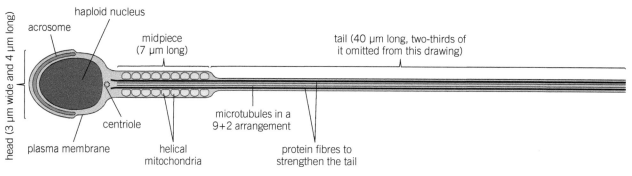

▲ **Figure 3** *The structure of a sperm cell*

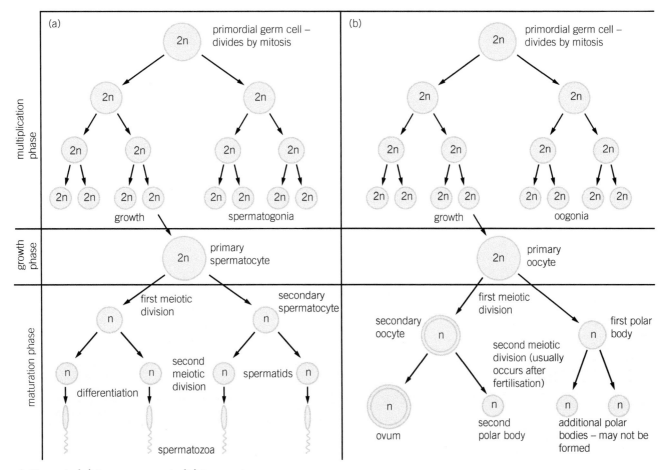

▲ **Figure 4** *(a) Spermatogenesis (b) Oogenesis*

Hormonal control of gametogenesis

A range of hormones control the development of gametes in both sexes. The initiation of gametogenesis is the same in both cases. The hypothalamus secretes gonadotrophin releasing hormone (GnRH). This hormone stimulates the anterior pituitary gland to secrete two more hormones: **follicle stimulating hormone** (FSH), and **luteinising hormone** (LH).

At this point, gametogenesis takes different courses in the two sexes. FSH and LH act on the ovaries and the testes, producing different effects (Table 1, overleaf) in the two reproductive systems.

Synoptic link

Gametogenesis involves both mitosis, which you learnt about in Topic 8.2 and meiosis, which was covered in Topic 9.1, Meiosis.

The hormonal control of gametogenesis relies on negative feedback, which you will learn about in Topic 29.1, Homeostasis – the key principles.

▼ **Table 1** *The roles of hormones in gametogenesis*

Hormone	Male	Female
LH	binds to cells in the testes (Leydig cells), triggering the release of testosterone	binds to follicle cells in the ovaries, causing them to mature and release oestrogen. A surge of LH causes a secondary oocyte to be released from the mature (Graafian) follicle
FSH	binds to cells in the testes (Sertoli cells) to make them more receptive to testosterone	binds to follicle cells in the ovaries, causing them to mature and release oestrogen
progesterone	-	inhibits LH and FSH
oestrogen	-	rising levels inhibit FSH and cause a surge in LH
testosterone	binds to Sertoli cells, triggering spermatogenesis. inhibits LH	-
inhibin	inhibits FSH	-

Summary questions

1 What type of cell division results in the formation of
   a oogonia?
   b sperm cells?
   c secondary oocytes?
   d primary spermatocytes? *(4 marks)*

2 Which structural features of a sperm cell enable it to
   a travel through the female reproductive system to reach a secondary oocyte?
   b penetrate into a secondary oocyte? *(4 marks)*

3 Outline three differences and three similarities between spermatogenesis and oogenesis. *(6 marks)*

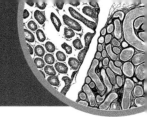

In the previous two topics, you have investigated the structure and function of human reproductive systems, and how these systems produce gametes. Women, unlike men, release their gametes in a well-defined cycle. This is called the **menstrual cycle** and is the focus of this topic.

The cycle

The menstrual cycle can be considered as two separate, yet synchronised, cycles: the ovarian cycle (which controls the release of secondary oocytes,) and the uterine cycle (which controls the development of the uterus lining). The same hormones regulate both, and together they are known as the menstrual cycle.

The menstrual cycle can be considered as four phases:

- proliferative phase – the lining of the uterus regenerates
- ovulation phase – the secondary oocyte is released into the oviduct from a mature follicle in the ovaries
- secretory phase – the uterus lining secretes nutrients to prepare for embryo implantation
- menstrual phase – the lining of the uterus is shed if implantation does not occur.

The cycle begins during puberty and continues until a woman reaches the menopause at approximately 50 years old. When successful fertilisation and embryo implantation occur, the cycle stops during pregnancy.

Hormonal control of the menstrual cycle

The menstrual cycle is controlled by four hormones (follicle stimulating hormone (FSH), luteinising hormone (LH), **oestrogen**, and **progesterone**), all of which you encountered in Topic 18.2 when considering oogenesis. These hormones interact and work in a set order. The process can summarised as follows:

1 The pituitary gland secretes FSH.

2 FSH is transported in the blood to the ovary, where it stimulates a follicle to develop.

3 The mature follicle releases oestrogen.

4 Oestrogen thickens the uterus lining, inhibits further FSH secretion, and stimulates the sudden secretion of LH from the pituitary gland, on approximately day 11 of the cycle.

5 LH causes ovulation (the release of a secondary oocyte from a follicle into the oviduct).

6 LH stimulates the follicle to develop into a structure called a corpus luteum, which secretes small amounts of oestrogen and large amounts of progesterone.

7 Progesterone inhibits FSH and LH, but stimulates the growth of blood vessels in the uterus lining.

8 With LH now inhibited, the corpus luteum degenerates and produces less progesterone, which in turn enables more FSH to be released.

9 If pregnancy does not occur, the fall in oestrogen and progesterone cause the outer layer of the uterus lining to be shed and lost from the body. This is menstruation. The cycle then begins again.

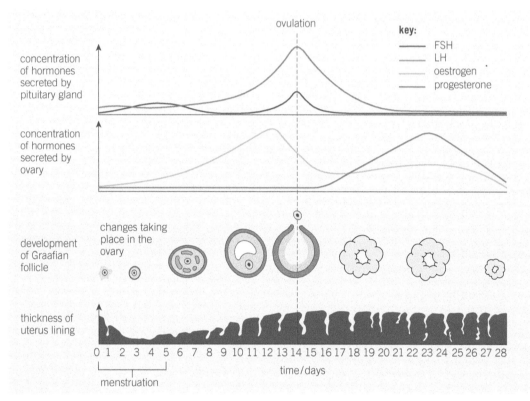

▲ **Figure 1** *The human menstrual cycle*

Synoptic link

You will learn about the principles of negative and positive feedback in Topic 29.1, Homeostasis – the key principles. The hormonal control of the menstrual cycle contains examples of both, as illustrated in Figure 2. For example, LH results in the secretion of progesterone, which inhibits further LH secretion. This is negative feedback.

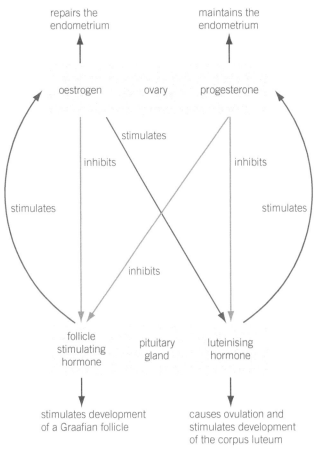

▲ **Figure 2** *The interactions of hormones in the menstrual cycle*

➕ Do other species have menstrual cycles?

Females of most mammalian species have an oestrous cycle rather than a menstrual cycle. In an oestrus cycle, the uterus lining is reabsorbed rather than being lost from the body, which happens in the human menstrual cycle.

Another significant difference is that females of other species are only sexually receptive during the time of ovulation (known as oestrous). During the rest of the oestrus cycle, females avoid sexual activity. The sexual activity of human females is not limited to the time of ovulation. They have the potential to be sexually receptive throughout the menstrual cycle.

Furthermore, human females have evolved concealed ovulation. In other words, they do not signal when they are ovulating, as other species do. This means that human males cannot be certain when sexual intercourse will result in fertilisation.

1   Suggest the types of signals that females of other species may use to signal their fertility during oestrous.
2   Suggest the possible benefits to female humans of concealing the timing of their ovulation.

Study tip

Remember that follicles and oocytes are present in the ovary prior to birth. FSH causes follicles to *develop*. It does not *produce* follicles.

Synoptic link

In Topic 10.3, Adaptation, you learnt about various adaptations that humans have evolved. Concealed ovulation and the menstrual cycle are other examples of human adaptations that have evolved from ancestral primate species with oestrous cycles.

Summary questions

1   Describe the role of oestrogen in the menstrual cycle.
    (3 marks)

2   Why do some cells respond to a hormone such as LH, whereas others do not?   *(2 marks)*

3   Using information from Topic 29.1 if necessary, describe the role of negative feedback in the menstrual cycle.   *(4 marks)*

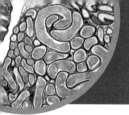

18.4 Fertility

Specification reference: 4.2.1

This chapter concerns reproduction. So far, you have studied the reproductive organs, the production of sperm and ova, and the physiological preparation for pregnancy in the menstrual cycle. The ultimate purpose of all that you have learnt in the preceding topics is now the focus of this topic: sexual reproduction. Here, you will look at the process of fertilisation and pregnancy. You will also learn about biological and ethical issues surrounding the numerous forms of infertility and their treatments.

Fertilisation

During copulation semen is released (or ejaculated) into the vagina of the female. Semen comprises fluids from the seminal vesicles (containing mucus and proteins to aid swimming, and fructose for energy), and from the prostate gland (an alkaline fluid to neutralise the acidic vagina). Approximately 200 million sperm are released from the epididymis during ejaculation and are present in the semen. Only one sperm, however, will fertilise the secondary oocyte (an immature ovum).

The sperm swim through the cervix, through the uterus, and up into the oviducts. If a secondary oocyte is present, fertilisation can occur. The mechanism by which a sperm cell fertilises a secondary oocyte is quite complex. The principal stages are as follows:

1 The acrosome (the large membrane-bound sac of enzymes in the head of a sperm cell) binds to the zona pellucida (a coat of glycoproteins) on the secondary oocyte.

2 The acrosome releases hydrolytic enzymes that digest the zona pellucida.

3 Proteins on the sperm's head bind to the plasma membrane of the secondary oocyte.

4 The sperm and secondary oocyte membranes fuse and the sperm nucleus enters the oocyte. This is fertilisation. The male and female nuclei then fuse, restoring the diploid number of chromosomes. The cell is now called a zygote.

5 The entry of a sperm activates cortical granules in the ovum. These granules release enzymes that digest the sperm-binding proteins on the ovum membrane. They also produce a hardening of the zona pellucida. This prevents other sperm from entering the ovum.

The importance of fertilisation

In Topic 9.1, you learnt about the importance of meiosis and subsequent fertilisation. The fusion of two haploid gametes in fertilisation restores the diploid number of chromosomes, which in humans is 46. Fertilisation is a random event, and only one sperm of the many millions that are released at ejaculation will fertilise an ovum. The DNA of each sperm is unique due to the occurrence of

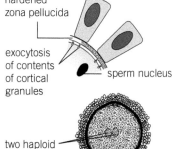

sperm try to push through the layers of follicle cells around the egg

follicle cell

zona pellucida

plasma membrane of egg

acrosome

tail and mitochondria usually remain outside

cortical granules

hardened zona pellucida

exocytosis of contents of cortical granules

sperm nucleus

two haploid nuclei from the sperm and the egg

▲ **Figure 1** *Stages in fertilisation*

crossing over and independent assortment in meiosis. Fertilisation adds another layer of genetic variation to the variation created during meiosis.

Pregnancy testing

Successful fertilisation produces a diploid zygote cell, which undergoes rapid mitosis to produce a bundle of cells called the morula. In the oviduct, the morula develops into a structure called the *blastocyst*, which moves into the uterus. If the blastocyst implants in the endometrium a pregnancy ensues. Part of the blastocyst begins to form a structure that will become the placenta, and this secretes a hormone called *human chorionic gonadotropin (hCG)*. Other regions within the blastocyst develop into the embryo.

Modern pregnancy testing kits detect hCG in a female's urine. The testing kits, in the form of a dipstick, contain monoclonal antibodies (mAbs) that are specific to hCG. hCG is produced from the earliest stages of pregnancy, so if it is present in a urine sample, it will bind to the mAbs. Figure 2 shows how a positive pregnancy test is confirmed by the appearance of two blue lines.

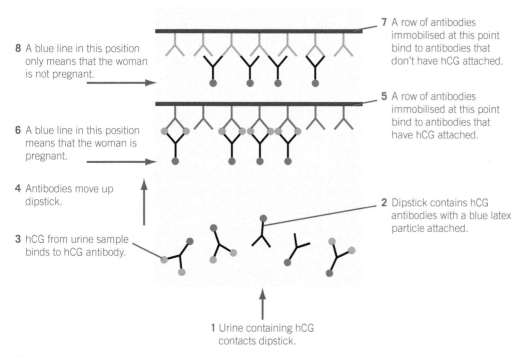

7 A row of antibodies immobilised at this point bind to antibodies that don't have hCG attached.

8 A blue line in this position only means that the woman is not pregnant.

5 A row of antibodies immobilised at this point bind to antibodies that have hCG attached.

6 A blue line in this position means that the woman is pregnant.

4 Antibodies move up dipstick.

2 Dipstick contains hCG antibodies with a blue latex particle attached.

3 hCG from urine sample binds to hCG antibody.

1 Urine containing hCG contacts dipstick.

▲ **Figure 2** *A pregnancy test kit*

Infertility

Approximately one in seven couples in the UK has problems conceiving.

Infertility has many potential causes, some connected to the female reproductive system and some to the male system. Table 1 outlines some of these possible causes of infertility.

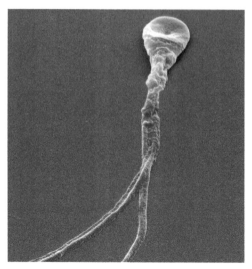

▲ **Figure 3** *Coloured scanning electron micrograph of an abnormal (double-tailed) sperm, ×2500 magnification*

▼ **Table 1** *Causes of infertility*

Males	Females
Abnormal sperm (e.g., lack of tails), due to failure to develop properly in the testes	Ovulatory disorders such as a failure to ovulate (e.g., because of insufficient hormone levels)
Production of antibodies that destroy the sperm in the semen	Production of antibodies that destroy the sperm in the uterus
Blockages in the sperm ducts A low sperm count	Blockages in the oviducts (e.g., from bacterial infection or from endometriosis, in which the uterine lining develops outside the uterus)

Treatments for infertility

Treatments are available for most forms of infertility. The development of technology that can assist fertilisation has raised ethical questions. For example, when a child is born from donated sperm, how much access should they be given to information about their genetic background? What time limits should be placed on embryo research into assisted fertilisation? Health considerations also exist. *In vitro* fertilisation, which is discussed below, creates a higher probability of multiple births (i.e., having twins or triplets). This carries an increased risk of stillbirth and maternal mortality.

Infertility treatments include:

● In vitro fertilisation (IVF) – oocytes are fertilised in a laboratory and the resultant embryo is transferred into the mother's uterus. Embryos from IVF treatment can be frozen and used at a later date. This is known as *frozen embryo replacement*. Sometimes, sperm struggle to fertilise the oocyte. In these cases, *intra-cytoplasmic sperm injection (ICSI)* can be used. ICSI is an IVF treatment in which a single sperm is injected directly into the oocyte, to increase the probability of fertilisation. Another modification to IVF is *gamete intra-fallopian transfer (GIFT)*. In GIFT, instead of fertilisation being carried out in a laboratory, sperm and oocytes are passed into the oviduct and allowed to fertilise naturally.

- Artificial insemination – any technique involving the introduction of semen into the female reproductive tract without sexual intercourse. Semen can be placed into the vagina or into the cervix without any special preparation. This can be semen from a male partner or semen from a donor (i.e., donor sperm insemination).

- Intrauterine insemination – sperm (rather than semen) are injected into the uterus. The sperm must be washed first or spasmodic uterine cramping can occur.

- Sometimes sperm are produced in the testes but are not ejaculated in a man's semen. It is possible to retrieve sperm from the epididymis in these cases. This is called *surgical sperm retrieval*.

- Ovulation induction – follicles in the ovaries are stimulated to develop and form viable secondary oocytes. Either drugs are taken that inhibit oestrogen, or FSH and LH treatments are provided.

Synoptic link

You looked at the idea of genetic variation being introduced during sexual reproduction in Topic 9.1, Meiosis. You have learnt here about the use of monoclonal antibodies in pregnancy testing. You discussed the structure and function of antibodies in Topic 12.2, Antibodies and immunity.

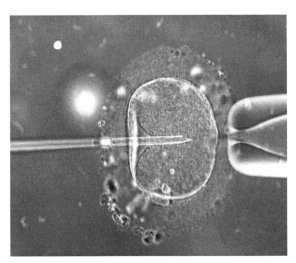

▲ **Figure 4** *Intra-cytoplasmic sperm injection*

Summary questions

1 Name the modification made to IVF treatment when sperm are unable to fertilise oocytes. *(1 mark)*

2 Explain why only one sperm can fertilise a secondary oocyte. *(3 marks)*

3 Describe and explain the role of antibodies in
   a detecting pregnancy
   b causing infertility. *(5 marks)*

Practice questions

1 Which of the following statements is/are true of luteinising hormone?

   Statement 1: Binds to Leydig cells in the testes.

   Statement 2: Binds to follicle cells in the ovaries.

   Statement 3: Is secreted from the pituitary gland.

   A   1, 2 and 3   B   Only 1 and 2

   C   Only 2 and 3   D   Only 1   (*1 mark*)

2 Which of the following statements is/are true of the seminal vesicle?

   Statement 1: Secretes fructose

   Statement 2: Transfers sperm

   Statement 3: Produces testosterone

   A   1, 2 and 3

   B   Only 1 and 2

   C   Only 2 and 3

   D   Only 1   (*1 mark*)

3 Which of the following cells is/are haploid?

   Statement 1: Oogonia

   Statement 2: First polar body

   Statement 3: Secondary oocyte

   A   1, 2 and 3

   B   Only 1 and 2

   C   Only 2 and 3

   D   Only 1   (*1 mark*)

4 Which form of infertility treatment involves both sperm and oocytes being passed into the oviduct to fertilise naturally?

   A   Artificial insemination

   B   Intrauterine insemination

   C   Gamete intra-fallopian transfer

   D   Intra-cytoplasmic sperm injection

5 a The process of spermatogenesis involves cell division by mitosis and meiosis.

   Identify the cells produced by the following types of cell division in the seminiferous tubule.

   (i)   Mitosis   (*1 mark*)

   (ii)   Meiosis   (*1 mark*)

b The diagram shows a human sperm cell.

   Name structure J.   (*1 mark*)

c The following passage about fertilisation appeared in a magazine for non-scientific readers. The terms that have been highlighted and numbered have not been written using the correct scientific terminology.

   > When the sperm reaches the **egg** (1) its head releases **chemicals** (2) which digest the **outer layer** (3). Vigorous movements of the sperm's **tail** (4) enable the head of the sperm to bind to the **egg** (1). Only one sperm can enter because a **barrier** (5) is formed to stop the other sperm. The two nuclei can now **join** (6) as fertilisation is completed.

   Write the correct scientific terms to replace the incorrect ones highlighted in the passage.   (*6 marks*)

d Most forms of contraception prevent ovulation or fertilisation from occurring. However, occasionally a woman needs to take an emergency contraceptive pill.

   Discuss two possible disadvantages of using an emergency contraceptive pill.

                            (*2 marks*)

   OCR F224 2010

6 a Most couples conceive a child within a year of trying, but some couples have difficulty conceiving. This could be due to a problem with either the male or female reproductive systems.

   IVF is a widely used treatment for infertility.

   At one IVF clinic, over 1000 treatment cycles were monitored. The number of live births was recorded as a percentage of the number of treatment cycles. The results were recorded against the age of the woman and are shown in the table.

Age of woman (years)	Percentage of live births per treatment cycle
Under 34	27.6
34 to 36	22.3
37 to 39	18.3
40 to 42	10.0
Above 42	Less than 5.0

The data in the table show that there is a decrease in the percentage of live births per treatment cycle with increasing age.

Explain this trend. *(3 marks)*

b Two other fertility treatments that use the same principles as IVF are GIFT and ICSI.

Outline how each of these treatments works. *(3 marks)*

OCR F224 2010

7 The theme of the completed crossword shown here is 'Functions of the Hormones of Human Reproduction'.

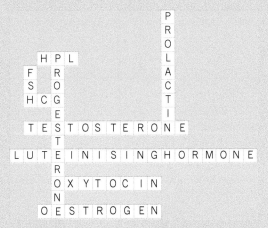

a Write a clue for each of the following crossword answers:

(i) Testosterone *(1 mark)*

(ii) Prolactin *(1 mark)*

b State which word or term in the crossword matches the following clues:

(i) Stimulates contraction of uterine muscles *(1 mark)*

(ii) Maintains uterine lining in the early stages of pregnancy *(1 mark)*

c Which two hormones can be used in the female contraceptive pill? *(2 marks)*

d State one other way contraceptive hormones can be administered to a woman other than by taking a pill.

(1 mark)

OCR F224 2010

8 A survey carried out on several IVF clinics in 2003 investigated the chance of having a multiple birth as a result of IVF treatment.

The results are summarised here.

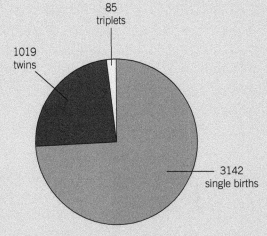

a Explain the difference between multiple pregnancy and multiple birth. *(2 marks)*

b Calculate, to the nearest whole number, the percentage of births that resulted in twins. *(2 marks)*

c During a multiple pregnancy, both mother and babies are deemed to be at a greater risk of complications.

Outline the increased health risks faced by the mother during a multiple pregnancy. *(3 marks)*

d Fertility treatment may be prescribed for some couples who have been unable to conceive.

Suggest three reasons why a man may be infertile. *(3 marks)*

e Some fertility treatments require sperm donation.

Discuss why it is becoming increasingly difficult to recruit sperm donors. *(2 marks)*

OCR F224 2010

9 Outline the structure and function of secondary oocytes and sperm cells. *(6 marks)*

19 THE EFFECTS OF AGEING ON THE REPRODUCTIVE SYSTEM

19.1 Effects of ageing on the female reproductive system

Specification reference: 4.2.2

In the previous chapter, you learnt about the biology of human reproductive systems. Here, you will look at the changes occurring in these systems as you age. This topic begins by examining the changes in hormone levels and fertility experienced by females.

How does ageing affect reproduction?

Ageing is the accumulation of physical and psychological changes that occur within a person over a period of time. Although ageing is inevitable, the deterioration of the human body (**senescence**) can be slowed by changes in lifestyle. Ageing causes many changes to the human body, such as decreases in basal metabolic rate (BMR) and cardiac output. For female reproductive capacity, the most significant changes are in hormone levels and the physiology of the ovaries.

Changes in hormone levels

Females are born with hundreds of thousands of primary oocytes in the follicles of the ovaries. Every month, starting from puberty, one of these follicles matures and ruptures, releasing a secondary oocyte. However, many of these follicles don't develop in this way. Instead, they disappear over time until there are no follicles or oocytes left. At this point various hormonal changes occur:

▲ **Figure 1** *Post-menopausal uterus tissue. The menopause is the end of the fertile period of a woman's life, caused by the cessation of oestrogen production, ×1650 magnification*

- secretion of oestrogen decreases
- secretion of FSH increases
- secretion of LH increases.

Levels of FSH and LH peak approximately 1 to 3 years after the last menstrual cycle, but the secretion of these hormones gradually falls to zero over time.

Changes in fertility

After puberty, female fertility increases and then decreases. Fertility levels usually peak between the ages of 20–25 years in humans and then decrease slowly until a sharper drop at the age of around 35 years.

The ability for an ovary to produce oocytes that can be fertilised (leading to pregnancy) is called the *ovarian reserve*. As the woman ages the number of oocytes that can be successfully fertilised decreases, this results in a fall in the woman's fertility level. This is a major factor in the inverse correlation between age and female fertility.

Ovarian follicles that are capable of further development secrete a hormone called anti-Müllerian hormone (AMH). The concentration of AMH in the bloodstream thereby indicates a woman's ovarian reserve. AMH concentration rather than a woman's age tends to be used as a measure of the potential for IVF treatment to succeed.

The menopause

The **menopause** is caused by a lack of ovarian follicles and is defined as the time the last menstrual cycle occurs. This usually occurs between the ages of 50 and 54 years, but can range from 35 to 59 years of age. Mothers and daughters tend to begin the menopause at very similar ages.

Prior to the menopause is the perimenopause. The perimenopause can occur several years before the woman's last menstrual period. At first the woman may experience more frequent periods as well as occasional missed periods. The length of each menstrual period may be longer or shorter than usual and there can be changes to the actual amount of menstrual flow. This will continue with longer times between each menstrual period until eventually they completely stop.

Once a woman has not experienced a menstrual period for one year the menopause is said to be complete. At this stage the woman is post-menopausal and she can no longer naturally become pregnant. Sometimes medical treatments, such as the removal of the ovaries or chemotherapy, cause a decrease in oestrogen (known as surgical menopause). You will learn in Topic 19.3, Effects of ageing on the male urogenital system, that men do not experience a decrease in levels of sex hormone to the same extent as women – they will not undergo a menopause.

Menopausal symptoms vary and may last five years or more. Other than changes in the regularity of periods, symptoms of menopause include:

- heart pounding or racing
- night sweats
- flushed skin
- insomnia
- hot flushes
- high heart rate.

Some women may also experience additional symptoms such as:

- a fall in libido (sex drive)
- dryness of the vagina leading to painful sexual intercourse
- irritability and anxiety
- depression
- headaches
- heart palpitations.

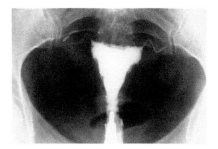

▲ **Figure 2** *X-ray of the uterus (white, centre) of a post-menopausal woman, which shows the degeneration of the uterine lining (mottled appearance)*

Surgical menopause

A hysterectomy is a surgical procedure to remove the uterus, after which the woman will no longer be able to become pregnant. A hysterectomy may be necessary for a variety of reasons, including:

- cancer of the uterus, ovaries, cervix or oviducts
- heavy periods (menorrhagia)
- long-term pelvic pain
- non-cancerous tumours (fibroids).

During a hysterectomy the ovaries can be removed which will result in the woman immediately going through the menopause, irrespective of her age (surgical menopause). In some cases one or both of the ovaries are left intact during a hysterectomy – this can lead to a 50% chance that the woman will experience the menopause within five years of her surgery. Hormone replacement therapy (HRT) is usually prescribed to women who experience a surgical menopause.

1 Suggest two reasons why a woman may go through the menopause even if her ovaries are left intact after a hysterectomy. *(2 marks)*

2 Suggest why women who have had a hysterectomy are sometimes offered testosterone supplements. *(3 marks)*

3 According to research by Prof. John Studd, hysterectomy in pre-menopausal women is associated with a three-fold increased risk of coronary heart disease. This is the case even when the ovaries are left intact. Suggest why the risk of CHD changes. *(2 marks)*

Synoptic link

You will need to understand the structure and function of the female reproductive system, which you learnt about in Topic 18.1, Human reproductive systems.

Summary questions

1 Why is it not scientifically accurate to refer to men experiencing a 'male menopause'? *(2 marks)*

2 Explain how changes in hormone levels are associated with three named symptoms after the menopause. *(3 marks)*

3 a Plot the most appropriate graph to present the data shown in the table. *(4 marks)*

Age (years)	Percentage of women who are postmenopausal (%)		
	20 cigarettes per day	10 cigarettes per day	Non smoker
44–45	13.9	11.3	8.7
46–47	31.3	20.9	16.5
48–49	50.0	33.9	26.9
50–51	80.1	73.0	56.5
52–53	90.2	83.5	70.0

b Suggest what factors, other than smoking, may affect the age of menopause onset. *(2 marks)*

19.2 Managing the effects of ageing on the female reproductive system

Specification reference: 4.2.2

In the previous topic you learnt about the changes that occur in the female reproductive system as it ages. Now you will examine the methods employed to manage these changes.

Hormone replacement therapy

Hormone replacement therapy (HRT) is the most noted treatment for relieving the symptoms of the menopause. It replaces female hormones (oestrogen and progesterone) that are at lower levels as the menopause approaches and as it progresses.

Different types of HRT

Different forms of HRT exist, including:

Unopposed oestrogen
This provides the woman with a supply of oestrogen only. It tends to be given to women who have undergone a hysterectomy at an early age.

Combined HRT
This involves administering oestrogen with progestin. Progestin is a synthetic molecule that has similar effects to progesterone. Oestrogen is known to increase the probability of uterine cancer – progestin reduces this risk.

The therapy can be administered using either a continuous or cyclic regime.

Continuous schedule
Women take the same hormones continually through the month, either in the form of an oral tablet or via a skin patch or under-skin implant. This form of HRT does not cause any withdrawal bleeds, which some women prefer.

Cyclical regime
In some forms of cyclical therapy, oestrogen and progestin are taken for three weeks and then stopped for the fourth week, causing the woman to experience withdrawal bleeding, which may be similar to a period. Another common regime is for oestrogen to be taken continuously, with progestin taken on days 10 to 14 each month. Withdrawal bleeding occurs after day 14.

Different delivery systems for HRT

There are several ways HRT can be taken, including:

- tablets, which can be taken orally
- skin patches

> ### Learning outcomes
> Demonstrate knowledge, understanding, and application of:
> → the advantages and disadvantages of hormone replacement therapy (HRT) to manage the effects of ageing on the female reproductive system
> → the use of treatments other than HRT to manage the effects of ageing on the female reproductive system.

> ### Study tip
> Progesterone and oestrogen are steroid hormones. They are able to diffuse through the phospholipid bilayers of cell membranes.

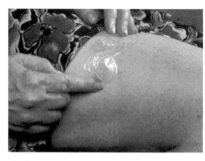

▲ **Figure 1** *A skin patch for HRT*

▲ **Figure 2** *Hormone replacement therapy pills*

- implants, which are inserted under the skin of the stomach, buttock, or thigh, under local anaesthetic
- vaginal rings
- oestrogen gel applied to the skin.

The choice of HRT delivery system is dependent on circumstances. For example, if a woman is only experiencing vaginal dryness, a form of HRT that can be applied directly to the vagina, such as a cream, gel, tablet or ring, may be prescribed. The hormones used in HRT will usually be prescribed at the lowest possible dose required to control the symptoms.

Advantages of taking HRT

The main advantages of taking HRT include:

- prevention of hot flushes – these symptoms usually improve within four weeks of HRT commencing
- reduced risk of developing osteoporosis
- prevention of vaginal dryness
- protective effect against the loss of connective tissue
- possible improvement in sleep and reduced muscle ache
- a reduction in mood swings and depressive symptoms.

Some studies have also shown a possible reduction in long-term risk of Alzheimer's disease and other forms of dementia. However, further studies are needed to confirm this.

Disadvantages of taking HRT

For most women, taking HRT for five years or less is safe and effective. However, both of the molecules used in HRT, oestrogen and progestin, can result in side effects. These side effects usually improve over time. Doctors recommend trying each treatment plan for at least three months before considering a change in dosage or a different form of HRT.

Serious health risks associated with HRT are rare but include the following conditions:

- Venous thromboembolism (VTE) – oral HRT increases the risk of VTE by a factor of two to three.
- Stroke – the risk of a stroke is slightly increased by most forms of HRT. Transdermal oestrogen, however, appears to be associated with a lower risk of stroke, compared to other forms of HRT.
- Breast cancer – Combined HRT increases the risk of breast cancer. However, the absolute risk is small at around one extra case of breast cancer per 1000 women per annum.
- Endometrial (uterine) cancer – Oestrogen-only HRT substantially increases the risk of endometrial cancer in women with a uterus. However, the use of cyclical progestin for at least 10 days per 28-day cycle appears to eliminate this risk.

▼ **Table 1** *Potential side effects of HRT*

Oestrogen	Progestin
fluid retention	fluid retention
bloating	breast tenderness
breast tenderness or swelling	headaches
nausea	mood swings
leg cramps	depression
headaches	acne
indigestion	backache
	pelvic pain

Data collected from studies over the past decade have shown that starting HRT when a woman is near to the menopause will probably provide a favourable benefit : risk ratio.

Alternatives to HRT

HRT can be started as soon as a woman begins to experience menopausal symptoms. However, HRT may not be suitable if the woman has:

- a history of breast cancer, ovarian cancer, or endometrial cancer
- a history of blood clots
- a history of heart disease or stroke
- untreated high blood pressure
- liver disease
- irregular periods.

In such cases, alternative medication may be prescribed. Phytoestrogens, for example, are chemicals found in plants, which either have a similar molecular structure to oestrogen or are converted to oestrogen-like compounds by bacteria in the gut. Phytoestrogens include:

- isoflavones – found in soya beans, and alfalfa
- lignans – found in cereal grains, most vegetables, and fruits
- flavones – found in cereals, celery, and herbs (e.g., parsley)
- coumestans – found in split peas, pinto beans, and alfalfa.

Phytoestrogens also seem to have antibacterial and antifungal properties and may reduce the effects of some virus infections. They can be consumed either as supplements or by increasing dietary intake. To rely on dietary intake alone would involve the ingestion of large amounts of leguminous food plants, such as peas and beans.

Some women have reported that relaxation therapies, such as yoga, aromatherapy and reflexology, reduce their menopausal symptoms. No scientific evidence exists to show that these activities are completely effective.

Menopausal symptoms can also be reduced by making lifestyle changes, such as:

- Taking regular exercise. This has been found to reduce symptoms of hot flushes and improve sleep as well as boost a woman's mood if she feels anxious, irritable, or depressed.
- Staying cool at night to relieve hot flushes and reduce night sweats.
- Reducing caffeine, alcohol, and spicy food intake.
- Reducing stress levels to improve mood swings.
- Giving up smoking, which reduces hot flushes.

Synoptic link

You will need to understand the role of female reproductive hormones. This is covered in Topic 18.2, Gametogenesis, and Topic 18.3, The menstrual cycle.

Summary questions

1 Describe the difference between continuous and cyclical HRT regimes. (*2 marks*)

2 In what circumstances might phytoestrogens, rather than HRT, be prescribed for menopausal women. (*2 marks*)

3 Evaluate the use of progestin in hormone replacement therapy. (*5 marks*)

19.3 Effects of ageing on the male urinogenital system

Specification reference: 4.2.2

Until now in this chapter, you have focused on the effects of ageing on the female reproductive system. Unlike women, men do not undergo a major, rapid change in fertility as they age, but ageing does result in gradual changes. You will learn about some of these changes to the male reproductive and urinary systems in this topic.

Changes in reproductive function

The male reproductive system may demonstrate changes in testicular tissue, sperm production, and erectile function.

Changes in the testes

There is a decrease in the mass of testicular tissue and there may be a slight decrease in the level of testosterone.

Changes in erectile function

An erection problem is when a man cannot get or maintain an erection that is firm enough to have intercourse. If the condition continues it is called erectile dysfunction (or impotence).

Erectile dysfunction may not be a direct result of physiological ageing, but may be caused by some medications, such as those for hypertension. Other disorders, such as diabetes, are often associated with the onset of erectile dysfunction.

Changes in fertility

Fertility varies between individuals and age is not a good predictor of male fertility. The testes continue producing sperm, but the rate of production goes down. The prostate gland, epididymis, and seminal vesicles lose some of their surface cells but continue to produce semen. Sclerosis can occur in the epididymis, which causes this tissue to become less elastic.

Male infertility is not the same as erectile dysfunction. A man who cannot maintain an erection may be able to produce sperm that can fertilise an ovum. Likewise, a man who is infertile can usually maintain an erection, but he may not be able to father a child due to problems with his sperm.

The decline in testosterone production with age can reduce a man's libido (sex drive).

Changes in urinary function

As a man ages some of the prostate tissue can be replaced with scar tissue causing the prostate gland to enlarge. This condition is called benign prostatic hypertrophy (BPH) and is thought to affect approximately 50% of men. BPH can lead to problems with ejaculation and slow down urination.

Approximately 60% of men over the age of 60 years have some degree of prostate enlargement. The cause of prostate enlargement is unknown, but most experts agree that changes in hormone levels play a role. Levels of dihydrotestosterone increase as men age. This may stimulate the growth of the prostate. Other scientists suggest that a decline in testosterone is responsible. The proportion of oestrogen in a man's blood increases with age, as testosterone declines. The relative increase in oestrogen may stimulate prostate growth.

BPH involves hyperplasia (an increase in the number of cells) rather than hypertrophy (an increase in the size of individual cells). BPH causes the formation of large nodules that can compress the urethra. This restricts the flow of urine from the bladder. Men with prostate enlargement, however, do not have a higher risk of prostate cancer compared to men without BPH. Signs of BPH can include:

- hesitancy in urination
- frequent urination
- irregular need to urinate
- a slow and weak urinary stream
- prolonged time taken to empty the bladder
- abdominal straining
- inability to completely empty the bladder
- post-urination dribble
- irritation during urination
- nocturia (need to urinate during the night)
- dysuria (painful urination)
- incontinence (involuntary leakage of urine)
- bladder pain
- difficulty ejaculating.

Diagnosis of BPH is usually the result of a rectal examination. Blood tests for prostate specific antigen (PSA) are carried out to exclude prostate cancer. In some cases, ultrasonography is also used.

BPH can be managed through lifestyle changes, such as reducing fluid intake at night and minimising alcohol and caffeine consumption. Medications are available that relax smooth muscle in the prostate, thereby easing the passage of urine. Surgery is a possibility when other treatments are ineffective.

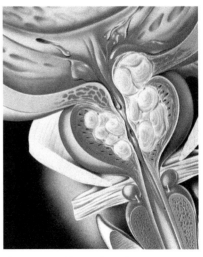

▲ Figure 1 Illustration of a prostate gland (red, centre right) showing hyperplasia. The nodules (cream swirls) are due to the excess growth of normal cells within the gland. The lower part of the bladder (pink) can be seen above the prostate, while part of the base of the penis (red/pink) lies below the gland

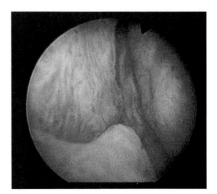

▲ Figure 2 Endoscope image from inside the urethra. This shows an enlarged prostate gland pushing on the urethral walls, causing them to bulge inwards

Summary questions

1 Why is prostatic hypertrophy described as benign? (1 mark)

2 Suggest why the symptoms of prostate cancer and BPH are similar. (2 marks)

3 Suggest why prostate function does not always affect fertility, and a man can father children even when his prostate gland has been removed. (3 marks)

Synoptic link

You learnt about the structure and function of the male reproductive system in Topic 18.1, Human reproductive systems.

Practice questions

1 Which of the following statements is/ are usually true of the hormonal changes experienced by women following menopause?

  1 Secretion of oestrogen increases

  2 Secretion of FSH increases

  3 Secretion of LH increases

  A 1, 2 and 3   B Only 1 and 2

  C Only 2 and 3   D Only 1

(1 mark)

2 Which of the following statements is/are true of combined hormone replacement therapy (HRT)?

  1 Lowers the risk of uterine cancer compared to unopposed oestrogen HRT

  2 It must be administered using a continuous schedule

  3 Oestrogen is administered with FSH

  A 1, 2 and 3   B Only 1 and 2

  C Only 2 and 3   D Only 1

(1 mark)

3 Which of the following is the most accurate explanation for the enlargement of the prostate gland in benign prostatic hypertrophy?

  A A tumour

  B Scar-like tissue

  C Bacterial infection

  D Viral infection   *(1 mark)*

4 Which of the following changes to lifestyle may reduce menopausal symptoms?

  A Maintenance of high room temperatures overnight

  B Reducing spicy food intake

  C Increasing caffeine intake

  D Reducing exercise levels   *(1 mark)*

5 Which of the following statements is/are true of the possible delivery systems for HRT?

  1 Skin patches

  2 Vaginal rings

  3 Progestin gel

  A 1, 2 and 3   B Only 1 and 2

  C Only 2 and 3   D Only 1

6 Hormone replacement therapy (HRT) is used to relieve menopausal symptoms.

a Describe the changes which occur during and after the menopause and explain how changes in hormone levels are associated with these changes. *(6 marks)*

b Some women use herbal remedies to treat menopausal symptoms.

From 2011, in the UK, all herbal products sold as medicines will need to be registered with the UK Medicines and Healthcare Products Regulatory Agency.

  (i) Suggest what evidence the regulatory agency will be considering before they register a herbal remedy for the treatment of menopausal symptoms. *(3 marks)*

  (ii) Suggest one possible active ingredient which may be found in a herbal remedy for menopausal symptoms. *(1 mark)*

c The use of HRT has been linked to an increased risk of developing some cancers, such as ovarian cancer. The graph shows the percentage survival at one and five years from diagnosis of ovarian cancer in women in England and Wales.

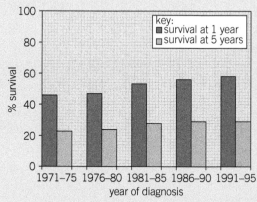

With reference to the figure, describe the changes in percentage survival for ovarian cancer between 1971 and 1995. *(3 marks)*

OCR F225 2011

7 Hot flushes, which are one symptom of menopause, can be embarrassing and uncomfortable and are one reason why some women choose to have HRT.

There are three main types of HRT: continuous combined, combined HRT, and cyclic HRT. The figure shows how hormones are given during two monthly cycles for the three types of HRT.

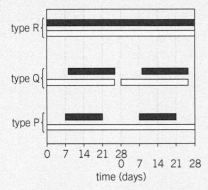

a Identify the two hormones represented by the dark and light shaded boxes. (*2 marks*)

b Identify which type of HRT is shown by the letters P, Q and R by copying and completing the table. (*3 marks*)

Type of HRT	Letter
Continuous combined HRT	
Combined HRT	
Cyclic HRT	

OCR F225 2012

8 Births in Japan were monitored over a one year period and the age of parents at the time of the child's birth was recorded.

The results are shown in the graph.

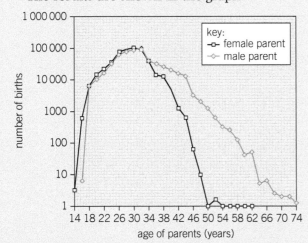

a Explain the difference in the pattern of data for male and female parents between the ages of 35 and 50 years. (*3 marks*)

b Suggest two reasons why no births were recorded for fathers under the age of 16. (*2 marks*)

OCR F225 2012

9 a Describe the difference between combined HRT and continuous combined HRT. (*3 marks*)

b The graph shows the results of one investigation into the effects of HRT on a range of possible conditions.

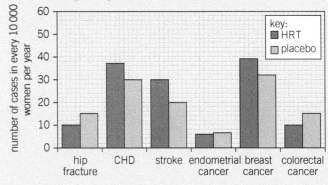

State one piece of evidence from the figure that indicates these results come from a clinical trial rather than an observational study. (*1 mark*)

c Using only the information in the figure, discuss the advantages and disadvantages of using HRT. (*5 marks*)

d Suggest why the following women were not included in the sample for this investigation:

• Women who had undergone a hysterectomy.

• Women whose blood pressure was routinely measured at or above 140/90 mmHg (18.6/12kPa). (*3 marks*)

OCR F225 2013

20 PHOTOSYNTHESIS AND ENVIRONMENTAL MANAGEMENT

20.1 Chloroplasts

Specification reference: 4.3.1

Learning outcomes

Demonstrate knowledge, understanding, and application of:

→ the ultrastructure of the chloroplast

→ the use of paper chromatography to investigate photosynthetic pigments.

This chapter examines aspects of food production. The focus will broaden as the chapter progresses. In subsequent topics, you will learn about nutrient cycling within ecosystems and how **photosynthesis** produces carbohydrates in plants. However, this topic will begin on a small scale, with the organelle responsible for photosynthesis: the **chloroplast**.

The structure of chloroplasts

Chloroplasts consist of two distinct regions: the **grana** (singular – granum) and the **stroma**. These are the locations of the two stages of photosynthesis, which will be covered in Topics 20.2 and 20.3. Chloroplasts range in length from 2–10 μm. They are encased in a double membrane, known as an envelope.

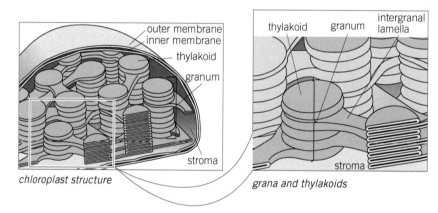

▲ Figure 1 *Chloroplast structure*

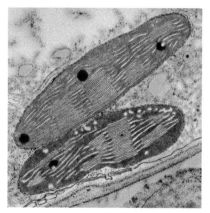

▲ Figure 2 *Electron micrograph of two pea chloroplasts, approximately ×16 000 magnification*

Grana

Grana are the structures in which the **light-dependent stage** of photosynthesis occurs. They are stacks of flattened membrane compartments known as **thylakoids**. Each granum stack is connected to another stack by intergranal thylakoids (also called lamellae). Embedded in the grana are all the structures required for the light-dependent stage of photosynthesis: photosystems (containing light-harvesting complexes), electron transport chains, and ATP synthase. You will examine this stage of photosynthesis in more detail in Topic 20.2.

68

Stroma

The stroma is the location of the **light-independent stage** of photosynthesis. It is a fluid-filled matrix that contains all the enzymes required for this part of photosynthesis. You will learn about these light-independent reactions in Topic 20.3, Photosynthesis – the light-independent stage.

The adaptations of chloroplasts for their role

- The inner membrane is embedded with transport proteins and is less permeable than the outer membrane. This enables control over which substances can enter the stroma from the cell cytoplasm.
- The stacking of thylakoid membranes produces a large surface area over which the light-dependent reactions can occur.
- The photosynthetic pigments are organised in photosystems, which maximise the efficiency of light energy absorption.
- As the grana are surrounded by the stroma, products from the light-dependent reactions can pass directly to the enzymes catalysing the light-independent reactions.
- Chloroplasts contain their own DNA (cpDNA) and ribosomes. This enables them to produce some of their photosynthetic proteins rather than importing them from the cell cytoplasm.

Study tip

The ability to convert measurements between decimal and standard form is an important skill. Placing small values into context is also a useful exercise. The following three examples provide good comparisons.

Palisade cell diameter

$40\ \mu m = 0.000\,040\ m = 4 \times 10^{-5}\ m$
(standard form)

Chloroplast length

$8\ \mu m = 0.000\,008\ m = 8 \times 10^{-6}\ m$

Glucose molecule diameter

$0.8\ nm = 0.000\,000\,000\,8\ m$
$= 8 \times 10^{-10}\ m$

Using chromatography to investigate photosynthetic pigments

Chloroplasts contain many pigments (e.g., chlorophyll and accessory pigments such as β-carotene and xanthophyll). Accessory pigments absorb wavelengths that chlorophyll absorbs poorly. These pigments can be separated and identified using a technique called chromatography.

In paper chromatography, the mixture of pigments is placed near the base of a paper strip. This point is called the origin. The paper is dipped in a solvent or mixture of two solvents (e.g., acetone and petroleum ether), which moves up the paper, carrying pigment molecules with it. The pigments will move at different rates, depending on their solubility in the solvent. This produces a chromatogram, which shows several separate spots or bands along the paper, each representing a different pigment.

The chromatogram can be analysed by calculating R_f values.

$$R_f = \frac{\text{distance moved by substance}}{\text{distance moved by solvent}}$$

R_f values will vary in different solvents, but each pigment has a standard R_f in a particular solvent, making identification possible. For example, Table 1 shows some R_f values for photosynthetic pigments using propane and ether as solvents.

Calculate the R_f values for pigments A and B in this chromatogram. Give your answers to three significant figures.

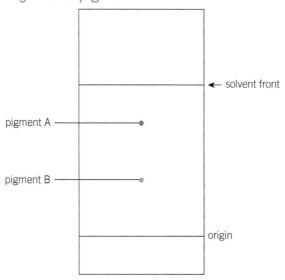

▲ **Figure 3** Chromatogram of chloroplast pigments

▼ **Table 1**

Pigment	Distance travelled (cm)	Solvent distance (cm)	R_f
carotene	4.75	5.00	0.95
chlorophyll a	3.25	5.00	0.65
chlorophyll b	2.25	5.00	0.45

What determines the colours of photosynthetic pigments?

Pigments vary in colour because they absorb different ranges of wavelengths of light. Wavelengths that are not absorbed will instead be transmitted or reflected by pigments. These wavelengths are detected by our visual systems and determine the pigment's colour. Figure 4 shows the wavelengths absorbed by three photosynthetic pigments. The colours of the three lines represent the colours you see. For example, chlorophyll a appears green because it absorbs a high proportion of blue/violet light (400–490 nm) and red light (620–700 nm), but very little green light (490–575 nm).

1 Explain the appearance of chlorophyll b and carotenoids in terms of their absorption spectra.
2 Estimate the peak absorption wavelength (m) of chlorophyll b in standard form.
3 Suggest the effect on growth of exposing a plant to blue/violet light (400–490 nm) rather than sunlight.

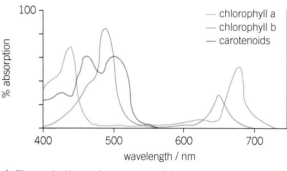

— chlorophyll a
— chlorophyll b
— carotenoids

▲ **Figure 4** *Absorption spectra of plant pigments*

Synoptic link

In Topic 1.6, Cell ultrastructure, you learnt about the ultrastructure of palisade mesophyll cells, which contain a high density of chloroplasts.

Study tip

You are not required to learn any structural details of the accessory pigment molecules found in photosystems.

Summary questions

1 Table 1 shows the results of a chromatographic analysis. Two other photosynthetic pigments, phaeophytin and xanthophyll, were separated in the same analysis.
  a Phaeophytin travelled 4.15 cm. What is the R_f value of phaeophytin?
  b Xanthophyll has an R_f value of 0.71. How far did it travel? (*2 marks*)

2 Describe two structural similarities between chloroplasts and mitochondria. (*2 marks*)

3 Figure 5 shows two chloroplasts from maize (*Zea mays*). Suggest how the different ultrastructures of the two chloroplasts affect their roles in photosynthesis. (*4 marks*)

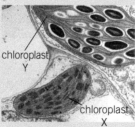

chloroplast Y

chloroplast X

▲ **Figure 5**

20.2 Photosynthesis – the light-dependent stage

Specification reference: 4.3.1

In the previous topic, you learnt about the structure of chloroplasts, the organelles of photosynthesis. This topic will focus on the first part of the photosynthetic process – the light-dependent stage – which takes place in the grana of chloroplasts. This stage is responsible for the conversion of light energy to chemical energy in plants.

Photosystems

Thylakoid membranes in chloroplasts contain photosystems, which are responsible for converting light energy into chemical energy. Photons of light are absorbed by chlorophyll b molecules and other accessory pigments arranged in light-harvesting complexes as illustrated in Figure 1. This energy is funnelled down to a reaction centre at the heart of a photosystem. Electrons in two chlorophyll a molecules in the reaction centre are excited and rise to a higher energy level. These high-energy electrons are transferred to nearby molecules known as electron acceptors.

The excited electrons are passed through a series of electron carriers in several oxidation–reduction reactions. This is known as an electron transport chain. Each new electron carrier occupies a lower energy level than the previous one, meaning that energy is released as the transport chain progresses. The energy is used to pump protons (H^+ ions) across the thylakoid membrane, into space inside the thylakoids. A proton gradient is established.

Two different photosystems are able to absorb light. Photosystem II is used before photosystem I in the series of light-dependent reactions. This numbering might seem odd, but it is simply a result of photosystem I being identified first.

Photophosphorylation

The protons that have accumulated inside the thylakoids move back across the membrane, down a concentration gradient. Significantly, however, the only pathway available to these H^+ ions is diffusion through the enzyme ATP synthase. This flow of protons through the ATP synthase channel is an example of chemiosmosis. The kinetic energy from the movement of protons is converted into chemical energy by ATP synthase. The enzyme bonds ADP and P_i (inorganic phosphate) to form ATP, which is then used in the light-independent reactions. The harnessing of light energy to produce ATP is called photophosphorylation.

Photophosphorylation has remarkable similarities to the production of ATP by oxidative phosphorylation in aerobic respiration.

Learning outcomes

Demonstrate knowledge, understanding, and application of:

→ the conversion of light energy to chemical energy in the light-dependent stage of photosynthesis

→ the role of water in photolysis.

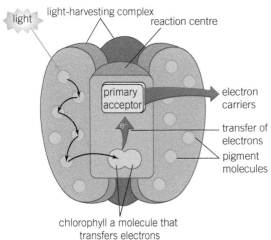

▲ **Figure 1** *A simplified light-harvesting complex in a photosystem*

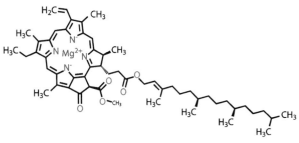

▲ **Figure 2** *A chlorophyll a molecule, which transfers electrons in photosystems*

Synoptic link

You learnt about aerobic respiration in Topic 16.3, Oxidative phosphorylation and the electron transport chain.

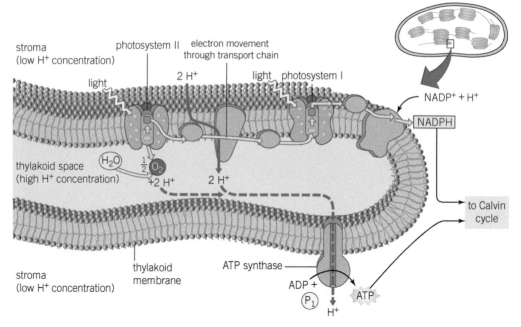

▲ **Figure 3** *The reactions of the light-dependent stage*

Photolysis

Water is essential for driving the light-dependent reactions of photosynthesis. An enzyme in photosystem II can split water in the presence of light. This is known as **photolysis**.

$$2H_2O \rightarrow 4H^+ + 4e^- + O_2$$

H^+ ions are pumped into the thylakoid space, enabling ATP synthase to produce ATP via chemiosmosis. The H^+ ions then reduce **NADP** in the stroma to reduced NADP, which can be used in the light-independent stage.

Electrons (e^-) replace those lost by the oxidation of chlorophyll a in photosystem II.

Oxygen (O_2) produced by photolysis will either diffuse out of the leaves through stomata or be used by plant cells in aerobic respiration.

The production of ATP and reduced NADP

By the end of the light-dependent reactions, molecules of ATP and reduced NADP have been produced on the stromal side of the thylakoids membrane. Both of these products are necessary for the production of carbohydrates in the light-independent stage of photosynthesis, which you can read about in the next topic.

Summary questions

1 Which product of the light-dependent reactions is not required in the light-independent stage of photosynthesis? *(1 mark)*

2 Outline the energy conversions that occur in chloroplasts to produce ATP. *(2 marks)*

3 Suggest three reasons why plants cannot use the ATP produced in the light-dependent stage of photosynthesis as their only source of ATP. *(3 marks)*

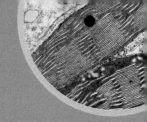

20.3 Photosynthesis – the light-independent stage

Specification reference: 4.3.1

The light-independent stage, also known as the **Calvin cycle**, is the final part of photosynthesis. It takes place in the stroma of chloroplasts, (as you learnt in Topic 20.1, Chloroplasts). As the name suggests, light plays no direct role in this stage. However, ATP and reduced NADP – two products from the light-dependent reactions, which you read about in Topic 20.2 – are essential for the Calvin cycle.

The roles of CO_2, ATP, and reduced NADP

CO_2

Carbon dioxide is fixed (i.e., converted from gaseous CO_2 into organic molecules) in the stroma of chloroplasts. Carbon dioxide is therefore the carbon source for all photosynthetic organisms.

ATP

The ATP that is generated in the light-dependent stage provides the energy to drive the carbon fixation reactions and the production of carbohydrates. Three molecules of ATP are used for every turn of the Calvin cycle, which is described below.

Reduced NADP

Reduced NADP provides reducing power (i.e., donating the hydrogen it acquired during the light-dependent stage) to enable the conversion of glycerate-3-phosphate (GP) to triose phosphate (TP) in the Calvin cycle. Hydrogen from reduced NADP is added to GP to form TP, and ATP provides the energy required for this reaction. Two reduced NADP molecules are used for every turn of the Calvin cycle.

The Calvin cycle

The following numbered stages correspond to those in Figure 1, which provides an overview of the reactions in the Calvin cycle.

1 Carbon dioxide diffuses through the stomata of a leaf, the plasma membrane of a cell, the cytoplasm, and across chloroplast membranes into the stroma.

2 The enzyme RuBisCO (ribulose bisphosphate carboxylase oxygenase) catalyses the reaction between ribulose bisphosphate (RuBP) and carbon dioxide.

3 The 5-carbon RuBP and the single carbon atom in CO_2 combine to produce two molecules of the 3-carbon GP.

4 ATP and reduced NADP are used to reduce GP to another 3-carbon molecule, TP.

Learning outcomes

Demonstrate knowledge, understanding, and application of:

→ the production of complex organic molecules in the light-independent reactions

→ the role of the products of the light-dependent reactions (ATP and reduced NADP) in the Calvin cycle

→ the role of CO_2 in the Calvin cycle.

Study tip

You might see the molecules of the Calvin cycle being referred to by a variety of names, depending on the book or article you are reading. Glycerate-3-phosphate (GP) is sometimes called PGA or 3PG. Triose phosphate (TP) is sometimes called G3P, GALP or PGAL.

Synoptic link

The reactions of the Calvin cycle are controlled by enzymes. You can remind yourself of the principles of enzymes and substrates in Chapter 3.

5 NADP has been regenerated and returns to the light-dependent stage, where it can be reduced again by accepting further H^+ ions.

6 Some TP is converted to molecules that can be used by the plant, such as glucose. You can read more about the products of the Calvin cycle in Topic 20.4, The products of photosynthesis.

7 The majority of TP is used to regenerate RuBP and to continue the cycle. Additional ATP is required for this reaction.

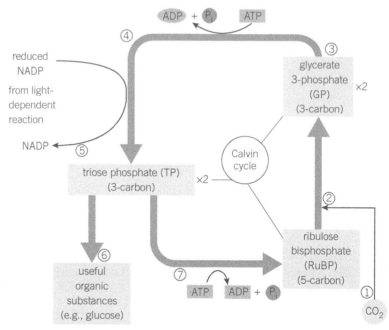

▲ **Figure 1** *A summary of the Calvin cycle*

Summary questions

1 Describe the role of RuBisCO in the Calvin cycle. *(2 marks)*

2 Explain why the Calvin cycle will stop after a plant has been placed in the dark for a period of time. *(2 marks)*

3 One glucose molecule can be produced for every six turns of the Calvin cycle. The mass of one molecule of RuBP is approximately 1.7 times greater than the mass of one molecule of glucose. Estimate the mass of RuBP required to produce 10 kg of glucose. Give your answer to three significant figures. *(2 marks)*

20.4 The products of photosynthesis

Specification reference: 4.3.1

In the previous topic, you learnt about the light-independent reactions of photosynthesis. CO_2 is fixed in these reactions. This enables a wide range of important organic (carbon-containing) molecules to be produced. This topic will look in more detail at the generation and fates of these molecules.

Products from GP

GP (glycerate-3-phosphate) is the 3-carbon molecule made from RuBP (5 carbons) and CO_2 in the Calvin cycle. GP can take part in three different reactions:

- Reduction to TP, which continues the Calvin cycle and, as you saw in Topic 20.3, requires both reduced NADP and ATP.
- Conversion to acetyl CoA, which is then converted to fatty acids.
- Formation of amino acids. This conversion requires plants to have absorbed mineral ions such as nitrates and sulfates. Nitrates provide nitrogen for the amine group (NH_2). Sulfates provide the sulfur required for the R groups in cysteine and methionine.

Products from TP

TP (triose phosphate) is the 3-carbon molecule formed by the reduction of GP in the Calvin cycle. It can be converted into a wide range of important products:

- Most of the TP is used to regenerate RuBP and continue the Calvin cycle.
- Two TP molecules can be joined to form 6-carbon sugars, such as glucose and fructose.
- Fructose and glucose can react to produce the disaccharide sucrose.
- Glucose molecules can also react to form polysaccharides such as amylose, amylopectin (starch), and cellulose.
- A single TP molecule is easily converted to glycerol, which can react with fatty acids to form triglycerides.

Respiration in both plants and animals is dependent upon the generation of carbohydrates from TP. So, for all organisms, TP is the source of the organic molecules that are used most commonly as respiratory substrates.

Learning outcomes

Demonstrate knowledge, understanding, and application of:

→ the production of carbohydrates, lipids, and amino acids from GP and TP

→ the dependency of respiration on the products of photosynthesis.

What changes when GP is converted?

Figure 1 shows some of the molecules that can be formed from GP. You can see that only small changes occur during these reactions. When GP is converted into the intermediate 1,3-diphosphoglycerate, the OH (hydroxyl) group is replaced by a phosphate group. When 1,3-diphosphoglycerate is converted to TP, this phosphate is substituted for a hydrogen atom. Figure 1 also shows the conversion of GP into another important biological molecule, serine.

1   Which molecules are required for the conversion of
    (a) GP to 1,3-diphosphoglycerate
    (b) 1,3-diphosphoglycerate to TP?
2   Serine is an amino acid. Draw the R group.
3   Suggest a mineral ion that would be required during the conversion of GP to serine.

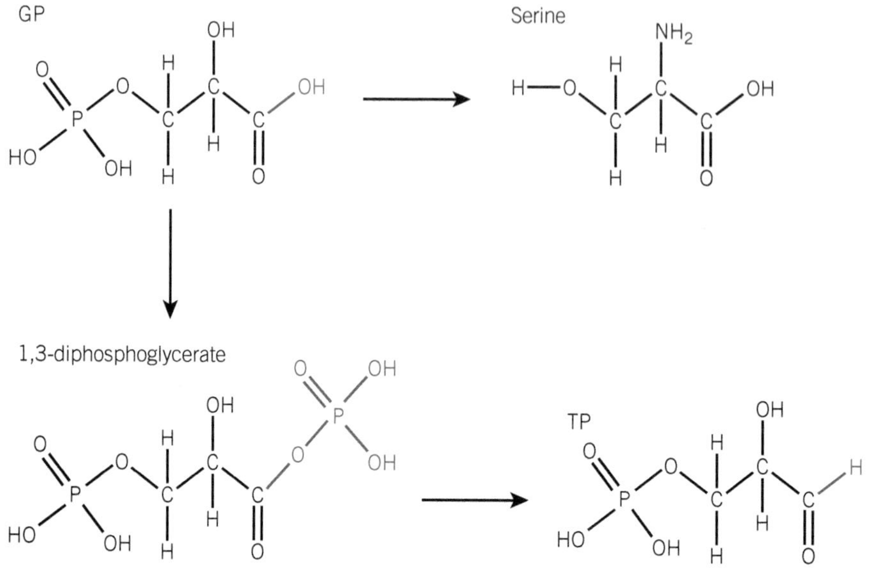

▲ **Figure 1** *An outline of molecules that can be produced from GP*

Synoptic link

You can remind yourself of the structures of the biological molecules discussed here in the following topics: Topic 3.1, amino acids – the building blocks of proteins, Topic 1.9, Cell membranes, and Topic 2.3, Carbohydrates in cells and biofluids.

Summary questions

1   Name two useful molecules that can be produced from GP and describe one way in which a plant can use each molecule. *(4 marks)*

2   State how many molecules of TP would be required to produce
    a   two molecules of glycerol.
    b   Six molecules of fructose.
    c   Four molecules of sucrose. *(3 marks)*

3   Outline the role of acetyl CoA in respiration and photosynthesis. *(4 marks)*

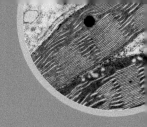

20.5 Factors affecting photosynthesis

Specification reference: 4.3.1

While studying the reactions of photosynthesis, you will have seen that several factors can affect photosynthetic rate. For example, in Topic 20.2 you looked at the light-dependent reactions, which are influenced by the intensity and duration of light to which plants are exposed. Topic 20.3 dealt with the Calvin cycle, which is reliant on the availability of CO_2. Here you will consider some of the factors affecting photosynthesis and discuss the methods that are available to study them.

Limiting factors in photosynthesis

Many variables affect the rate of photosynthesis, but the factor that is present at the least favourable level is known as the **limiting factor**. For example, if a chef has plenty of salt, yeast, and butter, but very little flour, the amount of bread that she can make is limited by the flour. Several abiotic variables have the potential to act as limiting factors in the rate of photosynthesis.

Light intensity

When light is a limiting factor, the intensity of light is directly proportional to the rate of photosynthesis. Light intensity limits the rate at which ATP and reduced NADP are produced in the light-dependent reactions. At low light intensities, GP concentrations will rise. RuBP and TP concentrations will decrease because insufficient ATP and NADPH are available to convert GP to TP and TP to RuBP.

Light wavelength

In Topic 20.1, you read about the different ranges of wavelengths absorbed by photosynthetic pigments. In direct sunlight, plants are exposed to the full spectrum of visible light wavelengths. In some environments (e.g., aquatic ecosystems) and in laboratory conditions, however, plants might not receive all wavelengths to an equal extent.

Light duration

The number of daylight hours will determine the amount of sunlight a plant is able to absorb. You will look at the effect of daylight duration when the idea of compensation points is introduced.

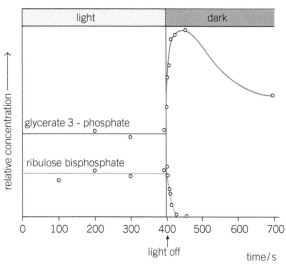

▲ **Figure 1** *The results of an experiment by James Bassham, which show the effects of light and dark on the concentrations of ribulose bisphosphate (RuBP) and glycerate-3-phosphate (GP) in plant cells*

Synoptic link

In Topic 10.4, Adaptations in plants, you learnt how species can evolve adaptations, such as a particular set of photosynthetic pigments, if they live in environments that receive a limited range of visible light wavelengths.

CO$_2$

The optimum carbon dioxide concentration for a high rate of photosynthesis is 0.1%. This contrasts with an atmospheric concentration of approximately 0.04%. Farmers and plant growers can increase the levels of CO$_2$ in greenhouses by burning methane, thereby boosting the rate of photosynthesis. At low CO$_2$ concentrations, RuBP will accumulate and GP formation will be limited.

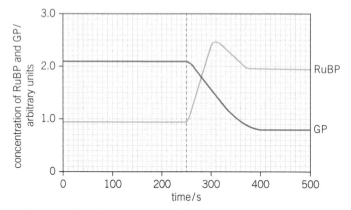

▲ **Figure 2** *The effect of carbon dioxide concentration on RuBP and GP concentrations. Left of green line = 1.0% carbon dioxide. Right of line = 0.003% carbon dioxide*

Temperature

Higher temperatures provide more kinetic energy for molecules involved in photosynthetic reactions, thereby increasing reaction rates. As a rule of thumb, photosynthetic rate approximately doubles for each 10 °C rise in temperature up to 25 °C. Temperatures higher than this will tend to slow the rate because enzymes such as RuBisCO are denatured.

Temperature has little effect on the light-dependent stage because few enzymes are required in these reactions, except for photolysis. Each of the light-independent reactions is catalysed by an enzyme, which means that temperature will have a significant impact on this stage of photosynthesis.

pH

Changes in pH can denature proteins that are essential to photosynthesis, including enzymes such as RuBisCO. This reduces the rate of photosynthesis.

Synoptic link

Changes of temperature and pH both alter the rate of photosynthesis by affecting how photosynthetic enzymes function. You can remind yourself of the principles of enzyme structure and function in Chapter 3, Proteins and enzymes.

Methods for investigating the factors affecting photosynthesis

Photosynthesis can be measured using a variety of approaches, including the use of a photosynthometer and experiments such as the Hill Reaction with DCPIP.

Several factors can act as independent variables. These include temperature, light intensity, CO_2 concentration. Remember, in a well-designed experiment only the independent variable will change. All other factors must be controlled. You will therefore be testing the effect of a single factor (the independent variable) on the rate of photosynthesis (the dependent variable).

Photosynthometer

Figure 3 shows how a photosynthometer is set up. It is designed to measure oxygen production in an aquatic plant such as *Elodea*. The rate of oxygen production is used as a measure of photosynthetic rate. Oxygen gas bubbles are collected in the capillary tube of the photosynthometer and the length of the bubble can be converted to volume (volume = length of bubble × πr^2, with r being the radius of the tube). The volume of oxygen evolved per unit time can then be calculated.

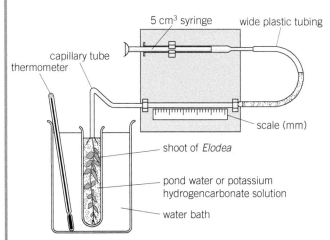

▲ **Figure 3** *A photosynthometer*

To investigate the effect of light intensity, the distance (d) of the light source from the equipment is varied. Light intensity is calculated as $1/d^2$. The *Elodea* is placed in a water bath at a constant temperature. A light bulb that generates little heat should be selected, otherwise the light source could change the water temperature as well as the light intensity. To ensure CO_2 concentration is not

a limiting factor, two drops of sodium hydrogencarbonate are added to the water. Sodium hydrogencarbonate generates CO_2.

Instead of light intensity, the effect of CO_2 can be tested by varying the concentration of sodium hydrogencarbonate added to the water. Temperature can be selected as the independent variable by varying the water bath setting and maintaining a constant light source. This method can be adapted to investigate other independent variables, such as pH and light wavelength.

DCPIP

DCPIP (2,6-dichlorophenol–indophenol) is a blue dye that acts as an electron acceptor. It becomes colourless when it accepts electrons and is reduced. This enables DCPIP to detect reducing agents produced by the light-dependent reactions of photosynthesis. An experiment can be conducted using the following steps:

1   Grind leaves (e.g., from spinach or lettuce) in ice-cold 2% sucrose solution.
2   Centrifuge the sample to obtain a pellet (high concentration of chloroplasts) and a liquid known as the supernatant (low concentration of chloroplasts).
3   Store the leaf extract in an ice-cold water bath and begin the experiment as soon as possible.
4   Set up the following tubes:
    i   DCPIP solution + leaf extract pellet (in light)
    ii   DCPIP solution + leaf extract pellet (in dark)
    iii DCPIP solution + supernatant (in light)
    iv   DCPIP solution + sucrose solution (in light)
    v   Distilled water + leaf extract pellet (in light)
5   Time how long it takes for DCPIP to change from blue to colourless.

The chloroplasts in tube (i) will generate electrons in the light-dependent reactions, causing DCPIP to decolourise and revealing the green colour of the chloroplasts. The supernatant in tube (iii) might contain a low concentration of chloroplasts, in which case it will decolourise, but more slowly than (i). The other tubes act as controls.

> Suggest why chloroplasts are isolated and stored in a 2% ice cold sucrose solution.

The compensation point

At night (in the dark), when no photosynthesis is occurring, plants continue to respire, taking in O_2 and giving out CO_2. As light intensity increases, photosynthetic rate increases. Eventually, the light **compensation point** is reached. This is when the rate of respiration matches the rate of photosynthesis. At the compensation point, the uptake of CO_2 for photosynthesis matches the volume released from respiration, and O_2 uptake and release are also equal. If light intensity is further increased, carbon dioxide uptake for photosynthesis outstrips its release from respiration.

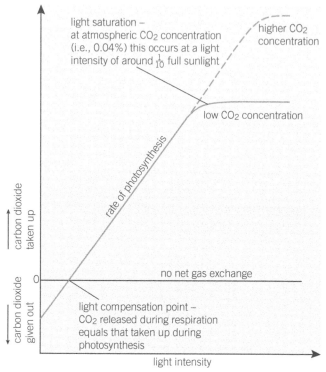

▲ **Figure 4** *A graph showing the light compensation point of a plant*

The compensation point is usually reached in the early morning (when photosynthetic rate is increasing) and in the evening (when photosynthetic rate is decreasing). In comparison, the rate of respiration in plants remains relatively constant. Plants will only increase their biomass when the rate of photosynthesis exceeds the rate of respiration. The compensation point will be exceeded for a longer period as day length increases. Therefore, an increase in daylight hours will increase plant growth rates. In the UK, in winter months, the lower light intensity and shorter day length means that the compensation point may not be reached so, although crops such as winter wheat will survive, they will not be growing.

The compensation points of different leaf samples can be investigated using hydrogencarbonate indicator solution. Hydrogencarbonate indicator is red at atmospheric CO_2 concentrations, turns yellow when pH drops and turns purple when pH rises. This means that below the compensation point, when more CO_2 is being produced in respiration than is being used in photosynthesis, pH will decrease and the indicator will turn yellow. Above the compensation point, when more CO_2 is being used for photosynthesis than is being produced during respiration, pH will increase and the indicator will turn purple.

The compensation points of plants can be estimated and compared by recording the effect of varying light intensities on the colour of hydrogencarbonate indicator.

▲ **Figure 5** *The range of colours of hydrogencarbonate indicator*

Summary questions

1 State and explain the effects on the rate of photosynthesis of the following temperature changes:
 a a decrease in temperature from 20 °C to 0 °C
 b an increase in temperature from 25 °C to 45 °C. *(4 marks)*

2 Describe and explain the shape of the graph in Figure 4. *(4 marks)*

3 A student used a photosynthometer to investigate the effect of light intensity on photosynthetic rate. When the light source was 0.5 m from the plant, an oxygen bubble 5.0 cm long was collected in the photosynthometer during a 2 minute period. A distance of 1.0 m produced 2.2 cm over the same time period. The diameter of the photosynthometer tube was 0.15 cm. Calculate the rate of photosynthesis for both measurements. *(2 marks)*

Synoptic link

In addition to its role in photosynthesis, light is also important in the control of flowering which you will learn more about in Topic 22.1, Plant reproduction – flowering and pollination.

Synoptic link

You learnt about the ultrastructure of bacterial cells in Topic 1.6, Cell ultrastructure, and about pathogenic organisms in Chapter 11, Diseases.

▲ **Figure 1** *A scanning electron micrograph of a root nodule, approx ×70 magnification*

You have already learnt about the ultrastructure of bacterial cells and some examples of pathogenic microorganisms. However, only a small proportion of bacterial species are pathogenic. Here you will learn about bacteria that are involved in nitrogen recycling, which play essential roles in maintaining ecosystems.

Nitrogen recycling

Ammonification

Saprotrophic microorganisms (also referred to as decomposers or putrefying bacteria) break down proteins, nucleic acids, and vitamins from dead organisms and nitrogenous waste such as urea. This process is known as ammonification, because it produces ammonia (NH_3) and ammonium ions (NH_4^+) and adds them into soils.

Nitrogen fixation

Another source of ammonium ions in soils is N_2 gas. Nitrogen gas constitutes 78% of the atmosphere, but most organisms are unable to use N_2 directly. Instead, atmospheric nitrogen is converted into nitrogen-containing compounds, such as ammonia, by nitrogen-fixing bacteria. This is known as **nitrogen fixation**.

Some nitrogen-fixing bacteria occur in soils and are free living (e.g., *Azotobacter*), whereas others have mutualistic relationships with plants. *Rhizobium* species, for example, live in nodules on the roots of leguminous plants such as beans. The bacteria fix nitrogen for the plants using the enzyme nitrogenase, and in return they receive carbohydrates from the plants. Nitrogen fixation requires a lot of energy. Approximately 16 moles of ATP are required to reduce 1 mole of nitrogen to ammonia.

 Culturing *Rhizobium*

Rhizobium bacteria can be cultured *in vitro* on agar plates using the following method:

1 Carefully cut a root nodule from the plant (a pink colour indicates a healthy nodule).

2 Wash and sterilise the nodule.

3 Crush the nodule and dilute with distilled water.

4 Prepare an agar plate with the following constituents:

Agar (20 g l^{-1}); mannitol (10 g l^{-1}) – this is an energy source; yeast extract (1 g l^{-1}) – a source of nitrogen; magnesium sulfate (0.2 g l^{-1}) – a source of essential ions; dipotassium phosphate (0.5 g l^{-1}), and sodium chloride (0.1 g l^{-1}) – for pH and osmotic buffering.

31.8 g of these constituents can be placed in 1 litre of distilled water, which is boiled, cooled, mixed, and then placed into sterile Petri dishes.

5 The Rhizobia are incubated on the agar plates at 25–30 °C for three days.

Suggest the reasons for Stage 2 and Stage 5 of the culture preparation.

The histology (microscopic tissue structure) of a root nodule can be studied using light microscopy. A thin section of root nodule can be treated with a suitable stain, such as crystal violet, and observed under a light microscope. Figure 2 shows a section of root nodule. The central region (stained purple) contains nitrogen-fixing bacteria and is surrounded by a cortex.

Nitrification

Plants absorb most of their nitrogen in the form of nitrate ions (NO_3^-). These NO_3^- ions are absorbed by root hair cells through active transport.

Nitrates are produced by nitrifying bacteria in a two-step process. Ammonium ions are first converted to nitrite ions (NO_2^-) by bacteria such as *Nitrosomonas*. Nitrite ions are then converted to nitrate ions by bacteria such as *Nitrobacter*.

Nitrifying bacteria carry out these reactions to obtain energy. Both **nitrification** steps are oxidation reactions that release energy.

Denitrification

Other bacteria in soil can carry out **denitrification**. This reaction involves the conversion of nitrates back into N_2 gas. Unlike nitrification, which involves oxidation, denitrification is a reduction reaction. Denitrifying bacteria are anaerobic and favour waterlogged soils that lack oxygen. This is a significant consideration for farmers. Soils that are poorly aerated, with a high concentration of water, will encourage denitrification, thereby reducing the availability of nitrogen-containing compounds. This limits the growth rate of crops.

Synoptic link

You learnt microscopy skills in Topic 1.1 Microscopy – the light microscope.

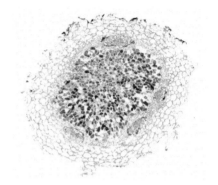

▲ **Figure 2** *A light micrograph of a broad bean root nodule, ×17.5 magnification*

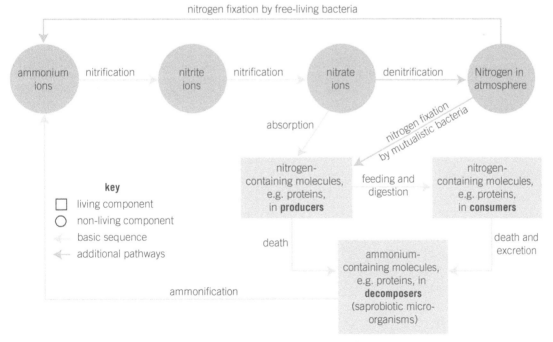

▲ **Figure 3** *The nitrogen cycle*

Synoptic link

You investigated the ultrastructure of bacterial cells and examples of pathogenic bacteria in Topics 1.6, Cell ultrastructure, and 11.1, Communicable diseases, respectively.

Study tip

It is easy to muddle the terms you will be using when describing the nitrogen cycle. Remember that 'nitrogen' is the element N, which is found in the atmosphere as a gaseous molecule, N_2. Compounds containing nitrate ions (NO_3^-) or ammonium ions (NH_4^+) can be described as 'nitrogen-containing compounds' (e.g., potassium nitrate (KNO_3) and ammonium sulfate $((NH_4)_2SO_4)$), as can ammonia (NH_3). If you are in doubt about any chemical formula you should use the full name.

 Luminescent microorganisms as biosensors

Bacteria involved in nitrogen recycling help to sustain ecosystems. Luminescent bacteria are another example of microorganisms that play an important ecological role, albeit a role created by scientists.

Bioluminescence is visible light generated by an organism. Scientists have genetically engineered some bacterial species to luminesce in the presence of particular pollutants. This enables luminescent bacteria to be used as biosensors to monitor environmental concentrations of certain chemicals.

Bacteria are genetically modified to produce the enzyme luciferase, which catalyses a reaction that generates visible blue-green light. The luciferase gene is linked with a promoter that responds to the presence of heavy metals, pollutants, or toxic conditions in an aquatic ecosystem. Depending on the design, the presence of pollutants will either cause luminescence to begin, or it will stop the bacterial cells from luminescing. This enables the water quality and safety of an environment to be monitored by recording increases or decreases in light levels.

Research one specific example of luminescent microorganisms that are used as biosensors.

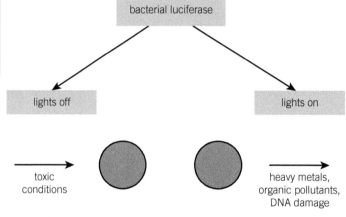

▲ Figure 4 *The two types of luminescent bacterial biosensors for environmental monitoring*

Summary questions

1 Add the most appropriate words to complete this paragraph, which describes the nitrogen cycle.

   Ammonium ions can be added to the soil through two processes, _____ (controlled by bacterial species such as *Azotobacter*) and _____ (controlled by saprotrophic bacteria). Nitrifying bacteria convert ammonium ions to _____ ions and nitrate ions. In anaerobic conditions, nitrate ions are converted to nitrogen gas in a process known as _____ . (*4 marks*)

2 Describe and explain the conditions required to culture *Rhizobium* in a laboratory. (*5 marks*)

3 A molecule called leghaemoglobin is produced in the root nodules of legumes only when they have been infected by nitrogen-fixing bacteria. Leghaemoglobin has a high affinity for oxygen. The nitrogenase enzyme, which is essential for nitrogen fixation, is sensitive to free oxygen in the root nodules. Use this information, knowledge from this topic, and your knowledge of haemoglobin in animals to suggest two reasons why leghaemoglobin is produced. (*4 marks*)

20.7 Food chains

Specification reference: 4.3.1

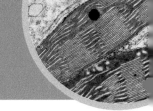

In the opening topics of this chapter, you examined how plants convert light energy to chemical energy. The chemical energy generated by photosynthesis and stored in plants is passed as food between other organisms. In this topic you will examine the efficiency of energy and biomass transfers in different food chains. You will also consider the role of ruminants, such as sheep, goats, and cattle, within the human food chain.

Food chains and webs

A food chain illustrates the transfers of energy between organisms within an **ecosystem. Producers**, which include plants, some bacterial species and algae, convert light energy to chemical energy, which is supplied to other species. Producers are eaten by **primary consumers**, which in turn are eaten by **secondary consumers**. **Tertiary consumers** then obtain their food from secondary consumers. In addition, decomposers obtain energy from dead organisms within a food chain, as you learnt in the previous topic. Each stage within a food chain is called a **trophic level** (from the Greek word 'trophe', meaning 'food').

Some food chains consist of only three trophic levels:

Maize   ⟶   Cattle   ⟶   Humans
(producer)   *(primary consumer)*   *(secondary consumer)*

Whereas others are longer and include quaternary consumers:

Phytoplankton ⟶ Krill ⟶ Penguins ⟶ Leopard seal ⟶ Killer whale
(producer)   *(primary*   *(secondary*   *(tertiary*   *(quaternary*
              consumer)   *consumer)*   *consumer)*   *consumer)*

Biomass and energy transfers

Some energy is lost at each trophic level of a food chain. Organisms require energy for metabolic processes such as respiration – some of this energy is eventually released as heat. As a consequence, less energy is available at each trophic level as you progress through a food chain.

The transfers of energy through a food chain can be represented as a pyramid of energy. An alternative approach is to use pyramids of biomass, which indicate the potential energy stored in organic molecules, such as fats and carbohydrates, at each trophic level. These two graphical representations both illustrate the loss of energy with each passing trophic level. Energy is measured in megajoules (or kilojoules) per square metre per year (MJ m^{-2} yr^{-1}).

The efficiency of energy transfers

Producers tend to convert less than 3% of the Sun's energy into biomass. Of the chemical energy stored in producers, approximately 10% is

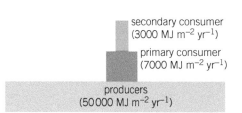

secondary consumer
(3000 MJ m^{-2} yr^{-1})

primary consumer
(7000 MJ m^{-2} yr^{-1})

producers
(50 000 MJ m^{-2} yr^{-1})

▲ **Figure 1** *A typical pyramid of energy for a grassland ecosystem*

Learning outcomes

Demonstrate knowledge, understanding, and application of:

→ biomass transfers through a food chain during food production

→ the comparative efficiency of biomass transfers in different food chains

→ the role of ruminants in the human food chain.

used by primary consumers for growth. The transfers to secondary and tertiary consumers tell similar stories, with less than 20% of the available energy being converted into biomass at the higher trophic level. The reduction in stored chemical energy can be attributed to some parts of organisms being inedible or indigestible, and to energy being released in excretory waste (e.g., urine), and as heat from respiration.

The efficiency of the energy transfer between each trophic level can be calculated as a percentage:

$$\text{Energy transfer} = \frac{\text{energy available after transfer}}{\text{energy available before transfer}} \times 100$$

Efficiency and sustainability in food production

In the food production industry, the efficiencies of energy and biomass transfers vary between food chains. This is a significant consideration for farmers and food producers because the efficiency of energy transfer plays a role in determining the sustainability of food production.

The energy transfer from maize to beef cattle is relatively inefficient – cattle usually require more than 10 kg of maize to produce 1 kg of biomass. In Topic 20.8, Farms as ecosystems, you will examine the methods that farmers adopt to improve this efficiency.

▲ **Figure 2** *A salmon farm in Scotland*

Fish are *ectotherms* (i.e., animals that rely on external sources of heat to regulate their body temperature). Fish consequently use less energy to generate heat. This increases the efficiency of energy transfer in their food chains. Usually less than 3 kg of food is required to produce 1 kg of fish biomass.

The demand for protein-rich food is increasing as the global human population rises. Collecting fish from the wild for food has depleted populations of many species and is not sustainable at the current rate of fishing. Fish farming, also known as aquaculture, is an alternative, which has greater sustainability than removing fish from the wild.

Another approach to improving the sustainability of the fishing industry is to promote the consumption of species that are low in the food chain. For example, tilapia, a group of cichlid fish, are primary consumers that feed on algae and plants. Farming species such as tilapia is efficient and sustainable because it involves only a single energy transfer between trophic levels. Unlike the farming of carnivorous, secondary and tertiary consumer fish (e.g., salmon), farming tilapia does not require the depletion of wild fish populations to act as food in their diet.

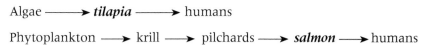

Algae ⟶ *tilapia* ⟶ humans

Phytoplankton ⟶ krill ⟶ pilchards ⟶ *salmon* ⟶ humans

Ruminants in food production

You might have been asked to ruminate on a question. In this context, 'ruminate' means to think about something, or 'chew it over' in your mind. The term derives from the rumination of food by certain species. Ruminants are animals that digest plant material slowly, in specialised stomachs, and regurgitate food to chew it a second time. This is also the source of the expression 'chewing the cud'.

The majority of farm animals are ruminants, including sheep, goats, and cattle. They have complex digestive systems, with four separate chambers in their stomachs.

The rumen contains microorganisms that are able to digest cellulose. Ruminants do not produce the cellulase enzyme that is capable of digesting cellulose, and they absorb little glucose directly from their diets. Instead, microorganisms in the rumen and reticulum break down cellulose and other carbohydrates into disaccharides and monosaccharides. Other bacteria convert these molecules into fatty acids. The abomasum acts like a human stomach by secreting hydrochloric acid and protease enzymes, which digest bacterial proteins into amino acids.

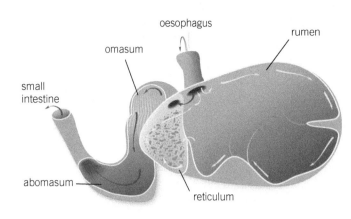

▲ **Figure 3** *The structure of a ruminant stomach*

Food passes from the rumen and reticulum into the omasum and the abomasum. The ruminant stomach is an ecosystem with its own specific abiotic conditions. For example, microorganisms in the rumen respire anaerobically, so the rumen harbours an anaerobic environment (i.e., there is a lack of oxygen).

Summary questions

1 Carp and salmon are two fish that can be farmed. Carp tend to be fed pellets that consist principally of plant material. Farmed salmon require food that contains protein from other fish species. State whether carp or salmon is more sustainable to farm. Explain your answer. (*2 marks*)

2 The trophic levels in a food chain can also be represented as a pyramid of numbers, which shows the relative number of individuals at each level. Explain why a pyramid of numbers illustrates the energy flows through a food chain with less accuracy than a pyramid of biomass. (*2 marks*)

3 The mean trophic level of species caught by the fishing industry indicates the health of an aquatic ecosystem. [1 = primary producer, 2 = primary consumer, 3 = secondary]. A decrease in the mean trophic level suggests that an ecosystem is in poor health and that the fishing practices lack sustainability. Suggest explanations for the trends shown in Figure 5. (*3 marks*)

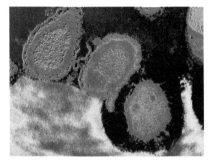

▲ **Figure 4** *Cellulose-digesting bacteria in the rumen of a cow, approximately ×35 300 magnification*

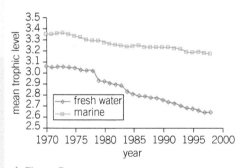

▲ **Figure 5**

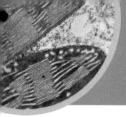

20.8 Farms as ecosystems

Specification reference: 4.3.1

Synoptic link

You learnt about ecosystems in Topic 10.6, Biodiversity.

As you have already learnt, an ecosystem comprises the complex interactions between the living and non-living elements within an area. A farm represents an ecosystem that is manipulated by humans. The productivity of a farm will therefore depend on both biotic (living) and abiotic (non-living) factors. In this topic, you will focus on the factors that affect productivity in a farming ecosystem and learn that the manipulation of these factors can result in conflict between agricultural and conservation needs.

Manipulating primary productivity

Primary productivity is the energy fixed by photosynthesis over the course of one year ($MJ\,m^{-2}\,yr^{-1}$). This is referred to as gross primary productivity (GPP). Not all of this energy is converted to carbohydrate stores in plants. **Net primary productivity** (NPP: mass per unit area per year) is defined as gross primary productivity minus the heat lost through respiration:

NPP = GPP − respiration loss

Many factors can influence productivity (Topic 20.5, Factors affecting photosynthesis). Farmers are able to manipulate some of these factors to increase the energy that is available to consumers.

Light

Plants can be grown under light banks in greenhouses to ensure constant, optimal light intensity and duration. The timing of sowing maximises the leaf area present during optimum conditions for photosynthesis over the summer months. The sowing density is chosen to prevent one plant overshadowing another.

Temperature

Greenhouses provide regulated, warm temperatures.

Water

Crops can be irrigated to ensure that water is not a limiting factor. Drought-resistant varieties are artificially selected. More recently, genetically modified (GM) drought-resistant crops have been developed. For example, GM oilseed rape can increase yields by up to 40% during droughts.

▲ Figure 1 *Irrigation of a Brussel sprout crop*

Nutrients

Farmers rotate crops (i.e., they alternate the type of crop grown in each field). Nitrate levels in the soil can be replenished by growing a nitrogen-fixing crop, such as beans, within the rotational cycle, or by the application of fertilisers in either organic or inorganic form.

Inorganic fertilisers are industrially produced chemicals such as ammonium or potassium nitrate. Organic fertilisers include plant waste materials and livestock waste such as manure. In addition to nutrient value, these have the advantages of adding organic material (humus) to the soil which improves its water-holding capacity and structure.

Competition

Herbicides are sprayed to kill weeds that compete with crops. Sometimes crops are genetically modified to be resistant to herbicides.

Pests

Pests, such as aphids and caterpillars, are killed by spraying pesticides. Control of pests and diseases maintains the leaf area for photosynthesis. GM pest-resistant plants, such as cotton, corn, and potatoes have been developed by inserting the bacterial Bt gene into their genomes. The Bt gene codes for a protein that is toxic to crop-eating insects.

Disease

Fungicides are sprayed on some crops to kill fungal pathogens. Several GM disease-resistant crops are being developed, such as GM oranges that resist citrus greening disease.

▲ Figure 2 *Oilseed rape* (Brassica napus), *has been genetically modified to be herbicide resistant*

Manipulating secondary productivity

Secondary productivity is the rate at which animals convert the chemical energy in the plants that they consume into their own biomass. Farmers are able to employ techniques to improve the efficiency of energy transfers from producers to consumers. The following techniques, for example, are widespread:

- Treatment with antibiotics reduces the energy expenditure of farm animals' immune systems by reducing bacterial infections.

- More energy can be channelled into growth by limiting the movement of farm animals. This is known as 'zero grazing'.

- The maintenance of a constant temperature also reduces energy expenditure for thermoregulation, allowing a greater proportion of energy to be used for growth.

- Selective breeding produces breeds of cattle, pigs, and sheep that are capable of increased productivity.

- Animals can be harvested before reaching adulthood. Juvenile animals invest a greater percentage of energy into growth than adults. Using the meat of young animals minimises energy loss from the food chain.

Many people think that these farming practices improve the efficiency of energy transfers at the expense of animal welfare.

Intensive rearing of livestock also means large quantities of waste both directly from the animals themselves or indirectly from, for example, the production of silage as food for cattle. This waste needs to be stored safely to prevent leakage into surrounding water courses leading to eutrophication. Many farms now process this material in 'digesters' to produce methane gas which can be used as fuel.

Conflict between agriculture and conservation

The farming practices that have been outlined are examples of intensive farming, which is characterised by the use of chemicals (e.g., fertilisers and pesticides), high costs, and a large input of labour. This contrasts with extensive farming, which requires fewer chemicals, and less money and labour.

▲ **Figure 3** *A hawthorn hedgerow*

Synoptic link

You have found out about a few examples of genetically modified crops during this topic. In Chapter 25, Gene technologies, you will learn more about the techniques used to create genetic modification.

The methods used during intensive agriculture produce higher yields but conflict with conservation strategies. Hedgerows, for example, which represent a food source and shelter to many species, tend to be removed during intensive farming, in order to maximise the space available for crops and large agricultural machines. Waste, fertilisers, herbicides, and pesticides from intensive farming can pass into watercourses and contaminate neighbouring ecosystems. Neonicotinoid insecticides have been blamed for a reduction in populations of bee species. These insecticides have consequently been banned in several countries.

The UK government offers payments to farmers who use methods that conserve the ecosystems on their land. This scheme was originally established in 1991 as the Countryside Stewardship Scheme and has since been renamed Environmental Stewardship (ES). Some of the practices encouraged by ES include:

- Growing crops that bear certain seeds, to attract bird species, and nectar, to attract bees and butterflies.
- Maintenance of hedgerows, which act as shelters and food sources for animals.
- Maintenance of dry stone walls. This provides a habitat for species of moss and lichen.
- Creating buffer strips. These are sections of land that have not been planted with crops or treated with fertilisers. These strips represent a buffer between intensive agriculture and natural habitats.

Summary questions

1   Outline, using examples, the role of the Environmental Stewardship scheme.   *(3 marks)*

2   Contrast the advantages and disadvantages of intensive farming with those of extensive farming.   *(3 marks)*

3   Over a 2 year period, a farmer grew 200 000 kg of a crop in a 1 km² field.
    a   Calculate the net primary productivity of this crop $(kg\,m^{-2}\,yr^{-1})$.
                                                                                                                  (2 marks)
    b   Explain how genetic engineering could increase the NPP of the crop.
                                                                                                                  (3 marks)

20.9 Land management
Specification reference: 4.3.1

You looked at techniques that can be used to manipulate farm ecosystems in Topic 20.8. Farming is an example of land management. The natural development of an ecosystem, known as **succession**, is interrupted by human management of land. This process of **deflected succession** due to land management affects the biodiversity within an ecosystem. In this topic, you will find out about deflected succession in managed areas, and examine the techniques that are available to investigate the biodiversity within such environments.

Learning outcomes

Demonstrate knowledge, understanding, and application of:

→ deflected succession as a result of land management

→ techniques used to measure biodiversity.

Deflected succession

Ecological succession is the change in the community of an ecosystem over time. Bare sand or rock is initially colonised by pioneer species that can survive in dry, exposed conditions, without soil. These pioneer species add nutrients to the ground as organisms decompose. The environment gradually becomes less hostile and is able to support other species. Soils develop and biodiversity increases. This process continues until a **climax community** is reached. This community has high biodiversity and is stable – no more succession occurs after this point has been reached.

Human activities, such as land management, can prevent a climax community from forming. This is known as deflected succession and results in the formation of a community known as a plagioclimax. Two examples of plagioclimax communities are found on agricultural land and in managed forests.

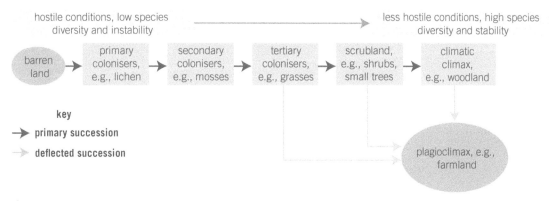

▲ Figure 1 *Ecological succession*

Agriculture

Deciduous woodland is the natural climax community over much of the UK. Over centuries, this woodland has been removed for timber and to create space for agriculture. Crops have been planted, whilst grazing by domesticated farm animals and mowing have maintained short grass and prevented tree seedlings from growing. For example, in the uplands of England, soils deteriorated as a result of deforestation, and heather, which grows well in poor soils, now dominates the plant community in this plagioclimax ecosystem.

▲ **Figure 2** *An agricultural plagioclimax – heather rather than trees dominates the plant community*

Forestry

Deciduous woodland represents the climax community in most parts of the UK, but many of the native deciduous trees have been removed. Non-native trees, including conifers such as spruces, pines, and firs, have been grown to meet the demands of the timber and fuel industries.

The remaining deciduous woodlands are often managed by processes such as coppicing and rotation, such that different stages of succession are always present in the woodland.

Measuring biodiversity

Plagioclimax communities, such as farmland, are likely to have a lower biodiversity than climax communities. You learnt about some common farming practices, including the use of fertilisers and pesticides, in the previous topic. As you read in Topic 20.8, farmers are now being funded to use methods that promote conservation and the maintenance of biodiversity.

Ecologists can monitor biodiversity by sampling parts of an ecosystem. Counting every single organism is impossible, but calculating the biodiversity in small sections of an ecosystem enables the overall biodiversity to be estimated.

The choice of sampling method is important, and this should be considered before the samples are taken and calculations are made. You should bear in mind a few key principles when designing your sampling method:

- *Larger sample sizes* are more representative of the whole ecosystem.
- A greater sample size increases validity, but the number of samples you can take will be limited by the time, energy, money, labour, equipment, and access available.
- *Avoiding bias* when choosing where to take samples will increase the validity of your results.
- Sampling, no matter how good, can only claim to represent a close estimate of an ecosystem's biodiversity. It is usually assumed that the population is normally distributed (follows the classic bell-shaped curve either side of the mean). This means that 95% of the sampled values will be within plus or minus two standard deviations from the mean.

The design of your sampling method will depend on the ecosystems being studied. Two methods for assessing biodiversity are **random sampling** and systematic sampling.

Random sampling

You can decide the location of sampling points in an ecosystem by generating random numbers. These random numbers are used as grid coordinates, and samples will be taken from these coordinates. You can place a square frame, called a quadrat, at each set of coordinates. The species present in the quadrat can be observed, identified with a key, and counted. Some quadrats are divided into a grid of smaller squares, which will help you to estimate the percentage cover of different species.

Synoptic link

You will learn how to calculate biodiversity in Topic 21.2, Biodiversity and food security.

Random sampling avoids bias, but it can produce an unrepresentative picture of an ecosystem. This is especially true if the surveyed area is large. In addition, some species may be unevenly distributed and found only in certain parts of the ecosystem. Random sampling could miss these species, especially if your sample size is small because of time constraints. Stratified sampling can overcome this problem.

Stratified sampling and transects

Other forms of sampling can be systematic rather than random. In stratified sampling the ecosystem being sampled can be divided into smaller areas, based on the distribution of habitats. This method ensures that sets of species are not overlooked. The sampling is more representative of the ecosystem.

For example, you might estimate that dense tree growth represents 80% of a woodland ecosystem and the remaining 20% consists of a shrubby clearing. These two areas are likely to contain different communities. If you plan to take 20 samples, 16 should be from the area with trees and four should be from the clearing. The number of samples is therefore proportional to the size of each area. Within each sub-region of your ecosystem, the location of each sample should be decided randomly.

Stratified sampling produces a more representative survey of an ecosystem's biodiversity by reducing sampling error. Bias can be avoided by using random sampling within each of your sub-regions. However, the relative size of each sub-region must be estimated accurately if stratified sampling is to provide a truly representative estimate of biodiversity.

A transect is used where a correlation may exist between an abiotic variable, such as soil pH or light intensity, and the distribution of organisms. This sampling method is illustrated in Figure 3.

Synoptic link

You learnt about the different definitions of biodiversity (genetic, species, and ecosystem diversity) in Topics 10.6, Biodiversity. You will learn how to calculate biodiversity from your sampling data in Topic 21.2, Biodiversity and food security.

Study tip

Calling a *quadrat* a *quadrant* is a common mistake in exams. Ecological sampling is carried out with a quadrat. Remember to leave out the 'n'.

▲ **Figure 3** *Students using a quadrat at intervals, along a line known as a transect, to sample biodiversity systematically. This is referred to as an interrupted belt transect*

Summary questions

1 Explain what is meant by a plagioclimax, and identify one example resulting from land management. *(2 marks)*

2 Describe the difference between succession and deflected succession, and explain why succession is likely to result in greater biodiversity. *(3 marks)*

3 Imagine you are planning to assess the biodiversity of an area of farmland. The farmland has an area of $1\,km^2$, comprising $7.00 \times 10^5\,m^2$ of grazed grassland, a shrubby area of $2.00 \times 10^5\,m^2$, and a small orchard of $1.00 \times 10^5\,m^2$.
   Outline the sampling method that you would use to produce the most valid representation of this ecosystem's biodiversity. *(6 marks)*

Practice questions

1   Which of the following statements is/are true of the changes that DCPIP undergoes when it reacts with products of the light-dependent reactions of photosynthesis?

    Statement 1: DCPIP is oxidised

    Statement 2: DCPIP becomes colourless

    Statement 3: DCPIP accepts electrons

    A   1, 2 and 3

    B   Only 1 and 2

    C   Only 2 and 3

    D   Only 1   *(1 mark)*

2   A pigment has an R_f value of 0.60. A student uses paper chromatography. The pigment travels 5.33 cm. How far (to two significant figures) does the solvent in which the pigment is dissolved travel?

    A   3.1 cm   **B**   3.2 cm

    C   8.8 cm   **D**   8.9 cm   *(1 mark)*

3   One glucose molecule is produced for every six turns of the Calvin cycle. The mass of one molecule of RuBP is approximately 1.7 times greater than the mass of one molecule of glucose. What is the mass of RuBP, to two significant figures, required to produce 5.75 kg of glucose?

    A   5800 g

    B   5900 g

    C   58 000 g

    D   59 000 g   *(1 mark)*

4   A photosynthometer was used to investigate the effect of temperature on photosynthetic rate. At 20 °C, an oxygen bubble 2.5 cm long was collected over a one minute period. The radius of the photosynthometer was 0.07 cm. What was the rate of photosynthesis to two significant figures?

    A   0.038 cm^3 min^{-1}

    B   0.040 cm^3 min^{-1}

    C   0.54 cm^3 min^{-1}

    D   0.55 cm^3 min^{-1} *(1 mark)*

5   A farmer grew 80 000 kg of a crop in a 480 m^2 field over a period of one year. What is the net primary productivity of this crop (units = g m^{-2} yr^{-1})?

    A 1.67×10^2

    B 1.67×10^3

    C 1.67×10^4

    D 1.67×10^5 *(1 mark)*

6   **a**   An experiment was carried out into the effect of different wavelengths of light on the rate of photosynthesis.

       Four sealed test-tubes were set up, each containing three leaf discs from the same plant suspended above hydrogencarbonate indicator solution. This solution changes colour at different pH values, as shown here.

Yellow	orange-red	purple
decreasing pH		increasing pH
←		→

       At the start of the experiment the contents of all four tubes were orange-red.

       Each tube was illuminated by a lamp with a coloured filter in front of it. The tubes were illuminated for the same length of time. The colour changes were noted and the results are shown here.

Colour of filter	Final colour of hydrogencarbonate indicator
colourless	purple
blue	purple
green	orange-yellow
red	red

       A fifth tube was set up in the same way as the other tubes. This tube was then covered in black paper before being illuminated for the same length of time. The final colour of the hydrogencarbonate indicator in this tube was yellow.

       (i)   State the purpose of the tube covered with black paper.   *(1 mark)*

       (ii)   State two precautions that need to be taken when designing and carrying out this experiment in order to obtain results from which valid conclusions can be drawn. Explain the need for each precaution.   *(2 marks)*

(iii) Name the pigment at the reaction centre of photosystems I and II.
(1 mark)

(iv) Explain the change observed in the tube exposed to green light. *(3 marks)*

b In order to maximise production, market gardens often grow plants in glasshouses. Light conditions can be controlled along with a number of other factors.

How can factors other than light conditions be controlled to increase the rate of photosynthesis and maximise production? *(4 marks)*

OCR F214 2010

7 A student carried out an experiment to investigate the effect of light intensity on the rate of photosynthesis in aquatic plants. The apparatus the student used is shown here.

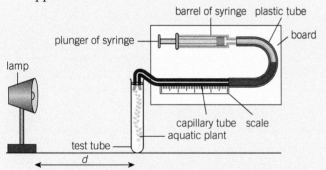

The student decided to measure the rate of photosynthesis by measuring the gas produced over a five minute period. The gas was collected in the capillary tube.

After five minutes, the length of the bubble was measured along the scale.

The light intensity was varied by altering the distance (*d*) between the lamp and the plant.

The student prepared the table below to calculate light intensity.

Distance (*d*) from lamp to plant (cm)	Light intensity $(\frac{1}{d^2})$
4	0.0625
8	0.0156
12	0.0059
16	0.0039
20	0.0025
24	
60	0.0003

a (i) Calculate the light intensity when the lamp was 24 cm from the plant.
(2 marks)

(ii) The length of the gas bubble was measured. State what additional information would be required to calculate the volume of gas produced.
(1 mark)

(iii) Suggest how the student supplied the aquatic plant with a source of carbon dioxide. *(1 mark)*

b (i) One assumption made during this experiment is that all of the oxygen produced by the plant is collected.

Suggest why not all of the oxygen produced by the plant is collected.
(2 marks)

(ii) Another assumption is that all of the gas collected is oxygen. Analysis of the gas collected reveals that 50% is oxygen, 44% is nitrogen and 6% is carbon dioxide.

Suggest a reason for the presence of nitrogen in the gas collected. *(1 mark)*

(iii) Comment on the percentage of carbon dioxide present in the gas collected and give reasons for this figure. *(3 marks)*

c Some aquatic photosynthetic organisms, such as seaweeds, contain pigments such as fucoxanthin and phycoerythrin, in addition to chlorophyll. These pigments give seaweeds a brown or red colour, and are produced in larger quantities in those seaweeds that live in deeper water.

Suggest why the presence of these pigments is an advantage to seaweeds that live in deeper water. *(2 marks)*

OCR F214 2011

8 Discuss the methods that can be employed on farms to increase primary and secondary productivity. *(6 marks)*

21 THE IMPACT OF POPULATION INCREASE

21.1 Human population growth and ecosystems

Specification reference: 4.3.2

In Chapter 20 you learnt about some of the ways by which humans affect ecosystems, such as agricultural and forestry land management. Our influence on ecosystems is a more significant issue now than at any time in human history because of the continuing rise in the global population. In this chapter, you will learn about some of the factors that influence the size of human populations and the impact of population changes on biodiversity.

Factors affecting birth and death rates

The size of a population is affected by four factors – immigration (the number of individuals entering the population from elsewhere), emigration (the number of individuals leaving the population), birth rate (the number of offspring born per year), and death rate (the number of deaths per year).

Immigration and emigration affect populations within an area rather than the global population. However, global birth rate currently exceeds death rate, and the global human population is undergoing exponential growth.

As countries develop economically, their populations tend to increase and then stabilise. This is illustrated in Figure 1, which is an example of a demographic graph. It shows the relationship between birth rate, death rate and population size.

Countries exhibit different birth and death rates. These rates are determined by a variety of factors, some of which are discussed below.

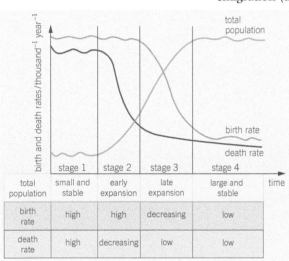

▲ Figure 1 A demographic graph showing the relationship between birth rate, death rate, and population size

	stage 1	stage 2	stage 3	stage 4
total population	small and stable	early expansion	late expansion	large and stable
birth rate	high	high	decreasing	low
death rate	high	decreasing	low	low

Food production

You have already learnt about some of the methods employed in modern food production. Intensive farming has resulted in excess food production in some regions, such as Europe and the USA. An increasingly secure food supply has meant that birth rates have increased and death rates have decreased in many nations.

Other countries, however lack the resources to carry out intensive farming. The United Nations estimates that more than 800 million people worldwide are undernourished. Advances in agriculture have yet to benefit everyone. In some nations, death rates are high because of malnutrition.

Medical technology

Medical technology has advanced rapidly during the last few decades. Vaccinations and antibiotics have been improved. New infertility treatments have been developed. These medical advances have boosted life expectancy and fertility in the most economically developed nations.

Huge inequalities, however exist in the availability of these treatments. Death rates remain comparatively high in economically underdeveloped regions such as sub-Saharan Africa and Central America. Birth rates in less economically developed nations are relatively high because birth control is less common. The economic prosperity of families depends on having large numbers of children who can earn and support the family.

Disease control

Diseases can be transmitted more easily in large populations. Disease vulnerability is therefore known as a density-dependent factor – its influence is dependent on population size.

Diseases such as HIV and tuberculosis continue to contribute to high death rates in many economically less developed nations. Conditions such as diarrhoea, which is controlled by water treatment and sewage disposal in economically developed countries, is a major cause of death in less economically developed countries.

How has the rising human population affected ecosystems?

Impact on abiotic factors

Remember that abiotic factors are the non-living aspects of an ecosystem, such as climate, soils, and water quality. Human populations have affected abiotic aspects of ecosystems in many ways.

Fertilisers can be washed from farmland and pollute aquatic ecosystems. This can cause eutrophication, which reduces the populations of many species.

The demand for water by rising human populations has resulted in rivers and lakes being drained, destroying aquatic ecosystems. The Aral Sea (which is really a saltwater lake) has shrunk to 10% of its original volume because the rivers that fed it have been diverted for irrigation projects. Its ecosystem has collapsed.

Salinisation (an increase in the salt content in soils) is one of the most widespread soil degradation processes on the Earth and can occur due to irrigation. Dissolved salts in the irrigation water are deposited in soils. Soil salinisation affects an estimated 1 million hectares in the European Union, mainly in the Mediterranean countries. It is a major cause of desertification. In Spain, 3% of the 3.5 million hectares of irrigated land is severely affected, reducing its agricultural potential, while another 15% is under serious risk.

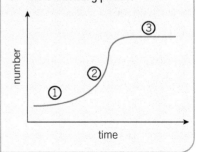

Study tip

The stages of a population growth curve are shown in the graph below. 1 = the lag phase, 2 = the exponential phase, and 3 = the stationary phase.

Synoptic link

You learnt about HIV and tuberculosis in Topic 11.1, Communicable diseases.

Study tip

Eutrophication results from an increase in nutrients in freshwater lakes and rivers. These nutrients can enter the water from sewage or fertilisers. The nutrients encourage the growth of algae, which prevents sunlight reaching deeper aquatic plants. When these plants die, decomposers deplete oxygen in the water, which further decreases biodiversity.

▲ Figure 2 *Eutrophication – an algal bloom resulting from nitrogen fertiliser being washed into a canal*

▲ **Figure 3** *The dried bed of the Aral Sea*

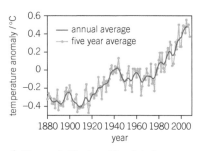

▲ **Figure 4** *The trend in global temperature anomalies (expressed as deviation from the mean global temperature between 1961 and 1990)*

Synoptic link

This topic links together threads from previous chapters. You have previously touched on some of the factors that influence the size of human populations, such as advances in agricultural methods (Chapter 20, Photosynthesis and environmental management) and medical technology (Topic 14.1, Non-communicable diseases).

Evidence suggests that emissions of carbon dioxide, methane, and other greenhouse gases from human activities have resulted in global climate change. Some climate scientists estimate that the mean global temperature could increase by up to 4.8°C by 2100, compared with pre-industrial levels.

Impact on biotic factors (biodiversity)

The abiotic changes already discussed can impact adversely on organisms within ecosystems. For example, global warming has been linked to the extinction of frog species. Frogs are sensitive to slight changes in temperature because of their thin skin, and temperature increases make them more vulnerable to the chytrid fungus.

Human activities can also damage organisms directly. Some estimates indicate that up to 50 000 species are lost each year because of deforestation. Tropical rainforests cover only 7% of the Earth's surface, but they are the habitat of half the world's species. By 2030, more than 80% of the rainforest in the whole world is likely to have been removed.

Agricultural monocultures, where only one crop species is grown and others are eliminated, reduces plant diversity and has the knock-on effect of removing food sources for birds and insects in the food web.

Pesticides such as insecticides and herbicides can kill organisms that are not being targeted. As you learnt in Topic 20.8, Farms as ecosystems, evidence suggests that neonicotinoid pesticides are toxic to bees, and these chemicals have been blamed for declining populations of bee species.

Summary questions

1   Describe how the rising human population has affected **a** soil quality **b** biodiversity in aquatic ecosystems.   *(4 marks)*

2   Examine the demographic graph in Figure 1. Describe and explain the change in population size during stages 2 and 3.   *(2 marks)*

3   Population pyramids illustrate the distribution of ages within populations. The two graphs in Figure 5 show age distribution in two countries with similar populations. State which country has the faster population growth. Explain your answer.   *(2 marks)*

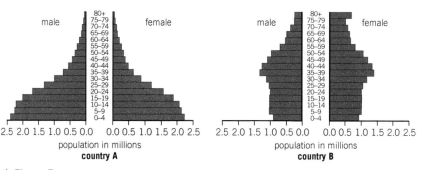

▲ **Figure 5**

21.2 Biodiversity and food security

Specification reference: 4.3.2

In the previous topic, you learnt about how the rising human population has affected ecosystems and biodiversity. Here, you will learn about how biodiversity is calculated using Simpson's diversity index. Another challenge resulting from the increasing human population is the pressure to produce sufficient food in a sustainable manner. In this topic you will examine the issues surrounding global food security, and the role of the United Nations Food and Agriculture Organisation in promoting food safety and sustainability.

Measuring biodiversity

The maintenance of biodiversity is important on ecological, scientific, and economic levels. High genetic diversity increases the chance of species adapting to future environmental change. Approximately half of our medicines contain a chemical derived from animal or plant species. Extinct species may have possessed genes that would have been useful in medical research or agriculture in the future. Of course, the aesthetic aspect of biodiversity is important as well.

Scientists monitor biodiversity to assess the impact of human activities on ecosystems. Species diversity can be calculated using the Simpson's diversity index (a number between 0 and 1).

Simpson's diversity index $(D) = 1 - [\Sigma \, (n/N)^2]$

where n is the number of individuals of a particular species and N is the total number of all individuals of all species. Remember, Σ is the symbol for 'sum'. In this case, you need to calculate $(n/N)^2$ for each species and then add the values together.

Learning outcomes

Demonstrate knowledge, understanding, and application of:

→ the ecological, scientific, and economic importance of species biodiversity

→ the use of statistical methods to assess species biodiversity

→ the importance of sustainable and secure food production and consumption.

Synoptic link

You learnt about about the animal and plant origins of some medicines in Topic 15.2, Medicinal drugs and clinical trials.

Synoptic link

In Topic 10.6, Biodiversity, you were introduced to the idea of species diversity, you also learnt how to calculate genetic diversity.

 Worked example: An example of the Simpson's diversity index calculation

	Field A (adjacent to intensively farmed land)			Field B (adjacent to extensively farmed land)		
	n	n/N	$(n/N)^2$	n	n/N	$(n/N)^2$
cocksfoot grass	60	0.60	0.3600	40	0.40	0.1600
timothy grass	29	0.29	0.0841	20	0.20	0.0400
meadow buttercup	4	0.04	0.0016	13	0.13	0.0169
cowslip	1	0.01	0.0001	8	0.08	0.0064
white clover	2	0.02	0.0004	13	0.13	0.0169
dandelion	4	0.04	0.0016	6	0.06	0.0036
Σ	-	-	0.4478	-	-	0.2438
$1-\Sigma$	-	-	0.5522	-	-	0.7562

In this example, n is the percentage cover of each plant species. N is therefore 100.

A high biodiversity is reflected by a high value of Simpson's diversity index. Low values indicate that an ecosystem is dominated by a few species and is unstable. In this example, field B has the higher Simpson's diversity index value and you can conclude that its biodiversity is greater than that of field A.

▲ **Figure 1** *Mystery shopping? Consumer trust in the content of food has decreased in the wake of recent scandals*

Global food security

Food security centres on the concept of ensuring that populations have access to sufficient amounts of safe and nutritious foods. However, global food security faces many challenges.

The rising human population is creating more demand for food. This has increased the need to employ sustainable practices in food and energy production. **Sustainability** involves meeting energy and food requirements today without compromising biodiversity or the ability to meet requirements in the future.

The Environmental Stewardship scheme in the UK promotes sustainable practices on agricultural farms. Fish farming and a greater reliance on primary consumer fish can improve sustainability. Populations of many fish species, notably cod and tuna, have been depleted to dangerously low levels. The European Union has established quotas that limit the numbers of fish caught per nation. However, some environmental scientists fear that without a complete ban on the fishing of some species, population numbers will fail to recover.

Food safety is another challenge facing the global food industry. There is a danger of hygiene and safety standards dropping as food production methods become increasingly intensive. In the UK, the Food Standards Agency works with businesses and local authorities to ensure that food safety regulations are enforced.

Food crime has increased to an alarming extent during the past decade. Food fraud, for example, involves food being sold with the intention of deceiving the consumer. Several examples have arisen in the last few years. In 2013, processed beef products in the UK were found to be contaminated with horsemeat. The chemical melamine was discovered in baby milk in China in 2008 and, in 2012, vodka in the Czech Republic was found to contain methanol. Both incidents resulted in deaths. 1700 tonnes of manuka honey, a product unique to New Zealand, is made each year. Yet 10 000 tonnes per year of honey branded 'manuka' is sold worldwide. Understandably, consumer trust in the food industry has ebbed in recent years.

The role of the UN Food and Agriculture Organisation

The United Nations Food and Agriculture Organisation (FAO) is in charge of protecting global food security. The FAO's main aims are to reduce hunger, poverty, food insecurity, and malnutrition, and to ensure that natural resources are managed sustainably. The FAO's Food Chain Crisis (FCC) Management Framework branch deals with food safety.

FCC management tackles threats to the human food chain, such as diseases, pests, and food safety emergencies. The Emergency Prevention System (EMPRES) is a fundamental component of the FAO's FCC management. Some of the recent global crises that the FAO has faced include outbreaks of avian influenza (H5N1) – linked to chickens, swine influenza (H1N1), locust infestations, and dioxin contamination of food. EMPRES activities include promoting research and assessment of potential crises in the food chain, and coordinating early warnings at the beginning of outbreaks. During epidemics of pests and diseases, EMPRES promotes containment and control of the outbreak.

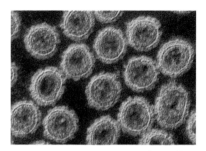

▲ Figure 2 *Avian influenza (H5N1) virus particles, ×113 300 magnification*

> ## Synoptic link
>
> You learnt about the different measures of biodiversity in Topic 10.6, Biodiversity.

Summary questions

1 Give one example of each of the following:
   a sustainable food production methods **b** a food safety crisis
   c a food crime. *(3 marks)*

2 Explain the importance of maintaining high species diversity. *(3 marks)*

3 Calculate the Simpson's diversity index for the two habitats in Table 1. Which habitat is more likely to be sensitive to environmental change? *(3 marks)*

▼ Table 1

Species	Habitat A	Habitat B
woodrush	1	4
holly (seedlings)	11	7
bramble	1	3
Yorkshire fog	1	2
sedge	2	3

Practice questions

1 Which of the following statements is/are a factor or factors that need to be known in order to calculate Simpson's diversity index?

Statement 1: The class of each species.

Statement 2: The number of individuals of each species.

Statement 3: The total number of individuals of all species.

A 1, 2 and 3

B Only 1 and 2

C Only 2 and 3

D Only 1 *(1 mark)*

2 Which of the following statements is/are an example or examples of the role played by the UN Food and Agriculture Organisation in protecting global food security?

Statement 1: Food chain crisis management.

Statement 2: Coordinating early warning systems during disease outbreaks.

Statement 3: Setting laws to regulate food safety.

A 1, 2 and 3

B Only 1 and 2

C Only 2 and 3

D Only 1 *(1 mark)*

3 Which of the following statements is true of eutrophication?

A A decrease in nutrients in aquatic habitats eventually results in the death of plant species.

B An increase in nutrients in aquatic habitats eventually results in the death of plant species.

C The death of algae depletes oxygen in aquatic habitats.

D The growth of algae increases oxygen content in aquatic habitats. *(1 mark)*

4 Which of the following statements represents a density-independent factor that can influence population size?

A Competition

B The spread of disease

C Climate

D Parasitism *(1 mark)*

5 a The graph shows the birth rate and death rate for the UK, Ghana and Nigeria for the year 2006.

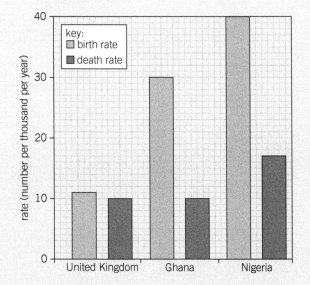

(i) Suggest reasons for the difference in birth rate between the UK and Nigeria. *(2 marks)*

(ii) Suggest why Ghana has the same death rate as the UK despite being a less economically developed country. *(2 marks)*

b Factors affecting the size of populations are termed density-dependent (DD) and density-independent (DI) factors.

Suggest whether each factor in the table is density-dependent or density-independent. *(2 marks)*

Factor	DD or DI
many people died due to an increase in malaria in Western Africa	
drought caused famine in parts of the Sudan	
a tsunami destroyed fishing villages on the east coast of India	

OCR F224 2012

6 Students collected some data about plant species in an area of ash woodland. Their results are shown in the table.

species	Number of individuals (*n*)
dog's mercury	40
wild strawberry	13
common avens	43
wood sorrel	4

Calculate the Simpson's index of diversity for the area of woodland. *(3 marks)*

OCR F212 2013

7 An ecologist collected data for butterfly species in an area of heathland. The data are shown in the table.

species	N
grayling (*Hipparchia Semele*)	3
large Heath (*Coenonympha tullia*)	11
gatekeeper (*Pyronia tithonus*)	6
green Hairstreak (*Callophrys rubi*)	2
silver-studded Blue (*Plebeius argus*)	2
small Heath (*Coenonympha pamphilus*)	7

a Calculate the Simpson's index of diversity for butterflies in this heathland ecosystem. *(3 marks)*

b Suggest the implications of a high value of Simpson's index of diversity in terms of plans to develop buildings on the heathland. *(2 marks)*

OCR F212 2010

8 The figure shows the relationship between annual deforestation rate and annual human population growth for six countries.

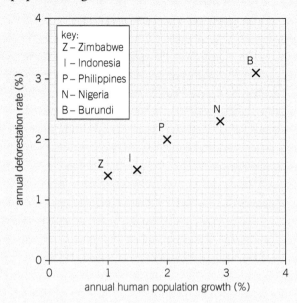

a Describe the relationship shown in the figure. *(2 marks)*

b Deforestation is one factor linked to increasing atmospheric carbon dioxide concentration.

Suggest ways by which individuals can reduce their impact on increasing atmospheric carbon dioxide concentration. *(3 marks)*

OCR F224 2010

9 Describe how the rise in human population has affected abiotic factors in ecosystems and explain the impact these changes have had on biotic factors. *(6 marks)*

22 PLANT REPRODUCTION
22.1 Plant reproduction – flowering and pollination

Specification reference: 4.4.1

You learnt about the human reproductive system in Chapter 18, Fertility and assisted reproduction. In this chapter you will learn about reproduction in flowering plants. You will first look at how plants control the development of their flowers and the strategies used by different species for pollination.

Pollination

Plants, like animals, can reproduce sexually, and they possess male and female gametes. The male gametes are found within **pollen** grains and the female gametes are located in **ovules**. In order for plants to reproduce, pollen from structures called **anthers** on one plant must be transferred to a structure called a **stigma** on another plant.

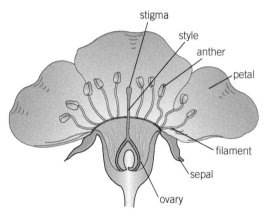

▲ **Figure 1** *The structure of a plum flower*

Study tip

Plant species such as wheat are hermaphrodites – their flowers produce both male and female gametes. These plants often self fertilise and are described as 'inbreeders'. Maize, in contrast, has separate male and female flowers and tends to 'outbreed'. Some hermaphrodite plants have adaptations that limit inbreeding.

Pollen can be transferred between flowers by either wind, insects, or other animals, such as birds or even bats, depending on the species of plant. Cereal plants rely on wind pollination, whereas legumes such as peas and bean plants use insect pollination. The type of pollination strategy used by a species determines the adaptations that it has evolved, such as the structures of the flowers and pollen grains.

Insect pollination adaptations

Species that rely on insects to transfer their pollen must have evolved a method of attracting these pollinators to their flowers. Many species, including legumes such as the pea plant (*Pisum sativum*), produce sugary liquid called nectar. This sweet liquid attracts insects such as bees and butterflies, and non-insect pollinators such as bats and hummingbirds. Pollen attaches to these animals when they feed on the nectar. The animals then deposit the pollen on the stigma of the next flower they visit.

Many species attract pollinators by using colourful and fragrant flowers, which exploit sensory preferences in insects. In addition to containing nectar with a high sugar content, the flowers of the crop plant oilseed rape are bright yellow, which attracts bee pollinators. Plants in the genus *Rafflesia* smell of rotting flesh to attract flies. Many species of orchid have unusual adaptations for ensuring pollen transfers. For example, lady's slipper orchids trap insects in pouches, forcing them to exit via the flower's anthers. Bee orchids (genus *Ophrys*) have flowers that mimic female insects. Pollen attaches to males when they attempt to mate with the flowers.

Wind pollination adaptations

Species that rely on wind to disperse their pollen grains do not require bright or scented flowers to attract insects. The most important adaptations for these plants are the size and quantity of their pollen. Light pollen grains are produced that can be carried through the air with ease. The pollen is produced in large amounts because many grains will fail to reach other plants. The anthers and stigma of wind-pollinated flowers hang outside the flower, exposed to air currents. Their stigmas tend to have large surface areas (Table 1).

Control of flowering

The growth and development of plants tends to be dependent on temperature. Agronomists use the concept of thermal time, measured as heat units, or growing day degrees. Temperature varies within a day and on successive days. The amount of heat a plant experiences in a day is its daily heat unit and a set number of heat units is required by the plant between developmental stages. Calculating heat units allows the timing of plant development to be predicted, which enables the application of fertiliser or other chemicals to be timed. Areas can be mapped in terms of their suitability for growing particular crops.

Many temperate plant species must be exposed to low temperatures (for example, 5 °C to 10 °C) for an extended period before they can flower. This is known as **vernalisation**. The cold temperatures during vernalisation promote the transcription of genes associated with flowering. Some species flower after vernalisation without further environmental signals. However, other plants require additional seasonal cues, such as changes in day length, following vernalisation.

Plants can measure changes in the length of night and day. This enables them to flower only when environmental conditions are favourable.

▲ **Figure 2** *Flowering couch grass (Agropyron repens) showing feathery white styles and yellow anthers exposed to the wind*

▲ **Figure 3** Ophrys *(bee orchid) flowers, which attract male bees by imitating females*

▼ **Table 1** *A comparison of features of wind and insect pollination*

Wind-pollinated flowers	Insect-pollinated flowers
often green or dull colours	bright colours
lack fragrance	often fragrant
small petals	large petals
no nectar	nectar (a sugar-rich liquid) produced in nectary
anthers usually outside the flower	anthers tend to lie deep inside the flower
light pollen grains	large pollen grains, often sticky
many pollen grains produced	few pollen grains produced
stigma is feathery and hangs outside the flower	stigma lies within the flower's petals

Investigating flowering – the long and the short of it

Cocklebur plants have been used to study the effect, on flowering, of exposure to light and dark periods. A scientist called Karl Hamner had been studying the relationship between nutrition and flowering in cockleburs. He had the idea to investigate the effect of light regimes only when he discovered that his plants had flowered after a power cut.

The table shows some results from experiments with cocklebur.

Period of darkness (hours)	Flash of light during the period of darkness?	Flower production
8.5	no	yes
6.5	no	no
12.5	red light after 6 hours	no
12.5	red light after 6 hours, followed by flash of far red light	yes
6.5	several flashes of far red light	yes

1 What effects do red light and far red light have on flowering in cocklebur?
2 Explain whether cocklebur is a long-day or short-day plant.

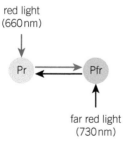

red light
(660 nm)

Pr ⇌ Pfr

far red light
(730 nm)

▲ **Figure 4** *Conversion between the two forms of phytochrome enables plants to measure day length*

Synoptic link

You learnt about aspects of human reproduction in Chapter 18, Fertility and assisted reproduction.

Long-day plants

Long-day plants flower during the late spring when day length rises above a critical value. Some plants (for example, oat and clover) cannot flower without experiencing this increase in day length. Other plants (for example, lettuce, barley, and peas) flower more quickly when they experience extended daylight hours, yet they will flower eventually regardless of the length of day.

Short-day plants

Short-day plants flower as nights grow longer and day length decreases below a critical value. They include species such as soya beans, rice, and cotton.

Experiments have shown that the length of the dark period is more important than the length of the light period in controlling flowering. For example, short-day plants will not flower if their dark period is interrupted by a flash of light, but interrupting their light period with darkness has no effect. It might therefore be more accurate for short-day plants to be called long-night plants, and long-day plants to be called short-night plants.

How do plants monitor photoperiods?

Flowering plants use a photoreceptor (a light-sensitive protein) called **phytochrome** to measure changes in day length. Phytochrome exists in two forms (known as isomers) – a form (Pr) that absorbs red light (660 nm) and a form (Pfr) that absorbs far red light (730 nm). Pr is converted to Pfr when a plant is exposed to red light. Exposure to far-red light (730 nm) or darkness converts Pfr back to Pr.

Pfr accumulates during the day, when plants are exposed to red light. Pfr is converted back to Pr during the night. The back-and-forth conversion between the two forms enables plants to measure day length. The relative proportions of Pr and Pfr are therefore thought to dictate when flowering is initiated.

Summary questions

1 Describe three ways in which plants can attract animal pollinators. (*3 marks*)

2 Describe how a decrease in night length would affect the following: a Pfr concentration in plant cells at dawn
b the probability of a short-day plant flowering. (*2 marks*)

3 Ragweed flowers when day length falls below 14.5 hours. In northern Maine, USA, day length drops below 14.5 hours in August, which can be as little as one month before frost first appears. Suggest why ragweed is not found in northern Maine. (*2 marks*)

22.2 Plant reproduction – seed formation and germination

Specification reference: 4.4.1

In the previous topic you learnt about how plants reproduce. You learnt how the development of flowers is controlled and how flower structure differs between species. In this topic you will investigate the next stages in the story of plant reproduction – fertilisation, seed formation, and seed **germination**.

Seed structure and formation

An endospermic seed, such as those of cereals, consists of three parts:

- the embryo (which is formed by successful fertilisation and will develop into an adult plant)
- the **endosperm** (a food source surrounding the embryo)
- the seed coat (a protective outer layer).

The fully developed seeds of some species, such as peas and beans, lack an endosperm. During the development of these seeds, the seed leaves, known as cotyledons, absorb food reserves from the endosperm. The cotyledons rather than an endosperm serve as the food source for the mature seed.

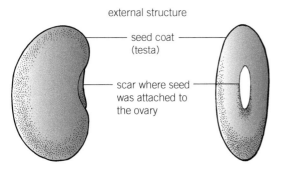

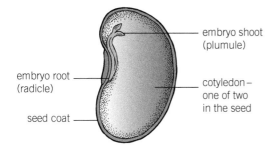

▲ **Figure 2** *The structure of a non-endospermic bean seed*

Fertilisation

Once a pollen grain has landed on the stigma of another plant, a pollen tube grows from the grain down towards the ovule. A pollen tube is a single cell, although it can grow longer than 30 cm in species

Learning outcomes

Demonstrate knowledge, understanding, and application of:

→ fertilisation and seed formation

→ the germination of seeds and investigations into factors affecting germination

→ the importance of cereals as staple foods.

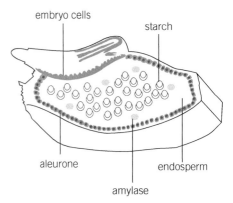

▲ **Figure 1** *The structure of a barley seed. The aleurone protein layer releases amylase during germination*

such as maize. The role of the pollen tube is to allow the passage of male gametes to the ovary, where they will fertilise the ovule.

The pollen tube delivers two sperm cells at the same time. One sperm cell fertilises an egg cell to form an embryo, which will develop into the future plant. The nucleus in the other sperm cell fuses with the two polar nuclei at the centre of the embryo sac to form the endosperm, which serves as the embryo's food supply. This is known as double fertilisation. The ovules develop into seeds, which contain the embryo and endosperm. The surrounding ovary can develop into a fruit.

Endosperm formation

The endosperm tissue surrounds the plant embryo inside a seed. It consists principally of starch, although it often contains oils and proteins.

Seed germination

Germination is the process by which a plant grows from a seed. **Gibberellins** are plant hormones that are essential for the germination process. They act as cell signalling molecules. At the beginning of germination, after a seed is exposed to water, starch in the endosperm is hydrolysed to glucose. Gibberellins initiate the breakdown of starch by inducing the synthesis of amylase.

Investigations into factors affecting germination

Seed germination is dependent upon both the conditions within the seed and external environmental conditions (for example, temperature, water, oxygen, and light levels). The conditions that encourage seed germination will differ between species.

The effects of variables on germination can be investigated experimentally. As with any process that is influenced by several factors, the effect of each variable should be analysed in isolation. All other variables (control variables) should be maintained at constant values, while only the value of the variable of interest (the independent variable) is altered.

Cress seeds are often used in germination experiments. The seeds can be placed in sets of Petri dishes containing solutions suitable for germination. One factor, such as temperature, is then selected as the independent variable. Seeds are assigned to groups exposed to different temperatures. Measurements of dependent variables such as root growth, shoot growth, and the number of seeds germinating during a set time period are easy to make. The age and variety of the seeds should be the same in each of the temperature groups – this adds a further element of control to the experiment.

Cereals as staple foods

The endosperm within a seed can act as a source of nutrition for humans. Endosperm is rich in starch and is therefore a good source of carbohydrate. The nutrition in many staple foods, such as rice, maize, and wheat is largely derived from the endosperms of the plants. Wheat endosperm can be ground into flour to produce bread.

▼ Table 1 *World cereal production, supply, and use*

Type of cereal	Production (million tonnes)		Supply (million tonnes)		Use (million tonnes)	
	2010/11	2014/15	2010/11	2014/15	2010/11	2014/15
all cereals	2253.8	2533.5	2777.1	3110.2	2271.0	2466.9
wheat	653.8	724.3	843.6	899.2	659.9	702.0
coarse grain (including maize)	1130.1	1313.1	1325.8	1533.7	1149.5	1264.8
rice	469.8	496.2	607.6	677.3	461.6	500.1

As Table 1 shows, the use of cereal crops is increasing, which means that annual production has risen to meet demand. However, in some years the global consumption of a crop may exceed its production (for example, more rice was used than was produced in 2014/15). In these cases, stocks remaining from the previous year are utilised to meet the demand. 'Supply' comprises the annual production of a particular crop plus the stockpiles that have been carried over from previous years.

Food security in relation to staple foods has become an important global issue. The world's population is increasing, raising the demand for food, but less land is available and access to water is limited in many regions. Furthermore, the costs of fertilisers and fuels are increasing. Producing sufficient food, using sustainable methods, is becoming increasingly difficult.

Synoptic link

You learnt about the ideas of food security and sustainability in Topic 21.2, Biodiversity and food security.

Summary questions

1 Describe how a plant embryo is formed. *(3 marks)*

2 Outline a method for analysing the effect of temperature on germination in a plant species. *(4 marks)*

3 Explain the role of gibberellins and endosperm during germination. *(4 marks)*

Practice questions

1   Which of the following statements is/are generally true of wind-pollinated flowers?

   Statement 1: Produce few pollen grains

   Statement 2: Have small petals

   Statement 3: Lack fragrance

   A   1, 2 and 3

   B   Only 1 and 2

   C   Only 2 and 3

   D   Only 1   (*1 mark*)

2   Which is the most accurate description of vernalisation?

   A   Exposure to warm temperatures triggers flowering.

   B   Exposure to cold temperatures triggers flowering.

   C   Exposure to warm temperatures triggers germination.

   D   Exposure to cold temperatures triggers germination.   (*1 mark*)

3   Which of the following scenarios is/are likely to result in flowering in short-day plants?

   1   An increase in night length beyond a critical period.

   2   Exposure to red light during the night.

   3   An increase in Pfr concentration within the plant.

   A   1, 2 and 3

   B   Only 1 and 2

   C   Only 2 and 3

   D   Only 1   (*1 mark*)

4   What is the role of the endosperm in a seed?

   A   Provides a protective layer.

   B   Develops into an adult plant.

   C   Fuses with a polar nucleus.

   D   Provides a food source.   (*1 mark*)

5   Which of the following statements is/are true of the Pfr form of phytochrome?

   Statement 1: It absorbs far red light.

   Statement 2: It is converted to Pr in periods of darkness.

   Statement 3: It accumulates during the day.

   A   1, 2 and 3

   B   Only 1 and 2

   C   Only 2 and 3

   D   Only 1   (*1 mark*)

6   Outline the importance of cereals as staple foods, including a consideration of sustainable farming practices.   (*6 marks*)

7   A student carried out an experiment to identify the content of two unlabelled bottles. One bottle contained auxin and the other contained gibberellin.

   In part 1 of the experiment, 30 seedlings had their shoot tips removed. The seedlings were divided into three groups of 10 and treated as shown in the table.

Group	Treatment
1	no treatment applied
2	solution J applied to cut stem at apex of seedling
3	solution of K applied to cut stem at apex of seedling

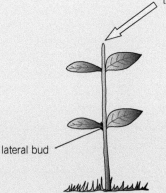

treatments applied to groups 2 and 3

lateral bud

   The seedlings were then exposed to light from all directions and left for seven days.

   In part 2 of the experiment, 30 coleoptiles had their tips removed. They were divided into groups of ten and treated as shown in the table.

Group	Treatment
4	no treatment applied
5	solution of J applied to cut tip of coleoptile
6	solution of K applied to cut tip of coleoptile

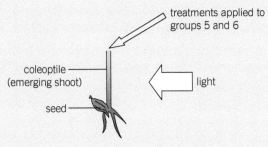

treatments applied to groups 5 and 6

coleoptile (emerging shoot)

light

seed

a Identify three variables that must be controlled in this experiment to produce valid results. *(3 marks)*

b Groups 1 and 4 were controls in this experiment. Explain why these controls were necessary. *(1 mark)*

OCR F215 2014

8 a Plant responses to environmental changes are coordinated by plant growth substances (plant hormones).

Explain why plants need to be able to respond to their environment. *(2 marks)*

b The following investigation was carried out into the effects of plant growth substances on germination:

A large number of lettuce seeds were divided into eight equal batches.

Each batch of seeds was placed on moist filter paper in a Petri dish and given a different treatment.

The different treatments are shown in the table. Each tick represents one of the eight batches.

Treatment	Concentration of gibberellin (mol dm^{-3})			
	0.00	0.05	0.50	5.00
A water	✓	✓	✓	✓
B abscisic acid	✓	✓	✓	✓

The batches were left to germinate in identical conditions and the percentage germination was calculated. The results are shown in the graph.

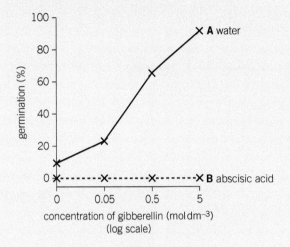

(i) Describe the effects of the plant growth substances on the germination of lettuce seeds. *(4 marks)*

(ii) Explain why all the lettuce seeds were kept at the same temperature (25°C). *(2 marks)*

(iii) State three variables, other than temperature, that needed to be controlled in this investigation. *(3 marks)*

OCR F215 2010

In this chapter you will learn about patterns of genetic inheritance. It draws together ideas you have already encountered, such as the nature of the genetic code (Topic 4.4, The genetic code) and the production of gametes via meiosis (Topic 9.1, Meiosis). For this topic you will need to start by familiarising yourself with some important terms that will crop up throughout this chapter. You will then learn about the pattern of inheritance for a single gene with two possible alleles.

Some important terms

In describing or explaining patterns of inheritance, it is essential that you use the correct terminology. Table 1 summarises the key terms with which you are expected to be familiar.

▼ **Table 1** *Important terms in genetic studies*

Genetic term	Definition
genotype	The genetic composition of an organism, which describes all the alleles it contains. **Genotypes** for a particular locus can be heterozygous or homozygous (later in table).
phenotype	An organism's observable characteristics. The **phenotype** results from the interaction of genotype and environment.
gene	A length/section of DNA that codes for the production of a particular polypeptide.
locus	The **locus** is the position on a chromosome of a particular gene.
allele	A gene variant. A gene can have many different variants. They occupy the same locus. Remember, a diploid organism has two copies of each gene. It could have two copies of the same **allele** or have two different alleles.
dominant	A **dominant** allele is expressed and affects an organism's phenotype even when present with the recessive allele.
recessive	A **recessive** allele is only expressed when the dominant allele is absent.
codominant	Two alleles that both contribute to the organism's phenotype to an equal extent are called **codominant**. (You will learn more about codominance in Topic 23.3, Codominance)
heterozygous	Having two different alleles of a gene.
homozygous	Having two identical alleles of a gene. Homozygous genotypes can be recessive or dominant.

An example – cystic fibrosis

Cystic fibrosis (CF) is a genetic disease that affects 70 000 people worldwide. It is possible to apply the terms in Table 1 to cystic fibrosis.

- The CF gene is known as CFTR and codes for a membrane transport protein that affects the salt composition of sweat and mucus.
- The locus of the gene is on chromosome 7 of the human genome.

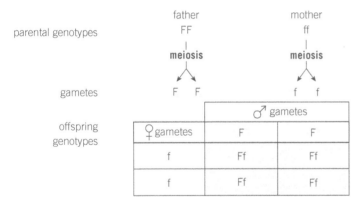

- Most people possess two copies of the correctly functioning CFTR gene variant (F). This is the **dominant** allele. However, this normal allele can be altered by one of hundreds of different mutations to produce a faulty variant (f). This is the **recessive** allele. You will look at examples of genetic mutations in the next topic.

- The **heterozygous** (Ff) and **homozygous** dominant (FF) genotypes code for the production of a functional transport protein (i.e., the healthy phenotype).

- The homozygous recessive (ff) **genotype** leads to a faulty transport protein, and this leads to symptoms such as the production of sticky mucus, infertility (especially in men), poor growth and salty skin (i.e., the CF **phenotype**).

Monogenic inheritance patterns

A monogenic trait is a characteristic controlled by a single gene. When two gene variants exist, the pattern of monogenic inheritance is relatively straightforward to interpret. In Topic 23.3, Codominance, you will learn about more complex patterns of inheritance involving multiple gene variants.

Cystic fibrosis can be used to illustrate monogenic inheritance. In reality, many different mutations can lead to a range of faulty alleles, each of which results in cystic fibrosis. However, for the purpose of illustration, these faulty alleles can be considered as a single CF recessive allele (f). The dominant healthy allele is represented by F.

Genetic diagrams can be drawn to illustrate the probability of inheriting a particular genotype. Here are a couple of examples for CF.

▲ **Figure 1** *Treatment of cystic fibrosis may include regular physical therapy to help clear mucus from the lungs*

▲ **Figure 2** *A genetic diagram showing the possible genotypes of offspring from a healthy father and a mother with cystic fibrosis*

In Figure 2, the father is homozygous dominant (FF) and all his gametes will contain F. The mother is homozygous recessive (ff) and produces only gametes containing f. Any offspring will be heterozygous (Ff) and will not have CF. They will be carriers, however. A carrier is an organism that has a recessive allele but does not express the trait for which it codes.

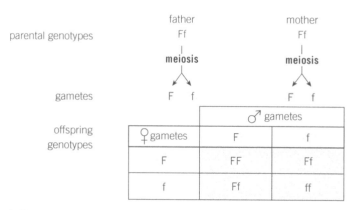

▲ **Figure 3** *A genetic diagram showing the possible genotypes of offspring from two parents carrying the CF allele*

Both parents in Figure 3 are heterozygous. There is a 75% chance that their offspring will be unaffected (i.e., those that have at least one dominant F allele). There is a 50% chance that their child will be a carrier of a cystic fibrosis allele (Ff), and a 25% chance that the child will have cystic fibrosis (ff).

Summary questions

1   Pea pods are either green or yellow. The genotypes GG and Gg code for green pods. The genotype gg codes for yellow pods. Choose two terms (other than genotype) from Table 1 to describe each of the following:
   a GG   b gg   c g   *(6 marks)*

2   Suggest why it is only necessary to consider one recessive allele (f) when analysing the inheritance of cystic fibrosis, despite there being many different mutations to the healthy allele (F) that cause gene variants.   *(2 marks)*

3   Paul and Sandra have a child together. Neither parent has cystic fibrosis, but Sandra is a carrier of the disease (genotype Ff). Paul has the genotype FF. Their child, many years later, has a baby whose other parent is a carrier of cystic fibrosis. Calculate the probability that Paul and Sandra will have:
   a   a child with cystic fibrosis
   b   a child who is a carrier of cystic fibrosis
   c   a grandchild with cystic fibrosis
   d   a grandchild who is a carrier of cystic fibrosis.   *(4 marks)*

23.2 Gene mutations

Specification reference: 5.1.1

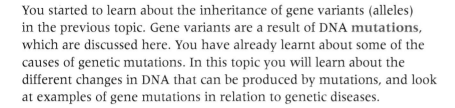

You started to learn about the inheritance of gene variants (alleles) in the previous topic. Gene variants are a result of DNA **mutations**, which are discussed here. You have already learnt about some of the causes of genetic mutations. In this topic you will learn about the different changes in DNA that can be produced by mutations, and look at examples of gene mutations in relation to genetic diseases.

Overview of mutations

Nucleotide sequences in DNA are altered by mutations. Genetic mutations in somatic cells are not inherited, but may result in cancer. Mutations in cells undergoing meiosis during gamete formation can be inherited.

Three types of mutation exist:

- Substitution (also known as a point mutation) – a nucleotide is exchanged for a different one.
- Insertion – an extra nucleotide or nucleotides are placed into the DNA sequence.
- Deletion – a nucleotide or nucleotides are removed.

Insertions and deletions are more likely than substitution mutations to result in major changes to the translated polypeptide. The removal or insertion of a nucleotide alters all subsequent triplet codes in the DNA sequence, not only the triplet in which the mutation occurs. This is known as a *frameshift,* and will usually result in a non-functional polypeptide.

The consequences of a substitution mutation vary. As Table 1 illustrates, sometimes a substitution results in the same amino acid appearing in the sequence after translation. This is a silent mutation and occurs because of the degeneracy of the genetic code. Some point mutations alter one amino acid in the primary structure of the polypeptide (missense mutations). This is likely to disrupt the tertiary structure of the protein and affect its function. Nonsense point mutations lead to stop codons in mRNA, which result in the length of the final polypeptide being shorter.

Learning outcome

Demonstrate knowledge, understanding, and application of:

→ types and examples of genetic mutations.

Synoptic link

This topic builds on your knowledge of DNA structure (Topic 4.2, Polynucleotides – DNA), protein structure (Topic 3.3, Protein structure – haemoglobin), and protein synthesis (Topic 4.7, Protein synthesis – translation.).

Synoptic link

You learnt in Topic 4.3, Semi-conservative replication, that a DNA mutation is a random change to an organism's genetic material. You learnt about genetic mutations in cancer in Topic 14.3, Cancer.

Synoptic link

You learnt about why the genetic code is called 'degenerate', in Topic 4.4, The genetic code.

▼ **Table 1** *Examples of substitution mutations and their consequences*

	DNA triplet	mRNA codon	Amino acid
before mutation	ACA	UGU	cysteine
silent mutation	ACG	UGC	cysteine
nonsense mutation	ACT	UGA (stop codon)	–
missense mutation	ACC	UGG	tryptophan

Synoptic link

You learnt about natural selection in Topic 10.5, Evolution.

A small proportion of mutations improve the function of the polypeptide for which the mutated gene codes. When a mutation benefits an organism, natural selection acts to maintain the new gene variant in a population. Many mutations are neutral and do not affect an organism – either the mutation is silent or it does not occur in a gene (i.e., the mutation is in a non-coding region of DNA).

A high proportion of mutations do have a negative impact on an organism. The next section describes a few examples of mutations that have persisted in human populations as heritable genetic diseases.

Mutations and genetic diseases

Cystic fibrosis

You have already learnt about the inheritance of cystic fibrosis (CF). CF can be caused by one of many hundreds of different mutations. The most common, which is the cause of 70% of CF cases worldwide, is the *deletion* of three specific nucleotides which prevents the insertion of an amino acid in the primary structure of the CFTR protein.

Normally, the CFTR mRNA is translated to produce a transmembrane protein that controls the movement of chloride ions. In CF this protein does not function, which leads to sticky mucus being produced. This thick mucus cannot be removed from the airways and can result in infections, scarring in the lungs and other problems.

Phenylketonuria (PKU)

Like CF, phenylketonuria (PKU) is caused by inheriting two recessive alleles. In PKU, the affected gene locus is on chromosome 12. A mutation prevents the enzyme that breaks down phenylalanine (an amino acid) from being made. A baby that is born with PKU will accumulate phenylalanine in the blood and tissue fluid. This results in brain damage.

As with CF, several different mutations can produce PKU and related conditions. These vary in severity. Some substitution mutations only change single amino acids in the enzyme and reduce its function. Other mutations, especially deletion mutations, stop any enzyme activity, preventing phenylalanine breakdown.

▲ **Figure 1** *A baby being tested for PKU*

Huntington's disease

Huntington's disease is caused by an accumulation of protein fragments in neurones in the brain. It reduces a person's ability to talk, think, and move.

The cause of the disease is an insertion mutation in a gene on chromosome 4. In healthy people, a cytosine–adenine–guanine (CAG) triplet is repeated between 10 and 26 times. People with Huntington's, however, have more than 40 CAG triplets in the gene – the greater the number of triplet repeats, the earlier that symptoms begin. Symptoms tend to begin after the age of 30 years.

Unlike CF and PKU, which are caused by inherited recessive alleles, Huntington's disease is the result of a dominant allele. This means that inheriting only a single copy of the faulty allele (i.e., being heterozygous) will cause the disease.

Sickle cell anaemia

Sickle cell anaemia is the result of a substitution mutation in the gene coding for the beta-polypeptides in haemoglobin. The gene locus is on chromosome 11. This mutation causes the amino acid valine to be added to the polypeptide's primary sequence instead of glutamic acid.

The FTO 'hunger' gene – an example of the subtle effects of gene mutation?

The FTO (fat mass and obesity-associated) gene codes for a protein that regulates appetite and feeding behaviour. Mutations have produced several known variants of the gene. One of these alleles has been linked to an increased risk of obesity.

In a study, people who are homozygous for the risk allele were found to be 3 kg heavier on average and were 70% more likely to be obese than people without a copy of this allele. A later study found that people with the risk allele have higher levels of the 'hunger hormone', ghrelin, in their blood, which means that they begin to feel hungry sooner than other people. Ghrelin is coded for by a separate gene.

1   Suggest the type of mutation responsible for the FTO gene variants. Explain your reasoning.
2   Suggest how the protein produced by the FTO gene might control feeding behaviour.

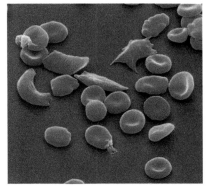

▲ **Figure 2** *Normal red blood cells (rounded) with distorted sickle-shaped cells (pink), ×3000 magnification*

Synoptic link

You will learn more about sickle cell anaemia in Topic 24.1, Factors affecting allele frequencies.

Study tip

Obesity has causes other than variants of the FTO gene. Some people who are homozygous recessive for the FTO risk allele do not develop obesity. Describing obesity as a 'genetic disease' would therefore be incorrect.

Summary questions

1   Look at the base sequence below. How many triplet codes in this sequence would be changed by **a** all guanine bases experiencing point mutations? **b** A point mutation of A? **c** A deletion of A?   *(3 marks)*

    G C A C T T G G G C C T C

2   Calculate the probability of two heterozygous parents (i.e., both having one healthy allele and one disease-causing allele) having a child with **a** Huntington's disease **b** PKU   *(2 marks)*

3   Aromatic rice is a popular cooking ingredient. A 'fragrance' gene has been sequenced in rice plants and found to have two variants – one that results in an aroma and one that does not. Sections of the two gene variants are shown here. The fragrant variant has four different mutations that make it different to the non-fragrant variant. Identify and describe the four mutations.   *(4 marks)*

    Non-fragrant variant: A A A C T G G T A A A A A G A T T A T G G C T T C A G C T G

    Fragrant variant: A A A C T G G T A T A T A T T T C A G C T G

23.3 Codominance

Specification reference: 5.1.1

Learning outcome

Demonstrate knowledge, understanding, and application of:

→ patterns of inheritance that show codominance and multiple alleles.

Synoptic link

This topic draws on your knowledge of blood groups (Topic 3.9, Blood donation), and the immune system (Topic 12.1, The immune system). You will learn about transplants in Topic 31.3, Kidney failure and its treatments.

The examples of inheritance that you have encountered so far in this chapter have involved genes with recessive and dominant variants. Heterozygous organisms that possess both variants will express only the dominant allele. Some traits, however, exhibit codominance, which means that both alleles are expressed in the phenotype. In this topic, you will learn about a few examples of codominant traits and their inheritance.

Codominance

Some genes have variants that, instead of being dominant and recessive, exhibit **codominance**. When two codominant alleles are present, both are transcribed and translated to produce polypeptides. This means both are expressed in the organism's phenotype. Often the phenotype takes an intermediate value between the phenotypes coded for by the two alleles. For example, snapdragon plants have three potential flower colours – red, white, and pink. One allele leads to the red phenotype, while one codes for white flowers. These two alleles are codominant. A heterozygous genotype therefore produces an intermediate pink colour.

▲ Figure 1 *Pink snapdragons – the pink phenotype is produced by two codominant alleles*

The examples of genetic inheritance that you have learnt about so far have involved genes with two variants. In many cases, however, a gene has multiple possible variants. Some of these alleles may be recessive, some dominant, and some codominant.

The next section will look at two examples of inheritance involving codominant and multiple alleles – blood groups and HLA antigens.

Blood groups

The inheritance pattern shown by the genes for human ABO blood groups is an example of both codominance and multiple alleles (multiple variants).

Four ABO blood groups exist – A, B, AB, and O. The existence of an AB group is a clue that codominance might be involved. Indeed, two of the possible alleles, I^A and I^B, are codominant. The third allele, I^O, is recessive to both I^A and I^B.

$I^A I^B$ heterozygotes produce both antigen A and B glycoproteins in the plasma membranes of their red blood cells. The homozygous recessive genotype ($I^O I^O$) results in no antigen protein being produced. Figure 2 shows the potential offspring genotypes and phenotypes of parents with $I^A I^O$ and $I^B I^O$ genotypes.

Synoptic link

You learnt about the importance of identifying the correct blood group for blood donations and transfusions in Topic 3.9, Blood donation.

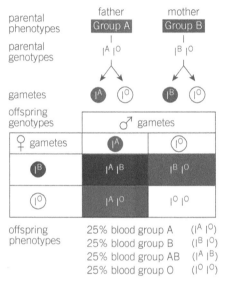

Study tip

As I^O is recessive to both I^A and I^B, the lower case symbol 'i' could also be used in genetic diagrams to represent this recessive allele.

offspring phenotypes

25% blood group A   ($I^A I^O$)
25% blood group B   ($I^B I^O$)
25% blood group AB ($I^A I^B$)
25% blood group O   ($I^O I^O$)

▲ **Figure 2** *A genetic diagram showing the possible offspring of a blood group A father and a blood group B mother*

HLA antigens

All cells, not only red blood cells, carry antigens in their cell surface membranes. Human leucocyte antigens (HLAs) are particularly important when considering organ and tissue transplants. A poor match between the HLA antigens of a transplanted organ and its recipient could result in rejection of the organ.

HLAs are the human version of the major histocompatability complex (MHC). MHC is a gene family found in many species. Humans have three main Class 1 MHC genes, known as HLA-A, HLA-B, and HLA-C, and three minor Class 1 MHC genes called HLA-E, HLA-F, and HLA-G. Despite the name, the proteins produced from these genes are present on the cell surface membranes of almost all cells. They display peptides, which have been made within the cell, to the immune system. Foreign peptides, such as those produced by an infected cell, are recognised and the cell is destroyed.

Three ABO blood group gene variants exist, but this level of variation is dwarfed by the number of gene variants that exist for HLA antigens. HLA antigens are coded for by six gene loci on chromosome 6. The HLA-A, HLA-B, and HLA-C genes, for example, have more than 2000 variants each, some more common than others. This range of alleles means that there are millions of potential combinations of the six HLA genes.

Each of the HLA genes exhibits codominance. Alleles inherited from both parents are expressed to the same extent.

The six HLA gene loci are positioned close together on chromosome 6. This means that they tend to be inherited together as a set, which is referred to as a haplotype. Inheritance of alleles on gene loci close together is called linkage. You will learn about other examples of linkage in the next topic.

Summary questions

1   Figure 3 shows an example of HLA inheritance. Suggest why tissue matches are more likely to be found between two siblings rather than between a parent and offspring.   (3 marks)

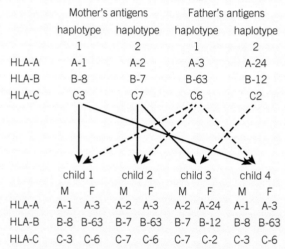

▲ **Figure 3**   An example of HLA inheritance

2   The colour of the four o'clock flower (*Mirabilis jalapa*) is controlled by two codominant alleles – C^R (red flowers) and C^W (white flowers). What are the genotypes of the two plants that would need to be bred to produce 50% red offspring and 50% pink offspring?   (2 marks)

3   A man claims not to be the father of a child who is blood group O. The father is blood group A and the mother is blood group B. Can you be certain whether he is telling the truth? Explain your answer.   (3 marks)

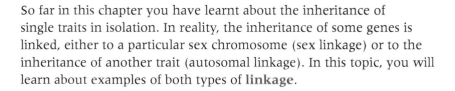

So far in this chapter you have learnt about the inheritance of single traits in isolation. In reality, the inheritance of some genes is linked, either to a particular sex chromosome (sex linkage) or to the inheritance of another trait (autosomal linkage). In this topic, you will learn about examples of both types of **linkage**.

Sex linkage

Most chromosomes occur in homologous pairs. Each of the two chromosomes in a pair carry loci for the same genes, in the same place along their length. Sex chromosomes, however, are different sizes and contain different gene loci. The X chromosome is significantly larger than the Y chromosome. As a consequence, most of the gene loci found on the X chromosome have no counterpart on the Y. Any gene present on a sex chromosome is known as sex-linked.

Some genes, such as SRY, are found only on the Y chromosome. These genes are termed Y-linked genes. The SRY gene triggers the activation of a non-sex-linked gene that is involved in the development of the testes. Many other Y-linked genes play crucial roles in sperm production.

Genes found only on the X chromosome are known as X-linked. Most of the genes on the X chromosome code for features other than female sex characteristics. For example, one X-linked gene codes for a blood clotting protein called factor VIII.

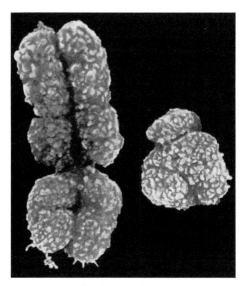

▲ **Figure 1** *X (left) and Y (right) chromosomes, ×10 200 magnification*

Haemophilia A

The blood of people with haemophilia clots slowly and they often suffer from persistent internal bleeding. Haemophilia A is an X-linked trait. The allele that produces a working copy of factor VIII is dominant (X^H). A mutated variant expresses an altered protein that does not function. This allele is recessive (X^h).

Learning outcome

Demonstrate knowledge, understanding, and application of:

→ patterns of inheritance that show sex linkage and autosomal linkage.

Synoptic link

Recapping the principles of meiosis and crossing over in Topic 9.1, Meiosis, might be useful before reading this topic.

If females inherit two recessive alleles, this produces a homozygous recessive genotype ($X^h X^h$) and results in haemophilia. Males, in contrast, have only one X chromosome and therefore need only a single recessive allele to have haemophilia ($X^h Y$). Consequently, the condition is found almost exclusively in males, although female haemophiliacs are known.

Study tip

When displaying sex-linked genotypes, choose an appropriate letter to represent the gene (H in the case of haemophilia). This is placed as superscript by the sex chromosome (X or Y), using capitals for dominant alleles and lower case for recessive alleles. In the case of X-linked conditions such as haemophilia, Y is written without any superscript because the gene is absent from the Y chromosome.

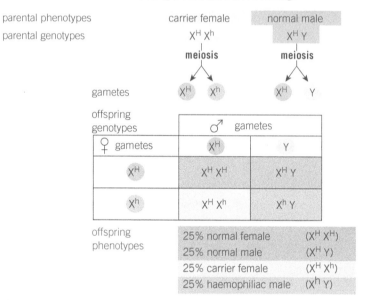

H = allele for production of clotting protein (rapid blood clotting)
h = allele for non-functional of clotting protein (slow blood clotting)

offspring phenotypes:
25% normal female ($X^H X^H$)
25% normal male ($X^H Y$)
25% carrier female ($X^H X^h$)
25% haemophiliac male ($X^h Y$)

▲ **Figure 2** *Inheritance of haemophilia from a carrier mother*

Synoptic link

You learnt about meiosis in Topic 9.1, Meiosis.

Autosomal linkage

Chromosomes other than the sex chromosomes are known as autosomes. Genes on autosomes can also be linked. Some gene loci are in close proximity on the same chromosome. Two genes positioned near each other are unlikely to be separated by crossing over during meiosis. An example of autosomal linkage is the inheritance of ABO blood groups and nail patella syndrome.

ABO blood groups and nail patella syndrome

Nail patella syndrome is a rare condition. Its symptoms include underdeveloped kneecaps (patellae) and nails, and an inability to straighten the elbows. Nail patella syndrome affects males and females equally (i.e., it is not sex-linked).

The allele that causes nail patella syndrome is dominant and is given the letter N. The recessive allele is n. The gene locus is found on chromosome 9, very close to the ABO blood group gene. This proximity means that blood group and nail patella syndrome alleles will probably be inherited together because crossing over during meiosis is unlikely to separate them. For example, if the I^A blood group allele and the nail patella syndrome allele are on the same copy of chromosome 9, they will almost certainly be passed into the same gamete following meiosis.

The inheritance of mitochondrial DNA

You previously learnt that organelles such as mitochondria and chloroplasts contain their own DNA (Topic 1.6, Cell ultrastructure and Topic 6.1, The transport system in mammals). Human mitochondrial DNA (mtDNA) could be thought of as an extra chromosome in the human genome, albeit a very small one, as it codes for 37 genes.

The inheritance of mtDNA does not show the type of sex-linkage that has been discussed in this topic. In virtually all species studied so far, however, mtDNA is inherited from the mother.

The rapid mutation rate of mtDNA enables it to be used to assess genetic relationships between species and between individuals of the same species.

1   Suggest why mtDNA is inherited from the mother and not from the father in most species, including humans.
2   Suggest why mtDNA is used to study genetic relationships.

Summary questions

1   Suggest why more X-linked traits exist than Y-linked traits.   (*2 marks*)

2   Draw a genetic diagram to show the possible offspring genotypes of a haemophiliac father and a mother who is homozygous dominant for the factor VIII gene. What is the probability of the parents having a daughter who carries the allele for haemophilia?   (*4 marks*)

3   A child's parents have the following genotypes for the ABO blood group (alleles I^A I^O I^B) and nail patella syndrome (alleles N – dominant, and n – recessive) traits: I^A N I^O n and I^O n I^B n.
   a   What is the probability of the child having nail patella syndrome?
   b   What are the possible blood group phenotypes if the child inherits nail patella syndrome? Assume that the two traits show autosomal linkage.   (*3 marks*)

23.5 Investigating patterns of inheritance

Specification reference: 5.1.1

▲ **Figure 1** Drosophila melanogaster

You have looked at several different patterns of genetic inheritance in this chapter including the inheritance of a single gene with two variants or several variants, sex-linked genes, and codominant alleles. Here, the discussion is extended to consider the patterns of inheritance when two genes are being studied. To study inheritance patterns, scientists employ statistical tests and often use **model organisms**. You will learn how these two tools help to reveal patterns of genetic inheritance.

Model organisms

A model organism is a species that is studied to help us understand biological processes and make discoveries that will have relevance to other organisms, including humans.

Model organisms are selected carefully and need to be suited to the questions that are being researched. A common model organism in genetic studies is the fruit fly (*Drosophila melanogaster*). Several characteristics make *Drosophila* a good model organism for genetic research:

● a short generation time (approximately 10 days), which means that many generations can be studied in a short space of time

● they breed in large numbers (females can lay 100 eggs in one day)

● small size and low costs of care mean that they are easy to breed in laboratory conditions

● scientists have detailed knowledge of the fruit fly's biology and genetics (e.g., the *Drosophila* genome has been sequenced)

● *Drosophila* has several morphological traits (such as body colour, eye colour, and number of wings) that have a range of gene variants and are easy to identify.

Dihybrid crosses

A dihybrid cross shows the pattern of inheritance for two traits at different gene loci. For example, the breeding of two individuals who are heterozygous (i.e., they each carry one dominant and one recessive allele) for two genes, produces offspring phenotypes in a 9 : 3 : 3 : 1 ratio. This can be illustrated by looking at the inheritance of two traits in *Drosophila melanogaster*.

Chi-squared test

The **chi-squared test** is used to analyse whether an inheritance pattern is statistically significant. More precisely, it tests whether there is a significant difference between the observed and expected numbers of offspring phenotypes.

χ^2 is calculated as:

χ^2 = the sum of (observed numbers – expected numbers)^2 / expected numbers

This is usually written as:

$$\chi^2 = \Sigma \, \frac{(O - E)^2}{E}$$

In the case of the *Drosophila* dihybrid cross, the expected results should be in a $9:3:3:1$ ratio. The observed results will rarely provide an exact match to the expected results. A chi-squared test value tells us whether there is a significant difference between the observed and expected results. If there is no statistical significance then our inheritance pattern theory for the traits being studied is correct. If there is a significant difference then we would need to adjust our model of inheritance.

A *Drosophila* dihybrid cross

Two of the observable traits in *Drosophila* flies are body colour and wing size. Several alleles for body colour exist, including a dominant allele for brown colour (B) and a recessive allele for black (b). Wing shapes can be normal (W, a dominant allele) or vestigial (w, a recessive allele). The possible genotypes of offspring from two heterozygous parents (BbWw) are shown in the genetic diagram.

Male gametes	Female gametes			
	BW	Bw	bW	bw
BW	BBWW	BBWw	BbWW	BbWw
Bw	BBWw	BBww	BbWw	Bbww
bW	BbWW	BbWw	bbWW	bbWw
bw	BbWw	Bbww	bbWw	bbww

The phenotypes of offspring will be present in the following ratios:

9 brown body, normal wings

3 black body, normal wings

3 brown body, vestigial wings

1 black body, vestigial wings

> When predicting a phenotypic ratio of 9:3:3:1 for a dihybrid cross, what assumptions do you have to make about the two gene loci being studied?

Calculating chi-squared

Results from a *Drosophila* genetic cross and the subsequent χ^2 calculation are shown in the table.

Phenotype	Observed (O)	Expected (E)	O – E	(O – E)^2	(O – E)^2 / E
brown, normal wings	87	90	−3	9	0.10
black, normal wings	31	30	1	1	0.03
brown, vestigial wings	35	30	5	25	0.83
black, vestigial wings	7	10	-3	9	0.90
					χ^2 = **1.86**

The calculated χ^2 value is compared with values in a χ^2 significance table.

Study tip

The use of χ^2 enables a scientific hypothesis to be tested. A null hypothesis states that no relationship or significant feature exists in the data. For example, a null hypothesis when analysing an inheritance pattern is that no difference exists between the expected and the observed results. If the χ^2 value is not significant then the null hypothesis can be accepted. If the χ^2 value is significant then the alternative hypothesis (i.e., a difference exists between the expected and the observed results) is accepted.

Study tip

A full table of chi-squared values can be found in the appendix.

Study tip

You will always be provided with the formula for χ^2 and relevant critical values. You should be able to apply the equation and understand what the test results mean. If the value of χ^2 is less than the critical value, the null hypothesis is accepted (i.e., the result is as expected). If the χ^2 value is greater than the critical value, the null hypothesis must be rejected. This means that the difference between the observed and the expected results is too great to be due to chance.

	Critical values of the χ^2 distribution									
	p									
df	0.995	0.975	0.9	0.5	0.1	0.05	0.025	0.01	0.005	df
1	0.000	0.000	0.016	0.455	2.706	3.841	5.024	6.635	7.879	1
2	0.010	0.051	0.211	1.386	4.605	5.991	7.378	9.210	10.597	2
3	0.072	0.216	0.584	2.366	6.251	7.815	9.348	11.345	12.838	3

The degrees of freedom (df) are calculated as the number of categories (phenotypes) − 1. In this example there are 3 degrees of freedom.

The P value tells us the probability of differences being the result of chance. Differences are statistically significant if $P = 0.05$ or less (i.e., this would mean that there is only a 5% probability that differences can be attributed to chance). In this case, the critical value of χ^2 (which represents a P value of 0.05) would be 7.815.

In the *Drosophila* example, the P value for a calculated χ^2 of **1.86** lies somewhere between 0.5 and 0.9. You can be very confident that the differences between the observed and expected results are down to chance. This means the model of inheritance for these two traits can be accepted.

Suggest the implications of a χ^2 value of 7.20 being calculated for a dihybrid cross.

Summary questions

1 State and explain two characteristics of a model organism suited to genetic research. *(4 marks)*

2 A scientist studied the inheritance of two traits in pea plants – seed colour and seed texture. The alleles for these traits are: R = round seeds, r = wrinkled seeds, G = yellow seeds, g = green seeds. R and G are the dominant alleles. What proportion of each phenotype would you expect to find in the offspring of one plant with the genotype RRGg and another plant that is RrGg? *(2 marks)*

3 Two other pea plants were bred together to produce 100 offspring. A scientist expected that 50 of the offspring plants would produce round, yellow seeds and the other 50 would produce wrinkled, yellow seeds. She expected no green seeds. The results were: 55 round, yellow seeds; 45 wrinkled, yellow seeds; 0 round, green seeds; 0 wrinkled, green seeds. Calculate whether the scientist's predictions can be supported by the chi-squared test. Suggest the likely genotypes of the two plants that were bred together. *(4 marks)*

So far in this chapter you have learnt about the consequences of DNA mutations (gene mutations) that alter the sequence of nucleotides. **Chromosome mutations** result in a change in the number of chromosomes or the structure of a chromosome. Here you will learn in more detail the causes of Down's syndrome, Turner's syndrome and Klinefelter's syndrome, along with the roles of genetic testing and counselling in cases of chromosome mutations and other genetic diseases.

Learning outcomes

Demonstrate knowledge, understanding, and application of:

→ chromosome mutations, including Turner's syndrome, Klinefelter's syndrome, and Down's syndrome

→ the use of genetic testing, pedigree analysis, and the role of genetic counsellors.

The processes that cause chromosome mutations

Translocation

Translocation is when a piece of chromosome breaks off and is transferred to another chromosome. Down's syndrome can be caused by translocation when the end of the long (q) arm of chromosome 21 breaks off and joins to another chromosome (usually chromosome 14). A gamete containing extra chromosome 21 genetic material is produced. Should this be inherited, a child would have Down's syndrome, which is characterised by learning disabilities and delayed development.

Non-disjunction

Non-disjunction involves a change in the diploid number of chromosomes. Either the homologous chromosomes fail to separate in meiosis I or sister chromatids fail to separate in meiosis II. The result is a gamete that is either missing a chromosome or has one extra chromosome.

Down's syndrome
A gamete is produced that contains two copies of chromosome 21. If this gamete is fertilised, the offspring will have three versions of chromosome 21, and 47 chromosomes in total.

Turner's syndrome
A gamete without a sex chromosome is produced – fusion of this gamete with a gamete carrying an X chromosome results in an embryo that has a single X chromosome. Symptoms of Turner's syndrome can include short stature, learning disabilities, and obesity.

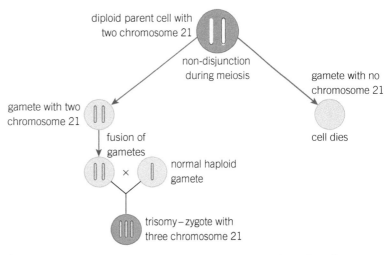

▲ **Figure 1** *Non-disjunction during meiosis, which gives rise to Down's syndrome*

Study tip

A chromosome comprises two arms – a long arm, referred to as q, and a short arm, referred to as p. These two arms are separated by a centromere. The q arm, because of its greater length, contains more genes than the p arm.

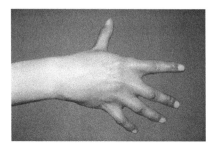

▲ **Figure 2** *The hand of a person with Klinefelter's syndrome*

Study tip

The X chromosome carries genes that are essential for various cellular activities. The zygote formed from the fusion of a gamete without sex chromosomes and a Y-bearing gamete will therefore fail to develop.

Synoptic link

In Topic 9.3, Monitoring fetal development, you learnt about techniques used in prenatal testing.

Klinefelter's syndrome

Klinefelter's syndrome only affects males. Most cases are caused by non-disjunction during sperm production, which results in the oocyte being fertilised with a sperm cell that is carrying both X and Y chromosomes. This means that a child with Klinefelter's syndrome has an extra sex chromosome (XXY). Symptoms can include infertility, weak bones, and less body hair than other males.

Genetic testing, analysis, and counselling

Genetic testing

The risk of a child inheriting genetic diseases can be assessed using a variety of genetic tests.

Parental carrier testing

Two people planning to have a child can be tested for the presence of recessive alleles associated with diseases. The couple's risk of having a child with a genetic condition is then assessed. The probability of cystic fibrosis can be analysed using carrier testing.

Prenatal testing

Embryos are tested for chromosome mutations and some heritable genetic diseases, especially if couples are believed to have a high risk of having a baby with an abnormality.

Newborn screening

Babies can be tested shortly after birth to identify genetic disorders that require immediate treatment, such as PKU which was discussed in Topic 23.2, Gene mutations.

Pedigree analysis

Another method for predicting the likelihood of a child being born with a genetic disorder is the use of pedigree charts. These charts resemble family trees and enable inheritance patterns within a family to be analysed. Figure 3 shows a pedigree chart for albinism. Albinos lack pigmentation in their eyes, hair, and skin. The convention in pedigree charts is to represent males with squares and females with circles.

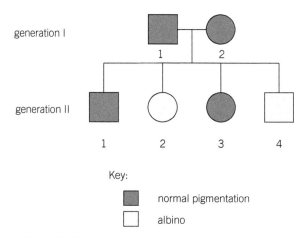

▲ **Figure 3** *Pedigree chart showing cases of albinism within a family*

Genetic counselling

Parents are offered genetic counselling when undergoing testing, or pedigree analysis, both of which can indicate a high probability of their child being born with a chromosome mutation or another genetic disorder. Genetic counsellors help people to understand the risks of children inheriting a particular genetic disorder, so that the prospective parents can make informed decisions (such as whether they should seek alternatives to natural conception).

Parents might be offered counselling if there is a known genetic condition in either family. It is also offered if there is an evident history of cancer in their families, or if testing confirms that both parents carry disease-causing recessive alleles, or if the couple have lost a baby during pregnancy or infancy.

Genetic counsellors must attempt to present their knowledge in a non-judgemental manner. Counselling has the potential to create ethical dilemmas:

- Parents, armed with more knowledge than ever before about the potential genetics of their children, often have to make difficult decisions. For example, should parents attempt to conceive if there is a high risk of their child inheriting Huntington's disease? (Topic 23.2, Codominance) Should a mother abort her embryo if she knows it has Down's syndrome?

- Genetic counsellors can learn sensitive information about a family. For example, should a counsellor inform a family if the screening process shows that the father is not a baby's biological father?

> **Study tip**
>
> You may be asked to evaluate the pros and cons of genetic testing and counselling. You will need to present a balanced account, outlining both sides of the argument.

> **Synoptic link**
>
> You can remind yourself about the process of meiosis in Topic 9.1, Meiosis. You also learnt about the techniques used to test for chromosome mutations in Topic 9.3, Monitoring fetal development.

Summary questions

1 Describe the difference between translocation and non-disjunction of chromosomes. *(2 marks)*

2 What methods can be employed prior to the conception of a child to assess the probability of parents having a baby with haemophilia? *(3 marks)*

3 Figure 3 shows cases of albinism in a family. Study the pedigree chart – state and explain whether a albinism is caused by a recessive allele b albinism is sex-linked. *(4 marks)*

Practice questions

1 Which of the following statements is/are true of human leucocyte antigens?

   Statement 1: HLA alleles exhibit codominance.

   Statement 2: The HLA genes are located on chromosome 6 and show genetic linkage.

   Statement 3: HLA genes code for proteins that are present only on the membranes of white blood cells.

   A   1, 2 and 3

   B   Only 1 and 2

   C   Only 2 and 3

   D   Only 1   (*1 mark*)

2 What is the name given to the type of mutation that results in the formation of a stop codon?

   A   Nonsense

   B   Missense

   C   Silent

   D   Insertion   (*1 mark*)

3 Two parents are both carriers of the recessive allele (f) that causes a rare condition called Friedreich's ataxia (FA).

   The parental genotypes are Ff. People with FA have the genotype ff.

   The two parents have a daughter. Their daughter, later in life, has a child with father who is a carrier of FA. What is probability that this child (the grandchild of the first couple) has FA?

   A   0%

   B   25%

   C   50%

   D   75%   (*1 mark*)

4 Which of the following statements is/are a reason or reasons for the selection of a model organism in a biological study?

   1   Small size.

   2   Morphological traits with a range of gene variants.

   3   Short generation time.

   A   1, 2 and 3

   B   Only 1 and 2

   C   Only 2 and 3

   D   Only 1   (*1 mark*)

5 Which of the following genotypes would belong to a person with Klinefelter's syndrome?

   A   XYY

   B   X0

   C   XXY

   D   XY   (*1 mark*)

6 Domestic chickens have been bred for many years to increase the number of eggs laid by the females. It is useful to be able to identify the young female chicks on the day after they hatch, as only the females need to be kept for laying eggs.

   Unlike mammals, where the sex chromosomes are known as X and Y, in chickens the sex chromosomes are known as Z and W.

   Male chickens have two Z chromosomes (ZZ). Female chickens have one Z chromosome and one W chromosome (ZW).

   a   Some genes for feather colour and pattern in chickens are carried on the Z chromosome but not on the W chromosome. One such example is the gene for striped feathers (barring).

   State the name given to this type of inheritance.   (*1 mark*)

   b   Inheritance of the barring pattern can be used to identify female chicks when they are one day old.

   The phenotypes associated with the two alleles of the barring gene are shown here.

allele	Adult phenotype	Day-old chick phenotype
dominant B	black feathers striped with white bars (barred)	black body with a white spot on head
recessive b	black feathers (non-barred)	black body and head

(i) State the adult phenotypes and sex of the following individuals: Z^BZ^b, Z^BW, Z^bW (*3 marks*)

(ii) A cross was carried out between a barred female and a non-barred male.

Copy and complete the genetic diagram to show the parental genotypes, their gametes, and the F1 genotypes. State the phenotypes of the offspring as day-old chicks.

Parent phenotypes	Barred female	Non-barred make
parent genotypes		
gametes		
F1 genotypes		

F1 day-old chick phenotypes

Male ....

Female .... (*5 marks*)

OCR F215 2013

7 In *Drosophila* flies, the gene controlling eye colour has two variants. The red eye allele (R) is dominant to the white eye allele (r).

A student crossed a red-eyed fly with a white-eyed fly. The results are shown in the table.

Phenotype of fly	Number of offspring
red-eyed female	27
red-eyed male	0
white-eye female	0
white-eyed male	23

a (i) Explain why these results cannot be explained merely by the red-eyed parent being heterozygous. (*2 marks*)

(ii) In *Drosophila*, the males are the heterogametic sex, possessing two different sex chromosomes, X and Y.

Draw a genetic diagram to show how the results shown above could have been produced. (*3 marks*)

(iii) Use the chi-squared test to analyse the results of the genetic cross. The expected ratio of red-eyed females to white-eyed males is 1:1. Calculate a chi-squared value for the test and state what this value shows based on the table of probabilities. (*4 marks*)

Degrees of freedom	Probability, *p*			
	0.90	0.50	0.10	0.05
1	0.02	0.45	2.71	3.84
2	0.21	1.39	4.61	5.99

OCR F215 2010

8 Huntington's disease is a genetic disease caused by an autosomal allele. If, at fertilisation, both gametes carry the mutant allele, the resultant embryo will not develop. The homozygous dominant genotype is described as lethal.

In some cases, Huntington's disease symptoms do not appear until an individual is aged 30 or over.

Use a genetic diagram to calculate the probability of a child developing Huntington's disease if both parents begin to show symptoms of the disease. (*5 marks*)

OCR F225 2011

9 Outline how non-disjunction can result in chromosome mutations, illustrating your answer with examples of these mutations. (*6 marks*)

24 POPULATION GENETICS AND EPIGENETICS

24.1 Factors affecting allele frequencies

Specification reference: 5.1.2

Synoptic link

This Topic builds on what you have learnt about DNA structure, transcription, and translation in Topics 4.2, 4.6, and 4.7, and about natural selection, in Topic 10.5.

Synoptic link

You learnt about the effects of natural selection on a gene pool in Topic 10.5, Evolution, and found out about the mutation that is responsible for sickle cell anaemia in Topic 23.2, Gene mutations.

You learnt about the principles of natural selection in Topic 10.5, Evolution. Variants of genes that benefit organisms are selected and their frequency in a population increases. In this topic, you will look in more detail at the role of natural selection in changing allele frequencies. You will investigate the links between gene variants, amino acid sequence and protein function, using haemoglobin variants as an example. You will also learn how factors other than natural selection can alter allele frequencies within populations.

The role of natural selection

All the variants of the genes in a population are collectively known as the population's gene pool. Natural selection acts to increase the frequency of beneficial alleles in a population and alter the gene pool. In some cases, however, the benefit of an allele is not immediately obvious. An example of this is the allele responsible for sickle cell anaemia.

Why is the allele for sickle cell anaemia higher in particular populations?

In general, you would not expect an allele responsible for a disease to increase in frequency. The allele that causes sickle cell anaemia, however, is a gene variant that produces both positive and negative effects in some human populations.

Sickle-cell anaemia is one of the most common genetic diseases in the world. A substitution mutation results in the production of abnormal haemoglobin. At low oxygen concentrations, the abnormal haemoglobin clumps together and deforms red blood cells into sickle shapes. This can cause blockages in capillaries and damage to organs.

The allele for normal haemoglobin production is H^A, the allele for abnormal haemoglobin is H^S. These two alleles are codominant. This means that people with the genotype $H^A H^A$ produce only normal haemoglobin. All the haemoglobin of people with the $H^S H^S$ genotype is abnormal and they have sickle cell anaemia. Heterozygotes ($H^A H^S$) have a condition called sickle cell trait – half their haemoglobin is normal, half is abnormal. People who inherit sickle cell trait are usually symptomless, but can experience symptoms when they are short of oxygen (e.g., at high altitude).

Children with the H^SH^S genotype have sickle cell anaemia and are unlikely to survive long enough to reproduce. The frequency of the H^S allele would therefore be expected to decrease. However, heterozygotes (H^AH^S) who have sickle cell trait are also protected against an infectious disease: malaria. Malaria is caused by a parasite called *Plasmodium*, which lives in red blood cells for part of its life cycle. One explanation for the protective effect of the H^S allele against malaria is that it slightly increases carbon monoxide production during the break down of red blood cells, which prevents the development of the disease.

There is a correlation between the distribution of malaria and the frequency of the H^S allele. In sub-Saharan African populations, where malaria is common, the H^S allele frequency is high. The mutant H^S allele provides an advantage here and natural selection maintains it at a high frequency in the population. The best genotype to have in populations where malaria is endemic is H^AH^S – this protects against malaria without people having sickle cell anaemia.

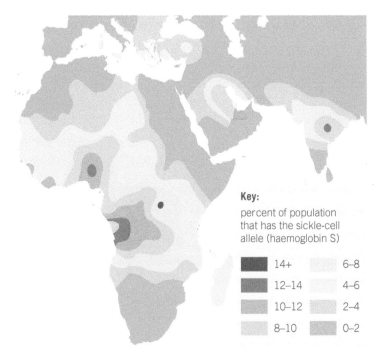

Key:

per cent of population that has the sickle-cell allele (haemoglobin S)

14+	6–8
12–14	4–6
10–12	2–4
8–10	0–2

▲ Figure 1 *The distribution of the H^S allele*

Linking allele variation with protein function

When a DNA mutation alters the amino acid sequence of a polypeptide, this is liable to change the tertiary structure of the protein. Remember that tertiary structures in proteins are held together by a variety of bonds between R groups. If a new allele causes even one different amino acid to be inserted in the chain, tertiary structure can be affected. A change in structure is likely to have a knock-on effect on protein function. For example, a different tertiary structure of an enzyme will produce a new shape in its active site. This will influence, often negatively, how well the enzyme can bond with its substrate.

Synoptic link

You learnt about how nucleotide sequence determines amino acid sequence in Topic 23.2, Gene mutations. You found out about bonding in proteins and the structure and function of haemoglobin in Topic 3.3, Protein structure – haemoglobin.

Many gene variants exist for the polypeptides in haemoglobin. Most people produce haemoglobin A, which consists of two α and two β polypeptide chains. Table 1 lists some other haemoglobin gene variants and shows the relationships between nucleotide sequences, amino acid sequences, and haemoglobin function.

▼ Table 1 *Types of haemoglobin*

Type of haemoglobin	Mutation	Change to protein structure	Change to protein function
haemoglobin S	substitution in β polypeptide gene	valine replaces glutamic acid	sickle cell disease: haemoglobin clumps at low oxygen levels
haemoglobin C	substitution in β polypeptide gene	lysine replaces glutamic acid	the effect is less severe than sickle cell disease, can cause some red blood cells to break down
haemoglobin H	severe mutations to α polypeptide genes	unstable haemoglobin constructed from four β polypeptides because of the lack of α polypeptides	higher affinity for oxygen than haemoglobin A, which results in little oxygen being released at tissues

The role of factors other than natural selection

In theory, the allele frequencies produced by natural selection will remain constant in a stable population. You will consider this idea further in the next topic. When population size decreases, however, allele frequencies are liable to change.

Genetic bottlenecks

A **genetic bottleneck** is a drastic reduction in population numbers. This can be caused by a natural event such as a volcanic eruption, habitat destruction, or hunting by human populations. The proportions of alleles in the surviving population could be very different to those in the original population. As survival is due to chance, some alleles might disappear if none of the individuals carrying those alleles survived. In this case genetic diversity would decrease.

The founder effect

The **founder effect** can be considered a type of genetic bottleneck. This occurs when a small group breaks away from the original large population to form a new colony (e.g., a group of birds migrating to a new island).

Examples of the founder effect in human populations

Ellis-van Creveld syndrome
When the Amish people migrated to Pennsylvania in 1744, two members of the new colony possessed the recessive allele for Ellis-van Creveld syndrome. The symptoms of this disorder include dwarfism, extra fingers, short ribs and cleft palates. After several generations of inbreeding within the new population, the Ellis-van Creveld allele had increased in frequency. It is now much more common among the Amish than it is among the general population.

original population has eight different alleles occurring at various frequencies

chance event reduces the size of the population significantly

the individuals that survive have fewer alleles (just four types) and with different frequencies (green twice as frequent as yellow)

as the population recovers, the number and frequency of the alleles are the same as those of the population that came through the bottleneck. This population is less diverse than the original population

▲ Figure 2 *An illustration of a genetic bottleneck effect*

Blood group distribution

Human global migration, over tens of thousands of years, resulted in many small populations being established. Each time a new population splintered off, this would have had the potential to filter the gene pool and change the proportion of each allele present.

The distribution of ABO blood groups represents a good example of the founder effect. The original human population that migrated to South America was almost entirely blood group O, for example. When humans first migrated into Asia, the original population had a relatively high proportion of the I^B allele.

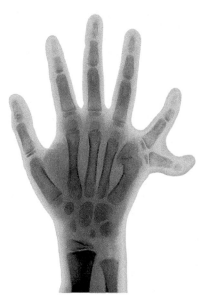

▲ **Figure 3** *Polydactyly (having extra digits) is a common symptom of Ellis-van Creveld syndrome*

Summary questions

1 How do genetic bottlenecks decrease genetic diversity?   (*2 marks*)

2 Why do Amish populations have a higher incidence of Ellis-van Creveld syndrome than the global population in general?   (*3 marks*)

3 Explain the difference in the frequency of the sickle cell allele (H^S) in Europe and Africa in terms of natural selection.   (*3 marks*)

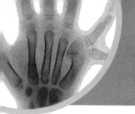

24.2 Analysis of allele frequencies

Specification reference: 5.1.2

You found out about some of the factors that affect allele frequencies in populations in the previous topic. Here you will learn that the frequencies of alleles can be estimated mathematically.

The Hardy–Weinberg principle

If you want to calculate the frequency of alleles in a population, you must make some assumptions about the population. The Hardy–Weinberg principle states that the proportion of alleles will remain the same from one generation to the next provided that the following conditions are met:

- no new mutations
- no migration in or out of the population, and therefore no flow of alleles
- no natural selection for or against alleles
- the population is large
- mating is random.

In reality, these conditions rarely exist. The Hardy–Weinberg principle nonetheless provides a basis for the study of gene frequencies.

Calculating allele frequency

Calculations can be carried out for genes with two variants (one dominant and one recessive allele). The first equation we need to understand is:

$p + q = 1.0$

Where p = the frequency of the dominant allele, and q = the frequency of the recessive allele. If only two alleles exist, their frequencies must add up to 1.0 (100%).

The second equation considers the frequencies of the possible genotypes in a population:

$p^2 + 2pq + q^2 = 1.0$

p^2 = the frequency of the homozygous dominant genotype; $2pq$ = the frequency of the heterozygous genotype; q^2 = the frequency of the homozygous recessive genotype. The frequencies of the three possible genotypes must add up to 1.0 (100%) since no other genotypes are possible.

Armed with these two equations, you will be able to calculate allele frequencies if you have knowledge of genotype frequencies. You can also calculate the proportions of each genotype in a population when you know the allele frequencies.

Worked example: Cystic fibrosis

Ireland has the world's highest rate of cystic fibrosis (CF). A recent study estimated that 1 in 1353 babies are born with CF. From this information alone, how do we calculate the proportion of unaffected carriers of the CF allele in Ireland's population?

1 You know from Topic 23.1 that CF is caused by a recessive allele (f). Only homozygous recessive genotypes have CF.

2 The homozygous recessive genotype = ff = q^2 in the Hardy-Weinberg equation.

3 q^2 = 1 in 1353 = 0.000739

4 To calculate the frequency of the recessive allele (q) we need to find the square root of q^2. $\sqrt{0.000739}$ = 0.027186

5 $p + q = 1.0$, so $p = 1 - q$. We now know $q = 0.027186$, therefore $p = 1 - 0.027186 = 0.972814$.

6 We are now able to calculate the frequency of heterozygous genotypes (Ff) in Ireland. Ff is represented in the Hardy–Weinberg equation by $2pq$.

$2pq = 2 \times 0.972814 \times 0.027186 = 0.052894$.

7 A frequency of 0.052894 means that 5.3% of Ireland's population are carriers of the CF allele. Another way of expressing the same value is to say that approximately 1 in 19 of the population in Ireland are carriers.

Summary questions

1 Suggest why the Hardy–Weinberg principle cannot be applied to populations of captive animals in zoos. *(2 marks)*

2 Albinism is a condition in which people lack melanin pigment in their hair, skin, and eyes. It is caused by a recessive allele, which means that only homozygous recessive genotypes produce albinism. 1 in 17 000 people worldwide have the condition. Calculate the allele frequency of the recessive allele for albinism. *(2 marks)*

3 You learnt about Ellis-van Creveld (EVC) syndrome in Topic 24.1, Factors affecting allele frequencies. The allele that causes the disease is recessive. In one Amish population, 5 in 1000 people have EVC. Calculate the percentage of the Amish population that carries one copy of the EVC allele. 1 in 123 people in the general population carry one EVC allele. Calculate the difference in the percentage of EVC carriers between the two populations. *(4 marks)*

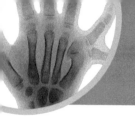

24.3 Speciation

Specification reference: 5.1.2

In AS Topic 10.4, you learnt that natural selection can lead to the formation of new species. For speciation to occur a population must be split into smaller isolated groups. Organisms can breed within the group but cannot breed with members of a different group – members of a species must be isolated from each other and unable to reproduce with each other. A variety of processes can result in isolation, and you will learn about some of them in this topic.

Mechanisms of speciation

Speciation is the formation of new species. This occurs when groups of individuals of the same species evolve in different ways. For this to happen, a form of isolation must exist. A species can be divided by a barrier into two separate populations (geographical isolation) or another factor can prevent members of the same species from successfully reproducing (reproductive isolation).

Geographical isolation

Geographical isolation usually involves a physical barrier that separates members of the same species. Geographical barriers include mountain ranges, rivers, distribution on different islands, and, in the present day, large-scale human activities such as agricultural and civil engineering developments.

Once two separate populations have formed they are likely to experience different selection pressures. This will drive their evolution in different directions, leading to the eventual development of separate species. Should the two new species encounter each other in the future, it is likely that their evolution will have diverged such that they are unable to breed together successfully. They will have become reproductively isolated from each other and will represent, effectively, two separate gene pools.

> ### Study tip
>
> Remember that the biological species concept defines a species as a population (or populations) of organisms that can breed with each other to produce viable offspring.

> ### Synoptic link
>
> You learnt about selection pressures in Topic 10.5, Evolution.

1 Species X occupies a forest area. Individuals within the forest form a single gene pool and freely interbreed.

species X lives and breeds in the forest

forest

2 Climatic changes to drier conditions reduce the size of the forest to two isolated regions. The distance between the two regions is too great for the two groups of species X (X_1 and X_2) to cross to each other.

forest A

group X_1

arid grassland

forest B

group X_2

▲ **Figure 1** *An example of speciation by geographical isolation*

3 Further climatic changes result in one region (Forest A) becoming colder and wetter. Selection pressures in the two isolated regions of forest are now different. Natural selection results in survival of group X_1 in forest A and group X_2 in forest B.

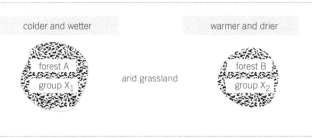

4 The differing selection pressures in the two areas result in the gene pools of the two groups becoming very different. There is no gene flow between the two groups.

5 A return to the original climatic conditions results in regrowth of forest. Forests A and B merge and the two groups of species are reunited. The two groups are no longer capable of interbreeding. They are now two species, Y and Z, each with its own gene pool.

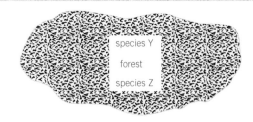

▲ **Figure 1** *Continued*

Reproductive isolation

Reproductive isolation arises eventually in geographically isolated populations. However, reproductive isolation can also evolve between organisms whose geographical distributions overlap. The isolation mechanisms include:

- temporal (e.g., different mating seasons among animals or flowering seasons in plants)
- behavioural (e.g., different mating rituals)
- mechanical/anatomical isolation (e.g., incompatible reproductive systems).

Sometimes none of these barriers exist and two species are able to mate, but a post-mating barrier stops successful reproduction. Post-mating barriers include gametes being unable to meet (e.g., sperm might be destroyed by chemicals in the female's reproductive system) or hybrid offspring being sterile.

Examples of reproductive isolation in primate speciation

Guenon species

Twenty-six species of guenon monkeys live in Africa. Several species of guenons have ranges with geographical overlap, but they do not interbreed. Some scientists have concluded that the strikingly different facial features of guenon species act to strengthen reproductive isolation.

Using facial recognition software, scientists found that guenon species that are likely to come into contact with each other are also those that show the greatest differences. Distinctive visual differences, such

▲ **Figure 2** *Three guenon species with very different facial features that reinforce reproductive isolation*

▲ **Figure 3** *Bonobos are thought to have been geographically isolated from chimpanzees by the Congo River*

Synoptic link

You learnt about primate classification in Chapter 10, Evolution and classification, and in Topic 10.5, Evolution, you considered the principles of natural selection.

as colour variations, ear tufts and mouth patches, appear to prevent breeding between species, reducing the risk of producing infertile offspring.

Hominids

Chimpanzees (*Pan troglodytes*) and bonobos (*Pan paniscus*) share a genus and have very similar genomes. They were once considered a single species. Similarly, humans and the extinct Neanderthals (*Homo neanderthalensis*) were thought for many years to be one species.

Bonobos and chimpanzees may have become separated and geographically isolated by the Congo River. Some scientists suggest that humans and Neanderthals experienced geographical isolation as well, perhaps being separated by a period of climate change.

Bonobos and chimpanzees will mate in captivity, and their hybrid offspring are thought to be fertile. Scientists have estimated that more than 1 million years on average is required for reproductive isolation in primates. Chimpanzees and bonobos probably diverged roughly 1 million years ago. The behavioural and anatomical differences between them, which have evolved during their geographical isolation, are insufficient to stop them mating. Nor have post-mating barriers evolved to render any hybrid offspring infertile.

Some scientists imagine a similar scenario existed with humans and Neanderthals, but that inter-species breeding would probably have been rare. Different behaviour patterns would have reduced the number of breeding opportunities.

Summary questions

1 Suggest the type of reproductive isolating mechanism that exists in the following examples:
   a The Kaibab squirrel and the Abert squirrel separated by the Colorado river.
   b The white sage plant has a large landing platform for pollinating insects, whereas the black sage has a small landing platform for pollinators.   *(2 marks)*

2 Describe and explain scientists' theories of reproductive isolation between species of guenon monkey.   *(2 marks)*

3 Evaluate the statement, 'bonobos and chimpanzees should be considered members of the same species'.   *(2 marks)*

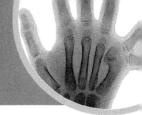

24.4 Epigenetics

Specification reference: 5.1.2

Scientists have realised only in recent decades the extent to which gene expression is modified during an organism's life. **Epigenetics** is the study of alterations in gene expression that are not a result of changes to DNA sequence. Here you will consider the role of the environment in modifying gene expression, and look at some of the studies into epigenetics within human populations.

What are epigenetic changes?

A DNA mutation is a change to the sequence of nucleotides in DNA. In contrast, epigenetic changes leave the structure of a gene intact. Instead, genes can be switched on and off by epigenetic changes – the extent to which they are expressed is altered. Some epigenetic changes can be inherited if they are present in gametes.

Two epigenetic mechanisms that have been studied in some detail are DNA methylation and histone modification.

DNA methylation

Methyl groups (CH_3-) can be added to DNA (specifically cytosine bases). This tends to decrease gene expression by preventing transcription.

Histone modification

In eukaryotes, DNA is associated with proteins called *histones*. The DNA is packaged and ordered by histones. DNA molecules are wrapped around the histones.

Histones can be chemically modified (e.g., by the addition of an acetyl, methyl, or phosphate group). These changes to histones either activate or deactivate the gene, depending on the modification. Modification can make the gene more or less accessible to transcription factors (proteins that control the rate of transcription). The addition of acetyl groups to histones, for example, increases transcription rates, thereby 'switching on' genes.

> **Learning outcomes**
>
> Demonstrate knowledge, understanding, and application of:
> → epigenetic changes to gene expression
> → the implications of human epigenetic studies.

> **Synoptic link**
>
> You learnt about the different types of DNA mutations, and about some human diseases that they cause, in Topic 23.2, Gene mutations.

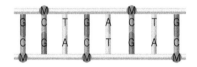

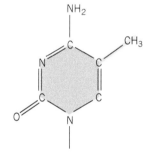

▲ **Figure 1** *DNA methylation is the addition of CH_3 (methyl; highlighted green on the base molecule) to cytosine bases*

gene "switched on"
• unmethylated cytosines (white circles)
• acetylated histones

transcription possible

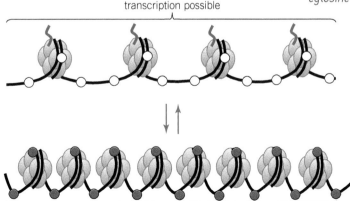

gene "switched off"
• methylated cytosines (red circles)
• deacetylated histones

transcription prevented

▲ **Figure 2** *Some possible epigenetic modifications and their results*

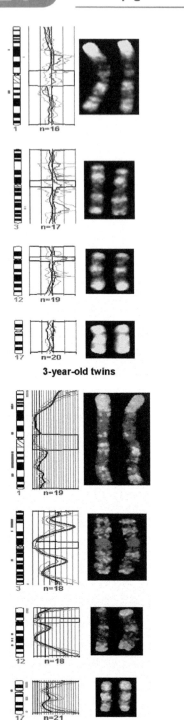

3-year-old twins

50-year-old twins

▲ **Figure 3** *Epigenetic changes in identical twins. Yellow = no epigenetic differences between twins, red and green = epigenetic differences between twins*

Epigenetic studies in human populations

Epigenetic changes have been linked with the development of several cancers, and epigenetics is now a major area of research. Previous studies have already demonstrated how a person's environment can bring about epigenetic changes that can be inherited.

Norrbotten studies

The people of Norrbotten, in northern Sweden, experienced famine in 1800, 1812, 1821, 1836, and 1856, interspersed with some years during which crops were abundant. Scientists discovered that the environment's feast and famine years impacted not only on the population at the time, but also on subsequent generations.

The descendants of women who experienced famine when they were fetuses had reduced life expectancies. The descendants of men who ate too much and were over-nourished during puberty, when their sperm were forming, also had reduced life expectancies. These discoveries show that the environment can produce epigenetic changes that are passed on to subsequent generations.

Dutch Hunger Winter

Children conceived during the 1944/45 wartime winter famine in the Netherlands have an increased risk of diabetes, obesity, and heart disease. This might be due to epigenetic alterations to genes associated with these diseases. The malnourishment of their mothers during early pregnancy appears to have produced an effect on children lasting their entire lives. Even more remarkably, the grandchildren of the malnourished women also tend to have the same health problems, regardless of the environment of their own early lives.

Twin studies

Twin studies are useful for teasing apart the influence of genetic and environmental factors on phenotypes. Identical twins have the same DNA sequences. However, if scientists are able to find twins that have been separated early in life and have been raised in different places, any differences between them will be a result of their environments.

Scientists have found that identical twins raised apart show few epigenetic differences in their early years but significant differences in middle age. Twins who spent less of their lifetime together showed the greatest epigenetic differences.

Of mice and memories?

Epigenetic changes related to fear have been shown to be heritable in mice. A study in 2013 exposed mice to electric shocks, which they learnt to associate with a cherry blossom smell. Eventually they exhibited a fear response to the smell alone, without shocks being administered.

The offspring of these mice appeared to be 200% more sensitive to cherry blossom odours. They showed fear when the smell was present. The scientists discovered that the gene coding for an odour receptor specific to the cherry smell had epigenetic tags that caused it to be expressed more. The offspring also possessed an enhanced section of the brain dedicated to smell.

The researchers suggest that a lack of DNA methylation in the odour receptor gene was responsible for the inherited effect.

This opens up the possibility that humans might inherit aversions for particular smells and tastes from their parents via epigenetic changes.

1   Explain how a lack of DNA methylation in the odour receptor gene might have caused the inherited fear of the smell of cherry blossom.
2   Some scientists suggest that the inheritance of the fear of the smell of cherry blossom needs more explanation. Suggest what aspects of the experimental results might require more testing and explanation.

Synoptic link

You looked at the principles of transcription in Topic 4.6, Protein synthesis – transcription.

Summary questions

1   Explain the difference between a DNA mutation and an epigenetic change.   (*2 marks*)

2   Descendants of people conceived during the Dutch Hunger Winter have an increased risk of heart disease and diabetes. What have scientists concluded is the explanation for these observations?   (*3 marks*)

3   Figure 4 shows some different conditions shared by twins and illustrates the relative influence of genes and the environment on each trait. Explain the evidence for the comparative influence of genetics and the environment on height and strokes.   (*3 marks*)

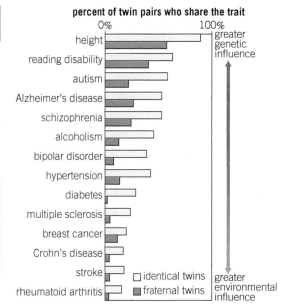

▲ Figure 4 *Different conditions shared by twins and the influence of genes and the environment on each trait*

Practice questions

1 Which of the following statements is/are true of the founder effect?

Statement 1: It can be considered a form of genetic bottleneck.

Statement 2: It often results in a change in allele proportions within a population.

Statement 3: It is usually a result of two populations merging.

A 1, 2 and 3

B Only 1 and 2

C Only 2 and 3

D Only 1 *(1 mark)*

2 Which of the following statements is/are true of the assumptions made when applying the Hardy–Weinberg principle to calculate the frequency of alleles in a population?

Statement 1: Natural selection is operating.

Statement 2: The population is large.

Statement 3: Mating is random.

A 1, 2 and 3

B Only 1 and 2

C Only 2 and 3

D Only 1 *(1 mark)*

3 Which of the following is an example of temporal reproductive isolation?

A Different mating rituals.

B Incompatible reproductive systems.

C Separation by a physical barrier.

D Different flowering seasons. *(1 mark)*

4 Which of the following statements is/are true of DNA methylation?

Statement 1: Methylation involves a CH_3 group being added to DNA.

Statement 2: Methylation tends to increase gene expression.

Statement 3: Adenine bases are methylated.

A 1, 2 and 3

B Only 1 and 2

C Only 2 and 3

D Only 1 *(1 mark)*

5 Which of the following statements is true of an epigenetic change?

A It occurs outside a cell.

B It alters the base sequence in DNA.

C It alters gene expression.

D It can never be inherited. *(1 mark)*

6 Albino rabbits have white fur as these individuals are unable to produce the pigment melanin. The ability to produce melanin is controlled by a gene with a dominant allele (B), resulting in brown fur, and a recessive allele (b), resulting in an albino.

Of the 60 rabbits in a pet shop, 45 are brown.

a Calculate the frequency of the dominant allele in this group. *(3 marks)*

b Give two reasons why it is not appropriate to use the Hardy–Weinberg principle to estimate the frequencies of alleles in this group of rabbits in the pet shop. *(2 marks)*

OCR F215 2014

7 The black form of ladybird is caused by an allele (B) that is dominant.

The red form of ladybird is therefore homozygous recessive at the same locus (bb).

A population contains 296 red ladybirds and 50 black ladybirds.

Calculate the frequency of the dominant allele and the recessive allele in this population. *(3 marks)*

OCR F215 2011

8 a The table shown here compares the frequency of certain traits in tame foxes and a control group.

Phenotypic trait	Animals showing trait (per 100 000)		Percentage increase in trait
	Control population	Tame population	
white patch of fur on head	710	12 400	1646
floppy ears	170	230	35
short tail	2	140	6900
curly tail	830	9400	1033

Suggest how genetic drift could account for the data in the table. *(2 marks)*

b The grey wolf species, *Canis lupus*, evolved into the domesticated dog species, *Canis familiaris*.

Suggest how different types of isolating mechanisms allowed dogs to evolve separately to wolves. *(3 marks)*

c Interbreeding between members of the wolf species and some dogs has been reported. However, there are some large breeds of dogs that cannot breed successfully with small dog breeds.

Use this information and your own knowledge to explain the problems of classifying wolves and different dog species according to the biological species concept and the phylogenetic species concept. *(4 marks)*

OCR F215 2012

9 The diagram and table show the behaviour patterns of workers of two species of bumblebee.

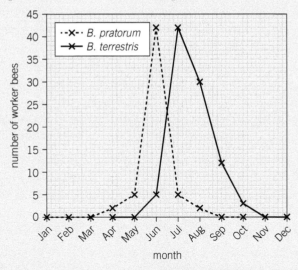

Species of bumblebee	Mean depth of flower visited (mm)	Visits to flowers when nectar only collected (%)	Visits to flowers when pollen only collected (%)	Visits to flowers when both nectar and pollen collected (%)
Bombus pratorum	7.4	23	10	67
Bombus terrestris	6.3	80	11	9

Using the information provided, suggest how an ancestral species might have evolved into the two species, *B. pratorum* and *B. terrestris*. *(3 marks)*

OCR F215 2014

10 Outline, with examples, how the founder effect can change allele frequencies in human populations. *(6 marks)*

25 GENE TECHNOLOGIES
25.1 DNA amplification – PCR

Specification reference: 5.1.3

In this chapter you will be studying a variety of genetic techniques. Many of these techniques require the amount of DNA available to be increased. For example, only small quantities of DNA are usually obtained from a crime scene or from a fossil. The amount of DNA needs to be amplified before it can be analysed. The method that is used to increase the amount of DNA available is called the **polymerase chain reaction** (PCR).

PCR

PCR is an *in vitro* method for copying and amplifying sections of DNA. The process is carried out in a machine called a *thermocycler*, which is a computer-controlled machine that varies temperatures at precise time intervals. The following molecules are placed in the PCR thermocycler:

- The DNA fragment that is being copied.
- Free phosphorylated nucleotides – which will be joined (bonded) together to form copies of the DNA fragment.
- Taq DNA polymerase – the enzyme responsible for bonding the free nucleotides together to form copies of DNA strand fragments. Human DNA polymerase is not used. Instead, the enzyme is obtained from a thermophilic bacterium called *Thermus aquaticus* (Taq). This form of DNA polymerase is tolerant to heat so does not denature during the PCR temperature cycling.
- **Primers** – short sequences (10–20 bases) of DNA, complementary to one end of a fragment. They provide the starting sequence for DNA polymerase to begin the copying process.

▲ Figure 1 *A PCR machine*

The stages

PCR involves three stages. The thermocycler switches between the three different temperatures required at each stage.

Separation of the DNA strands in the fragment – a temperature of 95 °C provides the kinetic energy required to break hydrogen bonds between the DNA strands and separate them.

Addition of primers – the thermocycler mixture is cooled to 55 °C. This enables primers to hydrogen bond to complementary bases at the end of each DNA strand.

DNA synthesis – the thermocycler is raised to 72 °C, which is the optimum temperature for Taq DNA polymerase to function. The polymerase joins free complementary nucleotides to each of the separated DNA strands, beginning at the primers.

Two copies of the original double-stranded fragment are produced by the end of one cycle. The cycle then begins again – two strands will result in four by the end of the second cycle. This means that more than one million DNA copies can be made in 20 cycles. A single cycle lasts two minutes.

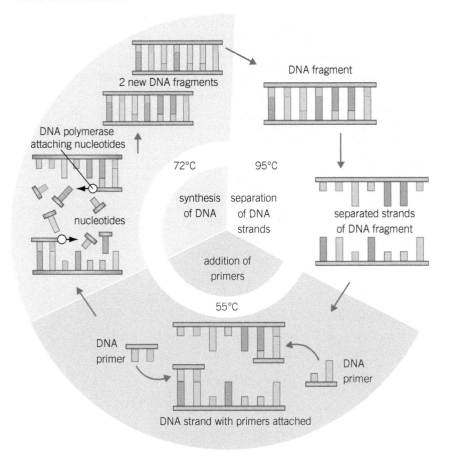

▲ Figure 2 *A cycle of the polymerase chain reaction*

Synoptic link

You looked at the structure and semi-conservative replication of DNA in Chapter 4, Nucleic acids.

Summary questions

1   Why is human DNA polymerase not used in the PCR process?
(1 mark)

2   If PCR begins with one fragment, calculate the number of fragments (represented on a log scale) produced after
    a   5 cycles
    b   12 cycles
    c   17 cycles.   *(3 marks)*

3   Describe two similarities and two differences between semi-conservative replication of DNA and PCR.   *(4 marks)*

The use of log scales to illustrate DNA amplification

The amplification of DNA is an example of exponential increase. Log scales can be used to show the relationship between cycles of heating and cooling and the increases in copy number.

After 10 cycles, a single fragment of DNA can, in theory, be amplified to 1024 fragments, which would be represented on a log scale as $10^{3.01}$.

Use the graph to estimate the number of DNA fragments that can be obtained after 20 cycles. Write your answer as a whole number rather than in standard form.

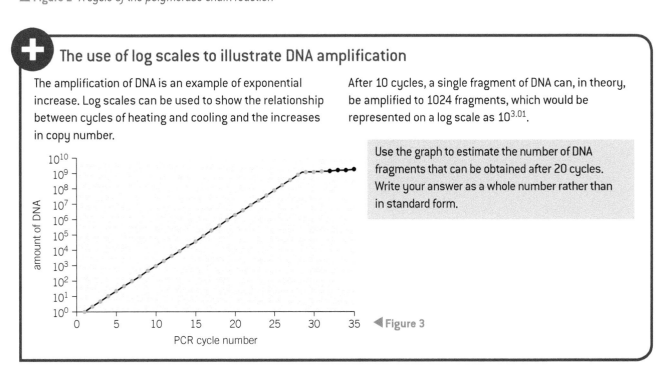

◀ Figure 3

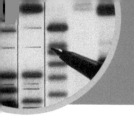

25.2 DNA analysis

Specification reference: 5.1.3

In the previous topic, you examined how the amount of DNA available for analysis can be increased by using a technique called PCR. Now you will look at some of the ways in which DNA can be analysed. First you will learn about an analysis technique, gel **electrophoresis**, which separates DNA by size. You will then learn about the types of DNA sequences that are commonly analysed during criminal investigations, paternity testing, clinical trials and investigations into human ancestry.

Electrophoresis

Gel electrophoresis is often performed after PCR has amplified the amount of DNA available. It can be used directly for analysis (e.g., in DNA fingerprinting), or to prepare DNA samples to be sequenced, or for cloning.

Electrophoresis separates DNA fragments for identification and analysis based on their size. Fragments that differ in length by only a single base can be separated.

The steps in gel electrophoresis

1 An electrophoresis gel plate consists of a gel called *agarose* covered in buffer solution. Electrodes are attached at both ends, which enables an electric current to be passed through the gel.

2 The DNA sample is cut into fragments using enzymes called *restriction endonucleases*.

3 The DNA fragments are placed in wells in the gel at the end closest to the negative electrode.

4 When a current is applied, DNA moves towards the positive electrode (*anode*) because it is negatively charged (due to the presence of phosphate groups).

5 Longer fragments do not move as far because they interact with the gel to a greater extent than shorter fragments. Shorter fragments will therefore be positioned nearer the anode once the procedure has ended.

6 The fragments appear as bands, which can be visualised using UV light and a fluorescent DNA dye.

7 DNA fragments can be extracted from the agarose gel for further analysis.

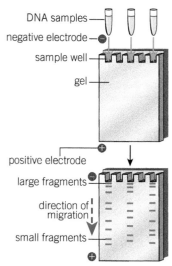

▲ **Figure 1** *The procedure for gel electrophoresis*

The buffer solution used in electrophoresis contains ions and maintains a constant pH. Suggest why both these aspects of the buffer solution are important in electrophoresis.

DNA sequences useful in genome studies

Variable number tandem repeats (VNTRs)

VNTRs are patterns of repeated nucleotides that are adjacent to each other in a DNA sequence (e.g., CAATT CAATT CAATT). There is variation in the VNTRs of different people, such that the probability of two unrelated people having the same VNTR pattern is very low. This allows the nucleotide sequences of VNTRs to be used as genetic fingerprints, for example, in forensic analysis.

Changes in VNTR nucleotide sequences alter the positions at which restriction endonucleases cut DNA. Gel electrophoresis can subsequently be used to produce VNTR banding patterns. These patterns are compared during paternity testing and criminal forensic investigations. In criminal investigations, forensic scientists look for a match between a suspect's DNA fingerprint and the specimen fingerprint taken from a crime scene. In paternity testing, all the child's bands will match bands from either the mother or the father. If the child's VNTR fingerprint contains a band without a match, then another man must have been the biological father.

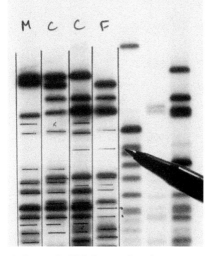

▲ **Figure 2** *DNA fingerprints from a mother (M), children (C) and father (F)*

Single nucleotide polymorphisms (SNPs)

SNPs are sequences of DNA that vary between people by a single nucleotide. They can occur in genes, but they are more common in non-coding DNA. When SNPs occur in genes we refer to them as *alleles*. They are caused by substitution mutations. Substitution mutations in a gene will not necessarily cause a change in the amino acid sequence of the protein being produced because the genetic code is degenerate.

▲ **Figure 3** *DNA fingerprints from VNTR sequences … but who is the criminal?*

Some SNPs can indicate a person's susceptibility to disease and how they respond to certain drugs, chemicals and vaccines. Knowledge of SNPs enables scientists to select particular individuals for clinical trials. Drugs may produce different effects in individuals with different SNPs.

Studies have been conducted in which the genomes of people with a disease are compared to the genomes of people who are free of the disease. Many non-infectious diseases are governed by several genes. SNPs that are more common in people with a disease can be identified. In this way, alleles are identified that indicate a greater risk of someone developing a particular condition. More than 200 traits and diseases have been analysed using this approach. For example, several SNPs at the *FGFR2* gene locus are associated with one form of breast cancer.

Haplotypes

A haplotype is a set of genes inherited together from one parent.

Haplotypes can provide insight into the history of human migrations. As a result of the founder effect, people with the same ancestral origins tend to share SNPs and have very similar haplotypes. The genetic

> **Synoptic link**
>
> You learnt in Topic 23.2, Gene mutations, about the effects on DNA sequence of different types of DNA mutation.

> **Synoptic link**
>
> You learnt about haplotypes in Topic 23.3, where human leucocyte antigens (HLAs) were discussed as an example. You found out about the process of DNA recombination in Topic 9.1, Meiosis.

sequences in haplotypes show clear geographical variation, which reflects the migrations of ancestral human populations. The set of similar haplotypes found in a particular ethnic group is known as a *haplogroup*.

The most common haplotypes for scientists to analyse when studying ancestral origins are those in mitochondrial DNA and the Y-chromosome. These forms of DNA are analysed because neither undergoes recombination – the only changes in their nucleotide sequences are from random mutations. Mitochondrial DNA passes solely from mother to offspring, whereas Y-chromosome DNA passes solely from father to son.

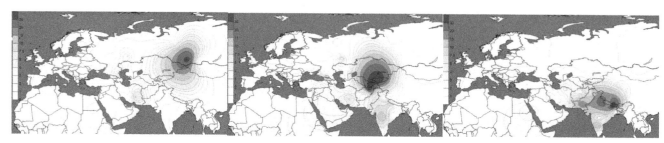

▲ **Figure 4** *Spatial frequency distributions of three sub-haplogroups from the human R1a haplogroup. The maps are based on analysis of the SNPs present on the Y-chromosomes of Eurasian men*

Synoptic link

Several of the ideas discussed in this topic tie in with ideas from earlier chapters. For example, the analysis of SNPs relates to substitution mutations, which you learnt about in Topic 23.2, Gene mutations, and geographical differences in haplogroups result from the founder effect, which you learnt about in Topic 24.1, Factors affecting allelic frequencies.

Summary questions

1 Which of the suspects in Figure 3 is likely to be the criminal? Explain your answer. (*2 marks*)

2 You learnt about chromatography in Topic 3.2, Separation and identification of amino acids. Suggest why electrophoresis is sometimes described as 'similar to chromatography'. (*2 marks*)

3 A person's ancestral origins can be assessed by analysing their haplogroups. Explain the principles behind this analysis. (*3 marks*)

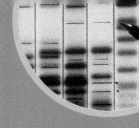

25.3 Genetic engineering – bacterial cells

Specification reference: 5.1.3

So far in this chapter, you have considered laboratory techniques for copying and analysing DNA. In this topic you will look at techniques for modifying an organism's DNA using recombinant DNA technology (**genetic engineering**). Scientists are now able to add or 'silence' genes from a range of species to alter their traits. This topic will examine the principles behind genetic engineering in bacterial cells.

General principles of genetic engineering

Genetic engineering combines DNA from two different organisms, often from different species. Three steps are required:

- obtaining the required gene
- placing the gene in a **vector** (a structure that carries the gene into the recipient cell)
- transporting the gene into the recipient cell.

A variety of approaches can be used at each stage. Once the gene has been placed in the target cells, the aim is for the recipient to express the gene and produce the protein for which it codes.

Genetic modification of bacteria

Bacteria can be genetically modified to produce a range of products, including human proteins such as insulin (to treat diabetes) and growth hormone (to treat dwarfism). Scientists can use two techniques to obtain the required gene: cutting out the gene using **restriction enzymes**, or producing the gene from an mRNA template.

Using restriction enzymes to extract a gene

Restriction enzymes (endonucleases) are found in bacteria. More than 50 different restriction enzymes are available for use in genetic engineering. Each enzyme cuts DNA at specific base sequences, known as recognition sites, which tend to be less than 10 bases in length. Many of the enzymes produce a staggered cut, which creates two short sequences of exposed, unpaired bases known as **sticky ends**. The enzyme recognition sequences are often *palindromic* (i.e., their base sequences are mirror images). An example of this is shown in Figure 1.

Plasmids can be used as vectors when genetically modifying bacteria. Plasmids are small, circular pieces of DNA taken from bacterial cells. The required gene and plasmids are cut with the same restriction enzyme. This creates two sets of sticky ends with complementary base pairings. The plasmid

Learning outcomes

Demonstrate knowledge, understanding, and application of:

→ genetic modification of bacterial cells to produce human proteins.

HindIII restriction endonuclease has the recognition site AAGCTT, which produces a staggered cut and therefore 'sticky ends':

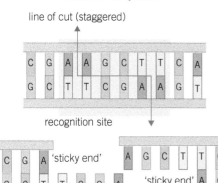

▲ **Figure 1** *The formation of sticky ends by a restriction enzyme*

Synoptic link

You learnt about HIV and reverse transcriptase in Topic 11.1, Communicable diseases.

and recombinant gene are *annealed* (hydrogen bonds form between complementary base pairs) and DNA ligase is used to join the sticky ends on the gene and the plasmid. The gene becomes integrated into the plasmid vector. The resultant DNA, which contains sections from two organisms, is known as a *recombinant plasmid.*

Using mRNA as a template to produce a gene

Scientists can extract a sample of transcribed mRNA for the required gene. The mRNA is used to synthesise the gene. Reverse transcriptase is an enzyme found in retroviruses such as HIV. The reverse transcriptase enzyme synthesises a single strand of DNA that is complementary to mRNA. This is known as cDNA. DNA polymerase is then used to join together free nucleotides and form a second DNA strand. The double stranded DNA copy is shorter than the original gene in the eukaryotic cell because it has been copied from mature mRNA, which has been edited. A plasmid vector is opened with a restriction enzyme. The gene is placed in the plasmid using DNA ligase.

Human insulin can be produced using transgenic bacteria. The human insulin gene is transferred into bacteria using the reverse transcriptase method. This process is illustrated in Figure 2.

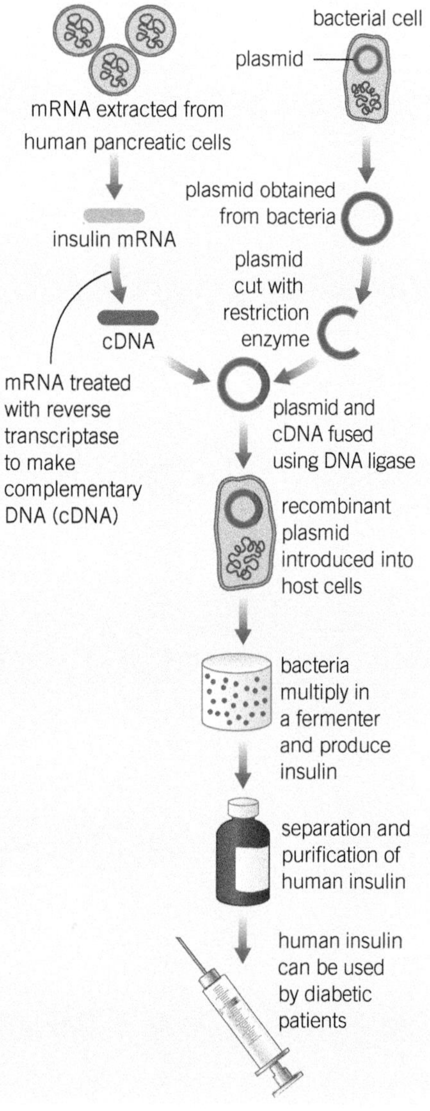

▲ **Figure 2** *The formation of genetically modified* E. coli *bacteria that can manufacture human insulin*

Recombinant plasmids are taken up by bacteria

Bacteria are mixed with recombinant plasmids, which are produced using one of the methods that have been described. Bacteria that take up the plasmids are known as *transgenic.*

However, the process is very inefficient. Not all plasmids take up the new gene, and less than 1% of plasmids are taken up by bacterial cells. A method is required to identify which bacteria have taken up the recombinant plasmid.

Identifying transgenic bacteria with reporter genes

Plasmids need to contain a reporter gene, which indicates whether the new gene has been taken up by bacterial cells. A reporter gene could be one that is inserted into the plasmid alongside the new gene. For example, a reporter gene that produces fluorescence would provide visual evidence that a recombinant plasmid has been taken up.

Alternatively, the required gene can be inserted into the DNA sequence of a reporter gene within the plasmid. This stops the reporter gene from functioning. As a consequence, only bacteria with the recombinant plasmid will have reporter genes that do not work.

Two reporter genes are used when inserting the human insulin gene into *E. coli* plasmids: the tetracycline-resistance (tet^R) gene and the ampicillin-resistance (amp^R) gene.

The amp^R gene is used to check which bacteria have taken up the plasmid. When bacteria are placed on an agar plate containing ampicillin, the only surviving bacteria are those that have taken up the plasmid and have the amp^R gene.

To assess which bacteria have taken plasmids that have incorporated the insulin gene, a restriction enzyme is selected that opens the plasmid in the DNA sequence of the *tet^R* gene. The insulin gene is integrated into the plasmid within the *tet^R* gene. Bacteria that take up the plasmid no longer have resistance to the tetracycline antibiotic and will fail to grow on agar plates containing tetracycline. We can now identify the original colonies of recombinant bacteria – these colonies can be grown on a large scale to produce insulin.

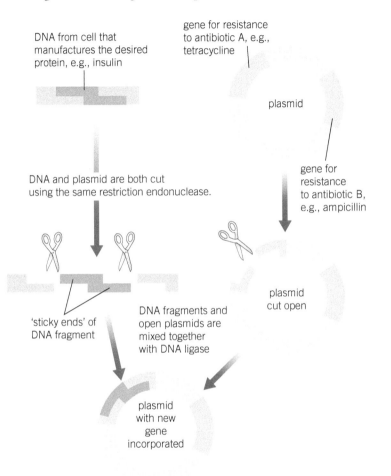

▲ **Figure 3** *Inserting the human insulin gene into a reporter gene in a plasmid vector*

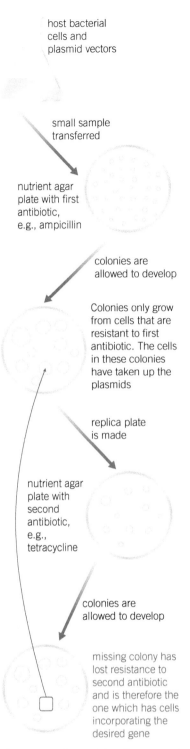

▲ **Figure 4** *How to identify transgenic bacteria using reporter genes*

Summary questions

1 Explain why, in genetic engineering, it is important to use the same restriction enzyme to cut both the gene being transferred and the plasmid vector. *(2 marks)*

2 Explain why mRNA is extracted from human pancreatic cells in the process shown in Figure 1. *(3 marks)*

3 Suggest why is it better to produce insulin from genetically modified bacteria than to extract insulin from animals. *(3 marks)*

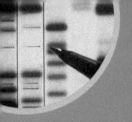

25.4 Genetic engineering – eukaryotic cells

Specification reference: 5.1.3

Learning outcomes

Demonstrate knowledge, understanding, and application of:

→ the uses of genetic engineering in eukaryotic cells.

In the previous topic you were introduced to the principles of genetic engineering and examined how bacterial cells can be engineered. Here you will learn about some examples of genetic modification in eukaryotic cells. The techniques used for introducing genetic material into eukaryotic cells are discussed in more detail in Topic 25.6, Gene therapy.

Genetic modification of eukaryotic cells

Producing human proteins in animals and crop plants

Many examples now exist of transgenic animals and crop plants. For example, goats have been genetically engineered to produce high-strength protein by incorporating a spider gene into their genome. Transgenic *Enviropigs* contain genes from mice and bacteria. These pigs are able to digest phosphorous-containing compounds in cereals more efficiently, thereby reducing feed costs and phosphorous pollution. Crops have been genetically engineered to produce many beneficial proteins, including a vitamin A precursor and insecticides.

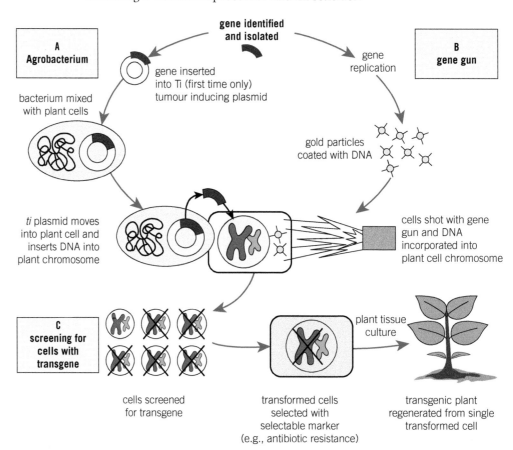

▲ **Figure 1** *A summary of genetic engineering techniques for plant cells. A and B are alternative methods of introducing DNA into plant cells*

154

Adding human genes to other eukaryotic species is a more controversial issue. Nonetheless, several experiments have produced transgenic species containing human genes, as Table 1 shows.

Synoptic link

You learnt about the principles of model organisms in Topic 23.5, Investigating patterns of inheritance.

▼ Table 1 *Transgenic organisms with human genes*

Transgenic organism	Human gene	Potential benefit
cows	lysozyme	antibacterial properties
goats	anti-thrombin	prevents blood clotting – it can be given to patients who cannot produce the protein naturally
tobacco	antibodies	protection from pathogens
rice	human serum albumin	blood transfusions

Knockout mice

You previously learnt about the principles of model organisms. The mouse is a common vertebrate model organism for studying disease development and treatment. Knockout mice are genetically engineered to have an inactivated gene. The function of the disabled gene can be assessed by comparing the phenotypes of knockout mice with those of normal mice.

Genes are knocked out by incorporating a nucleotide sequence resembling the target gene into an embryonic stem cell. The artificial DNA sequence replaces the functional gene. The stem cell is then fused with an embryo. The mouse develops with the artificial gene sequence instead of the functional gene.

Many diseases have been studied using knockout mice, including diabetes, heart disease and cancer. The gene that codes for a protein called *leptin* has also been studied in mice. By knocking out the leptin gene, scientists learnt that it plays a significant role in energy metabolism and fat deposition.

sperm from male goat → mature eggs from female goat's ovary → fertilised goat's eggs

human gene for anti-thrombin is inserted alongside gene for milk proteins → transformed goat's egg implanted into female goat → kids with the anti-thrombin gene are crossbred → milk from goat herd is rich in anti-thrombin → anti-thrombin is extracted from milk and purified → anti-thrombin is given to humans unable to produce it

▲ Figure 2 *Outline of the genetic modification of goats to produce anti-thrombin protein*

Summary questions

1 Describe how a plasmid is used to introduce a new gene into a plant cell. *(4 marks)*

2 Suggest the advantages of genetically modifying a crop to be herbicide resistant. *(2 marks)*

3 A scientist suspects that a particular gene may be associated with cancer. Explain how she could test her theory using knockout mice. *(4 marks)*

▲ Figure 3 *A normal mouse (right) and a mouse that has its leptin gene knocked out (left)*

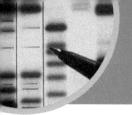

25.5 RNA manipulation

Specification reference: 5.1.3

You have previously learnt how a gene is transcribed into mRNA, which is translated into a polypeptide. However, the complete story is often more complicated than these two steps alone. Over the past few decades, scientists have discovered many ways in which mRNA can be altered following transcription. This topic will outline two of these processes: post-transcriptional mRNA splicing, and RNA interference. You will also consider the potential use of RNA interference in the treatment of diseases.

RNA splicing

Genes consist of sections called **exons** and **introns**. Exons code for sequences of amino acids, whereas introns do not. In eukaryotic cells, introns are transcribed into the initial mRNA strand (known as pre-mRNA) but are then removed. The exons are joined in a process called splicing. The mature mRNA, comprising only exons, can then be translated into a polypeptide.

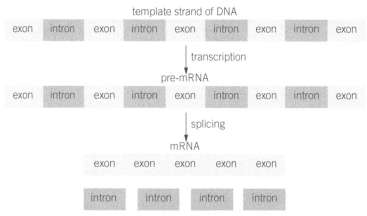

▲ Figure 1 *Splicing of pre-mRNA to form mature mRNA*

You might be questioning the purpose of introns. Why include sections of RNA that will then be removed? Splicing enables a range of different mRNA molecules to be formed from the same gene. Some introns can be retained and exons can be joined in different combinations. This is known as *alternative splicing* and means that a single gene can potentially code for many different proteins. Scientists have estimated that more than 90% of human genes are alternatively spliced. This may provide the genetic diversity needed to form complex structures such as the nervous system.

RNA interference

Many organisms use RNA interference (RNAi) to control genes. The process can also be used as a laboratory tool, and in the future it could be harnessed to provide treatments for diseases.

In RNAi, sections of DNA produce short RNA strands that bind to mRNA, effectively blocking protein synthesis. Gene expression is inhibited – this is sometimes referred to as *gene silencing*.

Two different RNA interference mechanisms exist, which involve two different forms of RNA – microRNA (miRNA), and small interfering RNA (siRNA). siRNA and miRNA are both derived from double-stranded RNA. An enzyme called dicer cuts the double-stranded RNA into short sections (usually 20–25 nucleotides in length). A protein called Argonaute then selects and binds to one of the two RNA strands, releasing the other.

The siRNA binds to a specific sequence of nucleotides on mRNA, which are complementary to its own nucleotides. The Argonaute protein then breaks the mRNA, rendering it non functional. miRNA is less precise than siRNA – only some of its sequence is complementary to mRNA. This enables miRNA to bind to many different mRNA molecules with the aid of Argonaute proteins, thereby blocking translation.

RNA interference in disease treatment?

In the future, it may be possible to exploit RNA interference in the treatment of disease. Liver failure in mouse model organisms has been reversed using RNAi. Cancers may be treated by silencing genes in tumour cells, such as those involved in cell division. Initial safety concerns were raised when mice with liver disease were given RNAi treatment. Almost half the mice died within two months of the therapy.

▲ Figure 2 *The molecules of RNA interference – an Argonaute protein (purple) holding siRNA (red and green)*

RNA interference in model organisms – genes down but not out?

In Topic 25.4, Genetic engineering – eukaryotic cells, you learnt about gene knockouts in model organisms. RNA interference can also be used to reduce gene activity in model organisms or in laboratory cell cultures. This enables the role of the gene product to be studied. Unlike gene knockouts, which eliminate the expression of a gene, RNA interference only reduces gene expression.

Scientists can either introduce double-stranded RNA, which will be cut into siRNA by the model organism's dicer enzyme, or they can introduce siRNA directly. siRNA tends to be used in mammalian model organisms because the larger double-stranded RNA molecules are less likely to be effective.

1 Why do you think RNA interference in model organisms is known as 'gene knockdown'?
2 Suggest why double-stranded RNA can be used successfully in less complex organisms such as invertebrates, but not in mammals.

Synoptic link

You learnt about the structure of mRNA and the process of transcription in Topic 4.6, Protein synthesis – transcription.

Summary questions

1 Explain how a gene consisting of 930 base pairs can be transcribed into mature mRNA consisting of only 615 nucleotides. *(2 marks)*

2 Explain the importance of alternative splicing in organisms. *(2 marks)*

3 Suggest why RNAi treatment for diseases is potentially dangerous, and explain why siRNA is more likely to be used in treatments than miRNA. *(3 marks)*

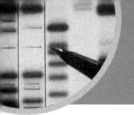
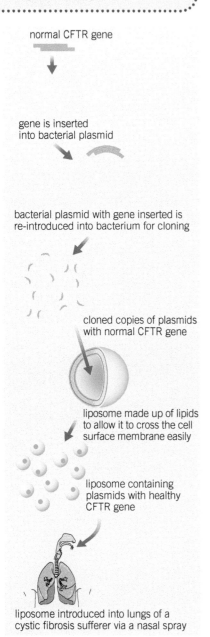

normal CFTR gene

gene is inserted into bacterial plasmid

bacterial plasmid with gene inserted is re-introduced into bacterium for cloning

cloned copies of plasmids with normal CFTR gene

liposome made up of lipids to allow it to cross the cell surface membrane easily

liposome containing plasmids with healthy CFTR gene

liposome introduced into lungs of a cystic fibrosis sufferer via a nasal spray

▲ **Figure 1** *Gene therapy for cystic fibrosis using liposomes for gene delivery*

So far in this chapter, you have examined methods for manipulating DNA and RNA. The final aspect of genetic technology we will cover is **gene therapy**. This is an application of genetic modification that treats genetic disorders by adding functional alleles to cells with defective alleles. We will focus on the ethical issues surrounding two diseases targeted by gene therapy: severe combined immune deficiency (SCID) and cystic fibrosis.

The principle of gene therapy

The underlying principle of gene therapy is to place working copies of genes into cells that lack them. Transcription and translation of the functional gene within cells may rid the patient of symptoms. If RNA interference, which you learnt about in the previous topic, can be thought of as silencing genes, then gene therapy gives them a voice.

Somatic cell gene therapy

This form of gene therapy targets non-gamete diploid cells (somatic cells) in tissue affected by a disease. The additional gene is taken up only by certain cells. For example, if cystic fibrosis (CF) is being treated then functional CF alleles will be taken up by lung cells. These new alleles will not be present in gametes and therefore cannot be passed to offspring. As targeted cells die and are replaced, the therapy needs to be repeated at regular intervals.

Vectors in somatic cell gene therapy

The therapeutic genes need to be delivered to the target cells. The structure that carries the genes that are to be delivered during therapy is known as a *vector*. Several types of vector exist.

A harmless virus can be employed as a vector. Adenoviruses, for example, cause respiratory diseases, such as colds, by inserting their DNA into cells in the lungs. This ability can be harnessed to transfer functional alleles into lung cells that have only faulty cystic fibrosis (CF) alleles.

The adenoviruses are made harmless by interfering with a gene controlling their replication. The functional CF allele is inserted into the viral DNA. The adenoviruses are delivered via nasal spray and are inhaled. In the lungs the viruses deliver their DNA, including the healthy CF variant, into lung cells.

SCID (a disease in which people do not show a cell-mediated immune response) was the first condition to be treated with gene therapy. A virus was used to introduce a healthy gene into the patients' cells. However, some of the patients were diagnosed with leukaemia as a result of the treatment. In more recent SCID gene therapy, stem cells were taken from a patient's bone marrow. A working version of the faulty gene was introduced into the stem cells, which were injected back into the patient.

An alternative delivery vector approach is for a gene to be wrapped in lipid molecules, known as a *liposome* (Figure 1). Bacterial plasmids are used to carry and copy the desired gene. Plasmids containing the gene are placed in liposomes. If CF is being treated, the liposomes are sprayed into the patient's nostrils and inhaled. Liposomes are able to pass across the cell surface membranes of respiratory cells in the lungs and deliver the new gene.

Germline gene therapy

Germline gene therapy involves adding a functional gene to a fertilised egg. This ensures that all an organism's cells will contain the added gene variant. This gene can be passed on to the treated organism's offspring.

Germline gene therapy in humans is prohibited, but research can be conducted on other organisms. The Clothier ethics committee, which was established in 1993, concluded that human germline gene therapy is unacceptable. The committee stated that inadvertent DNA modification could produce a new human disease, and permanent changes to the human genome creates ethical dilemmas that should be fully debated before committing to germline therapy.

Ethical implications of gene therapy

The potential future use of gene therapy has incredible scope. More than 4000 diseases have been identified as resulting from abnormal genes. Clinical trials have shown positive results for diseases such as cystic fibrosis, SCID, leukaemia and Parkinson's disease.

However, several ethical concerns remain, including:

- Health implications. Some people worry that new genes or vectors (either viruses or liposomes) could produce immune responses in patients. As we discussed earlier, in early gene therapy clinical trials for SCID, recipients developed leukaemia because the delivered genes were integrated in the wrong place within the genome.

- Is germline gene therapy ethical? Should we allow the genetic make-up of unborn children to be altered?

- Deciding which conditions are worthy of gene therapy could be difficult. Some people worry, for example, that gene therapy could eventually be used to alter height, hair colour, brain development, and other traits that are not considered to be disorders or diseases.

- Gene therapy research and treatments are expensive. Some people argue that money should be invested in proven treatments that are guaranteed to work.

- Future widespread use of gene therapy opens up the possibility of treatments only being available to those who can afford it, thereby creating inequality in healthcare.

▲ Figure 2 *A child with SCID must remain in a sterile environment. SCID is one condition that can be treated using gene therapy*

Study tip

Avoid writing that 'genes are replaced'. Defective genes are not removed. Cells that have undergone gene therapy produce both functional and non-functional proteins.

Summary questions

1 Why does gene therapy not provide a permanent cure for cystic fibrosis? *(2 marks)*

2 Why is gene therapy potentially most useful for treating diseases stemming from single-gene mutations? *(2 marks)*

3 Suggest why viruses and liposomes are used as vectors for delivering genes to target cells. *(2 marks)*

Practice questions

1 Which of the following statements is/are true of RNA interference?

Statement 1: Dicer is an enzyme that cuts double-stranded DNA into shorter sections.

Statement 2: RNA interference does not alter nucleotide sequences in DNA.

Statement 3: RNA interference, in general, reduces gene expression rather than eliminating gene expression.

A 1, 2 and 3

B Only 1 and 2

C Only 2 and 3

D Only 1   (1 mark)

2 During the polymerase chain reaction, why is the temperature lowered to 55°C?

A To enable DNA polymerase to function.

B To enable primers to attach to the DNA strands.

C To separate DNA strands.

D To enable free phosphorylated nucleotides to attach to DNA strands.   (1 mark)

3 After 12 cycles of PCR, a single fragment of DNA can, in theory, be amplified to how many fragments?

A $10^{3.01}$

B $10^{3.31}$

C $10^{3.61}$

D $10^{3.91}$ (1 mark)

4 Which of the following statements is/are true of the sticky ends produced during genetic modification of bacteria?

Statement 1: They are often palindromic.

Statement 2: They are produced by DNA ligase.

Statement 3: They are usually more than 20 bases in length.

A 1, 2 and 3

B Only 1 and 2

C Only 2 and 3

D Only 1   (1 mark)

5 a Genetic engineering uses the following:

A an enzyme that synthesises new DNA

B an enzyme that cuts DNA at specific sequences

C an enzyme that reseals cut ends of DNA

D small circular pieces of DNA found in bacteria – these pieces of DNA have antibiotic resistance genes

E an enzyme found in some viruses with an RNA genome – this converts RNA into DNA.

Name A to E.   (5 marks)

b Genes are cloned for a number of reasons. For example, one group of scientists wanted to sequence a disease-causing mutation to learn more about a human disease – these scientists started their research using white blood cells.

Another groups wanted to clone the insulin gene in order to manufacture its protein to treat diabetes – these scientists started their research using cells from the pancreas.

Suggest and explain the biological reasons why the two groups each started with a different cell.   (4 marks)

c A gene can be cloned *in vitro* by PCR or *in vivo* (in living cells) by introducing the gene into bacterial host cells. The table identifies some of the key steps in each process.

in vitro gene cloning (PCR)	*in vivo* gene cloning
At 95°C, DNA extracted from a cell separates into two strands.	A library of gene fragments is produced and introduced into host bacteria.
At 50°C, specially made primer sequences attach to the ends of the desired gene only.	Bacteria are screened for antibiotic resistance to identify those with recombinant DNA.
At 72°C complementary copies of both DNA Strands are made.	A gene probe is used to select the bacterial colony containing the desired gene.
The cycle of temperature changes is repeated and more copies of the gene are made.	This colony is grown on in nutrient broth and the DNA is then purified.

Compare the two processes of gene cloning by explaining the advantages of each. (6 marks)

OCR F215 2011

6 Some of the enzymes and vectors that are important in genetic modification are given an identifying letter in the table.

Enzymes	
A	Reverse transcriptase
B	DNA polymerase
C	DNA ligase
D	Restriction endonuclease
E	RNA polymerase

Vectors	
J	Plasmid
K	Virus
L	*Agrobacterium tumefaciens*
M	BAC
N	Bacteriophage

Select one correct letter from the table to fit each of the following statements.

An enzyme that cuts DNA

An enzyme that joins sections of DNA together

A vector to introduce foreign DNA into bacteria

A vector to introduce foreign DNA into plant cells

A vector to introduce foreign DNA into animal cells (5 marks)

OCR F215 2012

7 a Five goals that scientists would like to achieve are described here.

A producing large numbers of genetically identical model transgenic mice that show symptoms of disease.

B growing a replacement kidney identically tissue-matched to an individual patient.

C obtaining replacement hearts from transgenic pigs, partially tissue-matched to humans.

D genetically manipulating cells of one adult to cure a genetic disease in that individual.

E altering a prokaryotic pathogen for use as a vaccine.

Match the correct letters to the following procedures:

Xenotransplantation

Somatic cell therapy

Non-reproductive cloning

Animal reproductive cloning (4 marks)

b The table below shows four different combination of techniques used to achieve goals A to E. Copy the table and write the appropriate letters in the first column.

| Goal | Technique | | | |
	Vector used to transfer genes	Embryonic stem cells manipulated	Non-embryonic stem cells manipulated	Tissue designed for use in a different species
	✓	✗	✓	✗
	✓	✓	✗	✗
	✗	✓	✗	✗
	✓	✓	✗	✓
	✓	✗	✗	✗

(5 marks)

OCR F215 2013

In chapter 17 you learnt how the body responds to exercise both in the short and long term. In this chapter you will learn how many of the body's responses are brought about by the nervous system. A nervous system is an essential part of any advanced organism because it allows an organism to react to environmental and internal changes and to make appropriate behavioural modifications.

Organisation of the nervous system

The nervous system in humans is organised into a central nervous system – the brain and spinal cord – and the peripheral nervous system. Neurones in the peripheral nervous system carry impulses both into and out of the central nervous system (CNS). The peripheral nervous system can be further subdivided into the sensory and motor systems. In turn, the motor nervous system is divided into the somatic and the autonomic nervous systems.

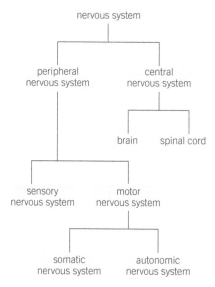

▲ Figure 1 *The organisation of the nervous system*

In the somatic nervous system, motor neurones carry impulses from the CNS to skeletal muscles, which are mostly under voluntary control.

In the autonomic nervous system motor neurones carry impulses from the CNS to muscles in the internal organs such as the smooth muscle of the gut and blood vessel walls, and to other effectors such as endocrine glands, and the sinoatrial node in the wall of the right atrium.

Organisation of the autonomic nervous system

The autonomic nervous system plays a key role in the homeostatic mechanisms that regulate the internal environment of the body.

There are a number of differences between this system and the somatic nervous system.

- Unlike the somatic nervous system, the autonomic nervous system consists of two 'branches', called the sympathetic and the parasympathetic nervous systems. The parasympathetic differs from the sympathetic in structure and function and the two branches are antagonistic, meaning that they have opposite effects.

- In the autonomic nervous system, there are two neurones linking the CNS to the effector. In the somatic system there is a single neurone linking the two. The two neurones in the autonomic nervous system connect via a synapse situated in a structure called a ganglion, which is outside the CNS.

- Somatic neurones are myelinated along the length of the axon or nerve fibre. Autonomic neurones are myelinated between the CNS and the ganglion (the pre-ganglionic neurone), but between the ganglion and the effector (the post-ganglionic neurone) they are unmyelinated. The lengths of the pre- and post-ganglionic fibres also differ between the sympathetic and parasympathetic branch.

- The neurotransmitter released by somatic motor neurones at their effector is always acetylcholine.

- In the autonomic nervous system, the pre-ganglionic neurotransmitter in both branches is acetylcholine – however the post-ganglionic neurotransmitter is different. The parasympathetic branch uses acetylcholine, whereas the sympathetic branch is adrenergic, relying on noradrenaline as the neurotransmitter.

> **Synoptic link**
>
> The structure and function of neurones and synapses are studied in Topic 26.2, Establishment and transmission of nerve impulses. You will need to refer to this topic to understand how neurones and synapses interact.

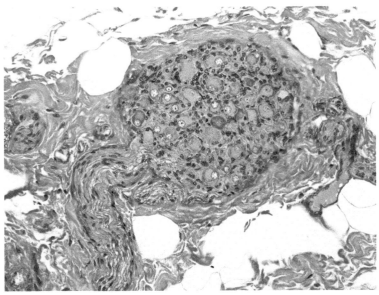

▲ **Figure 3** *A ganglion in the sympathetic nervous system, with a nerve fibre. The ganglion contains several individual cell bodies, which have pink stained cytoplasm, ×50 magnification*

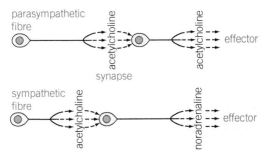

▲ **Figure 2** *Comparison of the sympathetic and parasympathetic nervous systems. The synapses are found in structures called ganglia and the dotted lines refer to neurotransmitters*

The effects of the parasympathetic and sympathetic nervous systems

The **parasympathetic** system is active under normal relaxed conditions, so it is often described as the 'rest and digest' system.

The **sympathetic** system is active under stressful situations, such as excitement and danger, so may be described as the 'fight, fright, flight,

> **Synoptic link**
>
> You will study the hormone adrenaline further in the Topic 29.3, Regulation of heart rate.

or excite' system. In this case the neurotransmitters are the chemical noradrenaline, which is similar to the hormone **adrenaline** and acetylcholine.

▼ Table 1 *Comparison between parasympathetic and sympathetic nervous systems*

Parasympathetic nervous system	Sympathetic nervous system
operates during rest and relaxation and so allows digestion and increases food movement through the gut	operates during stress-related activities and slows down movement through the gut
slows heart rate and dilates arteries and arterioles	speeds up heart rate and constricts arteries and arterioles
constricts bronchioles	dilates bronchioles
contracts circular muscles in iris so pupil constricts	contracts radial muscles in iris so pupil dilates
stimulates tear production and saliva	no effect on tear production and can make saliva production thicker (hence dry mouth when you are scared)
relaxes bladder and anal sphincters, and no effect on erector pili muscles in skin so hairs lie flat	contracts bladder and anal sphincters, and erector pili muscles in skin so hairs stand on end
no effect on sweat production	increases sweat production
ganglia linking the neurones are within the target organ	ganglia are close to spinal cord, not within the target organ

The structure and function of the human brain

The human brain consists of several distinct regions, each with its own function. Figure 4 shows the location of some of these regions.

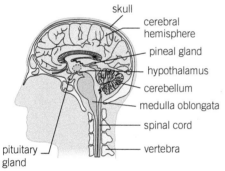

▲ Figure 4 *The human brain is composed of several distinct regions*

- The **cerebrum** – the largest part of the brain, which is formed as two hemispheres. Its surface is covered by a layer of nerve cell bodies known as the cerebral cortex, which is extensively folded. This is the region that enables our conscious thought processes, emotional responses, understanding, and use of language and intelligent thought, such as reasoning and the ability to draw conclusions. The left cerebral hemisphere receives sensory inputs from receptors on the right side of the body and controls voluntary muscles in the right side from the motor cortex. The right hemisphere carries out the same functions for the left side of the body.

- The **cerebellum** – controls the coordination of the muscles and non-voluntary movement, balance, and posture, so activities such as walking are almost automatic. It takes in sensory inputs from the retina and other receptors such as the spindle fibres in muscles and organs in the inner ear.

- The **medulla oblongata** – controls the autonomic activities in the body such as heart rate, breathing rate, blood pressure and peristalsis in the gut muscles.

- The **hypothalamus** – controls body temperature, osmoregulation, and secretion of hormones via the pituitary gland. The hypothalamus produces the hormones that are secreted by the posterior pituitary, and also produces releasing factors that stimulate the secretion of hormones produced by the anterior pituitary. Most homeostatic mechanisms are controlled here, with the notable exception of blood sugar control.

- The **pituitary gland** – the posterior lobe stores and secretes hormones produced by the hypothalamus, but the anterior lobe produces and secretes hormones.

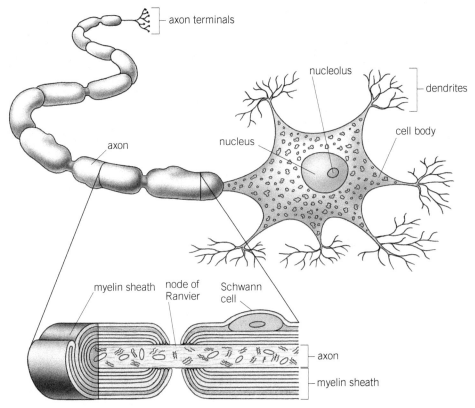

▲ Figure 5 *A motor neurone showing the myelin sheath*

The structure of the neurones

Nerve cells, or **neurones**, are adapted to transmit nerve impulses. They consist of a cell body, which contains the nucleus and other cell organelles, such as mitochondria and endoplasmic reticulum. Dendrites are short branched extensions from the cell body, which provide an increased surface area to receive the impulses from other nerve cells into the cell body. The axon is a single long extension from the cell body that carries the impulses away from the cell body. The axon is usually surrounded by **Schwann cells**. These are flattened cells that are rolled around the axon as they develop. The plasma membrane of the Schwann cell forms many layers surrounding each axon, forming the **myelin sheath**.

There are three types of neurones:

- *Sensory neurones* carry impulses from receptors to the central nervous system. The dendrites join together into a single process, a long myelinated dendron, which carries the impulses to the cell body. The axon carries the impulse from the cell body, so the cell body is found part way along the neurone and to one side.

- *Motor neurones* carry impulses from the central nervous system to effectors. There are many short dendrites surrounding the cell body, and a single long axon that connects to the effectors or to other neurones.

- *Relay neurones* lie within the central nervous system and transfer impulses from sensory to motor neurones. Relay neurones have short processes with no myelin sheath.

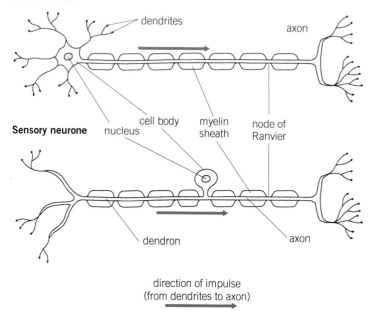

▲ **Figure 6** *A motor neurone and a sensory neurone*

Summary questions

1 State one structural and one functional difference between a sensory and a motor neurone.   (*2 marks*)

2 The drug atropine blocks the action of acetylcholine on structures innervated by post-ganglionic cholinergic neurones.
   a State precisely which part of the nervous system will be affected by atropine.   (*1 mark*)
   b Suggest why atropine might be used during anaesthesia.   (*2 marks*)

3 The graph shows the relationship between body mass and brain size in a range of mammals.
   Discuss the evidence for the hypothesis that humans have the largest brain mass of all primates.   (*4 marks*)

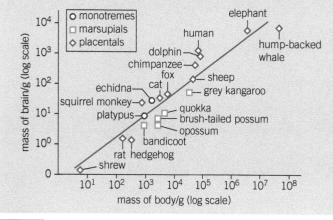

26.2 Establishment and transmission of nerve impulses

Specification reference:5.2.1

A nerve impulse is caused by temporary changes, mainly in the concentrations of sodium and potassium ions, across the axon membrane in a neurone. The changes in ion concentrations result in a change in the electrical potential difference across the membrane, causing a polarised membrane to become **depolarised**. The **resting potential** is re-established once the wave of depolarisation (the **action potential**) has passed.

The establishment of a resting potential

- At rest, the inside of the axon membrane has a negative charge relative to the outside of the membrane. The potential difference is about −65 mV. The negative sign indicates that it is the inside that is more negative. The membrane is said to be **polarised**.

- This **potential difference** is caused by a carrier protein that uses ATP to actively pump sodium ions (Na^+) out and potassium ions (K^+) into the axon across the membrane.

- The active transport of Na^+ out of the axon is greater than the transport of K^+ in, since three sodium ions are moved out for every two potassium ions moved in. So there are more positive ions outside than inside the cytoplasm.

- The axon membrane is also more permeable to K^+ than to Na^+, so some K^+ diffuses out.

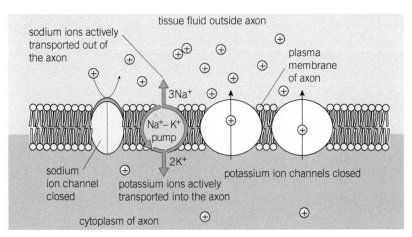

▲ Figure 1 *The ion distribution at resting potential*

The action potential

- An **action potential** arises due to changes in membrane permeability to both K^+ and Na^+. Changes to the electrical membrane potential cause voltage-gated ion channels, in the axon membrane, to open or close, resulting in the movement of K^+ and Na^+ ions across the membrane. This movement further affects membrane potential, stimulating a continuation of the action potential, as a positive feedback mechanism.

Synoptic link

You will need to review Topics 1.9–1.11 (in Chapter 1, Cells and microscopy) which covered the structure of the cell plasma membrane, and the roles that protein channels and carrier proteins play in transport across membranes.

- In an action potential, voltage-gated Na^+ ion channels open, allowing Na^+ to diffuse in down the concentration gradient and causing the membrane to depolarise (become more positive on the inside than the outside).

- When the inside of the axon membrane reaches $+40\,mV$ the Na^+ ion channels close and the K^+ ion channels open.

- K^+ ions diffuse out of the axon down a concentration gradient (**repolarisation**). This causes the inside to become even more negative (**hyperpolarisation**) than it was at the resting potential.

- The K^+ gates now close and the sodium–potassium pump re-establishes the resting potential.

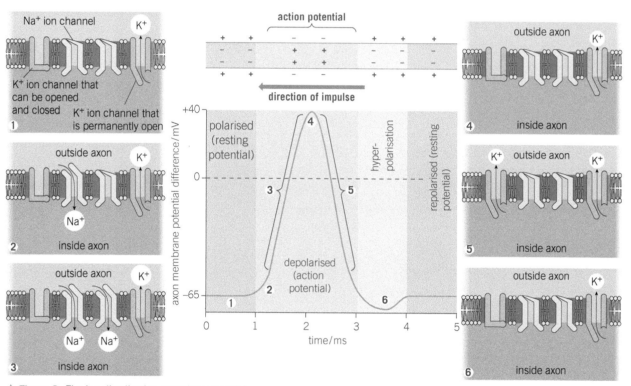

▲ Figure 2 *The ion distribution at action potential*

Study tip

The opening of the voltage-gated Na^+ ion channels is caused by a positive feedback mechanism. A stimulus causes a disruption to the axon membrane, allowing some Na^+ ions to enter – a generator potential. As more Na^+ enters it causes more gates to open, so more Na^+ diffuses in and even more gates open, causing depolarisation and an action potential. You may be asked to compare negative and positive feedback.

The propagation of a nerve impulse

This is the transmission of a nerve impulse along an axon. As one region becomes depolarised it stimulates the region just ahead to depolarise. Each region is depolarised for only about three milliseconds.

Following an action potential the depolarised region of the axon becomes repolarised, returning to the resting potential.

However, since repolarisation results in the membrane becoming hyperpolarised, no further action potentials are possible until the ions are re-balanced. This takes approximately one millisecond and this period is called the **refractory period**. As a result, action potentials can be triggered in the region ahead of an 'active zone' but not behind – meaning the impulse can only travel forwards.

a The resting state

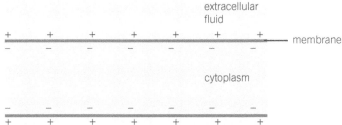

b Depolarisation of the membrane causing the flow of sodium ions into the neurone

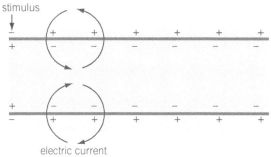

electric current

c The inflow of sodium ions causes a wave of depolarisation to spread along the membrane

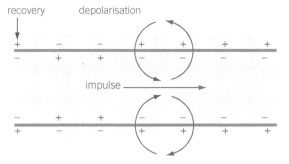

d The process repeats itself along the axon

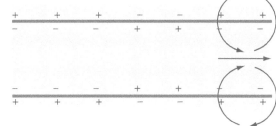

▲ Figure 3 *Propagation of an action potential*

The 'all or nothing' rule

A stimulus has to be of a particular intensity in order to generate an action potential. This is referred to as a threshold. A stimulus below the threshold will not result in an action potential. A stimulus above the threshold will generate the same size of action potential regardless of the size of the stimulus. The larger the intensity of stimulus, the more frequent the action potentials will be.

The role of the myelin sheath

In an unmyelinated neurone, the nerve impulse will travel along the whole length of the axon as a wave of depolarisation.

In myelinated neurones, the ions cannot pass through the myelin sheath. Instead, depolarisation occurs at the gaps between Schwann cells, where the voltage-gated sodium and potassium ion channels are found. These gaps are called the **nodes of Ranvier**. The action potential jumps from one node to the next in a process called **saltatory conduction** – speeding up the transmission of nerve impulses in myelinated neurones.

The structure and function of synapses

Synapses occur where two neurones almost connect, but with a small gap between them called the synaptic gap. The neurones are referred to as the pre- and postsynaptic neurones. In the cytoplasm of the synaptic knob of the presynaptic neurone there are many vesicles containing a neurotransmitter chemical – most commonly acetylcholine. The synaptic knob also contains smooth endoplasmic reticulum and many mitochondria. The plasma membrane of the synaptic knob has voltage-gated calcium ion channels. The postsynaptic membrane has chemical-gated sodium-ion channels that carry receptor sites for acetylcholine.

Study tip

Axons with a wider diameter will transmit impulses more rapidly than narrower axons, because they have a smaller surface area to volume ratio and so lose fewer ions by leakage. The giant axons of squid, used to trigger muscle contractions for water jet propulsion as an escape mechanism, show this feature, and this accounts for their rapid movements.

Figure 4 shows how impulses are transmitted across the synaptic gap in a fraction of a millisecond by diffusion of the neurotransmitter chemical.

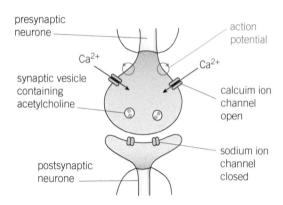

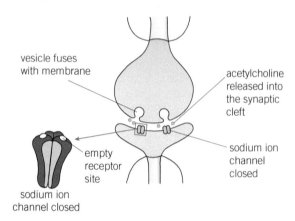

1 An action potential arrives in the presynaptic neurone. This causes calcium ion channels to open and calcium ions (Ca^{2+}) enter the synaptic knob.

2 The calcium ions cause the vesicles of acetylcholine to fuse with the presynaptic membrane. Acetylcholine is released into the synaptic cleft and diffuses across to the postsynaptic membrane.

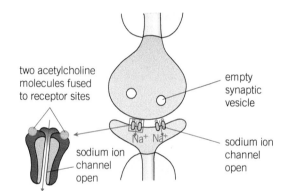

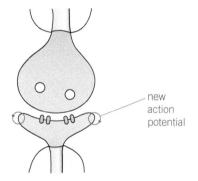

3 Acetylcholine binds to specific receptor sites on the sodium ion channels in the postsynaptic membrane. This causes the sodium ion channels to open, allowing sodium ions to enter the postsynaptic neurone.

4 If enough sodium ions enter to overcome the threshold value, an action potential is generated in the postsynaptic neurone.

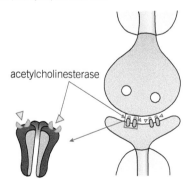

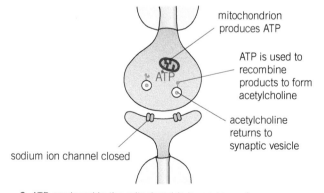

5 An enzyme called acetylcholinesterase, hydrolyses acetylcholine. The breakdown products diffuse back across the cleft and back into the synaptic knob. The sodium ion channels close once the acetylcholine is broken down.

6 ATP produced in the mitochondria is used to re-form acetylcholine from the breakdown products. This is stored in vesicles in the synaptic knob until it is needed again.

▲ **Figure 4** *The impulse crossing the gap of a cholinergic (acetylcholine) synapse*

Excitatory and inhibitory postsynaptic potentials can occur at synapses and are due to different neurotransmitters. **Excitatory postsynaptic potentials** (EPSP) increase the likelihood of an action potential occurring in the postsynaptic neurone because the neurotransmitter causes the membrane potential to become less negative and closer to threshold value. One example is glutamate, which opens Na^+ ion channels in the postsynaptic membrane, allowing Na^+ to enter. **Inhibitory postsynaptic potentials** (IPSP) decrease the chances of an action potential occurring in the postsynaptic neurone because the neurotransmitter causes the membrane potential to become more negative. One example is GABA which, depending on the receptor, either causes K^+ to leave the post synaptic neurone or causes chloride ions to enter.

A reflex arc

Reflec arcs are the simplest type of nerve pathway. They are important in many of the protective mechanisms of the body as they allow extremely rapid responses to stimuli without involving the conscious part of the brain.

A spinal reflex is one that involves the spinal cord as the coordinating part of CNS. Examples include the knee jerk reflex and moving a hand from a hot surface.

A cranial reflex operates in the same way but involves the lower centres of the brain, and not the spinal cord. Examples include the iris reflex, blinking of the eye and salivation.

Doctors use a reflex known as the plantar reflex as a diagnostic tool. This is the reflex that occurs when the sole of the foot is stimulated using a blunt instrument or a finger. In a normal adult or a child from at least a year old, the plantar reflex causes the foot to flex downwards. However, in some individuals who are suffering from a disease of the spine, the reflex causes the foot to flex upwards. The same reflex is used in the diagnosis of brain or spinal cord damage, or damage to specific nerves in the leg. The reflex responses of pupils can also indicate nervous system damage. Both pupils should constrict in response to light shone in one eye, and differences in pupil sizes in these tests could indicate damage to the brain or optic nerve. The blink reflex is one of the last to be lost as a person becomes unconscious and, if absent, indicates that the person is in a coma.

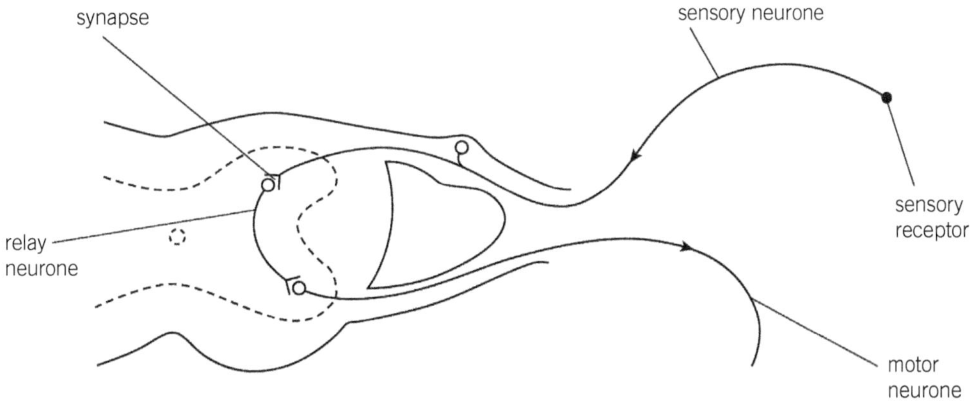

> **Study tip**
>
> The action potential is only generated if the threshold is reached, so the gap acts as a filter for preventing the onward transmission of impulses from small stimuli. The enzyme acetylcholinesterase breaks down the transmitter so that it can be released from the receptor, and diffuse back to re-enter the synaptic knob, where it is re-synthesised into acetylcholine.

> **Synoptic link**
>
> Synapse activity is one example of cell signalling. You have come across other examples of cell signalling previously in Chapter 12, Immunity and in Topic 14.3, Cancer. The interaction of drugs with cell signalling receptors is covered in Topic 26.3, Brain and spinal cord damage.

▲ Figure 5 *A reflex arc*

Practical investigation into reflexes

One quantitative method for investigating reflex time is to determine reaction times by catching a falling half metre ruler. The 50 cm ruler is held at the 50 cm mark, by the person timing, with the zero point between the open finger and thumb of the catcher. The aim is for the ruler to be caught as soon as possible after it is dropped. Repeat three different times for each person using their right hand, and record the average for 10 people. Test the left hand on 10 different people and record the results. The equation can now be used to calculate the reaction time for each person on both the right and left hand.

Y = distance in metres,

t = time in s,

g = acceleration due to gravity (9.8 m/s^2)

$$t = \sqrt{\left(\frac{2Y}{g}\right)}$$

▼ Table 1 *Results*

Right hand for person	Average distance / cm	Time / s	Left hand for person	Average distance / cm	Time / s
1	11.5	0.15	11	21.0	0.21
2	20.0	0.20	12	13.0	0.16
3	21.0	0.21	13	11.0	0.15
4	17.0	0.19	14	25.0	0.23
5	12.5	0.18	15	26.0	0.23
6	16.0	0.16	16	25.5	0.23
7	12.5	0.16	17	22.0	0.21
8	16.0	0.18	18	25.0	0.23
9	15.0	0.17	19	26.5	0.23
10	12.0	0.16	20	20.5	0.20
mean	—	0.172	mean	—	0.205
SD					

An unpaired Learner's t-test can then be carried out to compare the values of the right- and left-hand reaction times for the group studied in this experiment, using the following equation:

$$t = \frac{\bar{x}_1 - \bar{x}_2}{\sqrt{\frac{S_1^2}{n_1} + \frac{S_2^2}{n_2}}}$$

Where $\bar{x}_1 - \bar{x}_2$ is the difference between the means of the two sets of data.

S_1^2 is the square of the standard deviation for set 1 (right hand) and S_2^2 is the square of the standard deviation for set 2 (left hand).

n is the number of measurements collected, so for both n_1 and n_2 this is 10.

$t = (0.033) \div (0.00034)$

$t = 3.00$

Based on your calculated value of 't', what do you conclude about the difference in reaction times for the right and left hand?

Summary questions

1   Describe the pathway taken by a nerve impulse in a reflex, and
    explain the importance of a reflex arc.   (3 marks)

2   State how, and explain why, the transmission of an action potential
    in a myelinated neurone is different from that of a non-myelinated
    neurone.   (3 marks)

3   Suggest how, and why, the pupil response may differ from normal
    in someone with glaucoma in the right eye. You may need to read
    Topic 28.2, Effects of ageing on the nervous system.   (2 marks)

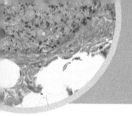

26.3 Brain and spinal cord damage

Specification reference: 5.2.1

Learning outcomes

Demonstrate knowledge, understanding, and application of:

→ the use of brain scans in assessing brain and spinal cord damage

→ the consequences of brain and spinal cord damage

→ the use of drugs to modify brain activity and function

→ psychological and physical drug dependency.

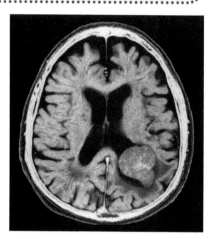

▲ Figure 1 MRI image of the brain showing a lesion

Study tip

MRI scans, unlike CT scans, do not use X-rays, which means that they can be carried out during pregnancy. Powerful magnets line up the protons in the hydrogen atoms in water molecules. Radiowaves then 'knock' the protons out of line. This is carried out in short bursts, and in between each burst the protons fall back into line, emitting radiowaves that are analysed to form an image. Remind yourself of the structure of the water molecule, and that some tissues contain more water than others.

Brain damage and brain scans

Acquired brain injury (ABI) refers to any brain injury that occurs after birth, such as a stroke, an infection, haemorrhage or anoxic injury. There are two main types of brain damage:

- Traumatic brain injury (TBI) occurs following a severe blow to the head, such as may be experienced in a traffic accident, a fall or an assault.

- Non-traumatic brain injuries do not involve a severe blow but are caused by stroke, infection, or anoxic injury.

In addition there are progressive neurodegenerative diseases such as Alzheimer's and Parkinson's.

There are a number of neuro-imaging techniques that have become very important in assisting the diagnosis of brain damage. They determine the location and extent of any damage and so indicate the necessary treatment, surgery or rehabilitation.

- MRI – **magnetic resonance imaging**. This can be used to detect tumours, strokes, areas of infarction and also areas of demyelination in the CNS.

- fMRI – **functional MRI** allows identification of activity in specific areas of the brain due to increased blood flow. It relies on the fact that the magnetic properties of haemoglobin change depending on whether it is oxygenated or deoxygenated. fMRI can be used when planning brain surgery, as important areas such as speech and language can be clearly identified and so be kept as intact as possible.

- CT scans – computed tomography or CAT scans use X-rays and a computer to construct images of the internal structures. Further detail can be obtained by using CT perfusion, where a substance is injected to show which areas are adequately supplied (or perfused) with blood.

- PET (positron emission tomography) scan – a radioactive form of glucose (FDG) is introduced into the vein, and is taken up by tissues – the scanner then detects the positrons released from the glucose, to form an image. As cancerous cells process glucose differently, compared with normal cells in the same tissue, this can be detected and is then displayed on the image as different colours. Because PET scanning relies on the functional aspect of the tissue it is not routinely used in diagnosis of TBI, but it is useful for monitoring long term care and therapy, and indicates prognosis.

- EEG – **electroencephalogram**. Small electrical impulses sent from brain cells to other brain cells are detected by electrodes on the scalp. These impulses are recorded and monitored for any abnormal readings, which may indicate conditions such as epilepsy, dementia, brain inflammation, head injury or coma.

The effect of brain damage on the individual

The effect of brain damage will depend entirely on the region of the brain affected and the degree of damage.

For example, strokes can vary in severity, from minor strokes, where the patient may be unaware of the problem, through to strokes following which a full or good recovery occurs, and then major strokes, which cause continued disability or death. A stroke occurs when one of the blood vessels supplying blood to the brain becomes blocked, so that the blood supply is interrupted. The blockage is usually caused by a blood clot becoming lodged in the blood vessel. In some cases strokes result from bleeding into the brain from a ruptured or damaged blood vessel. The type of damage depends entirely on the region of the brain affected by the interruption in blood flow. For example, if the speech centre is affected then the ability to speak will be impaired, whereas damage to the motor area of the cerebral cortex could affect the ability to control certain voluntary muscles.

Rehabilitation can help patients to regain some or many of the skills lost or affected by the brain injury, by compensating with new neural pathways.

TBI may cause damage to the hypothalamus or to the pituitary gland. These structures control the body's hormones, and damage to them will cause the production and release of a large number of hormones, including those controlling the homeostatic mechanisms of the body, such as ADH, thyroxine, and also the sex hormones.

Extensive brain damage can result in a vegetative state, but the brain stem is still functioning so there is still a slim chance of recovery. A person is confirmed dead when their brain stem function is permanently lost. This may happen even when their heart beat and oxygen supply are being artificially maintained. After confirmation of death, it may be possible for organs to be removed for donation if consent can be obtained, so establishing brain death is crucial.

The use of drugs to modify brain activity

Some brain diseases such as Alzheimer's disease may be treated in the future using drugs to reverse the condition. Currently, drugs taken early enough in the course of the disease will prevent it from getting worse, but there is only limited success in reversing the damage. A group of drugs known as cholinesterase inhibitors provide some benefit by inhibiting the enzyme cholinesterase from hydrolysing the neurotransmitter acetylcholine. The result is an increase in acetylcholine, which seems to allow the restoration of a limited level of lost function.

Some brain damage may be prevented by providing anti-inflammatory drugs, such as aspirin and ibuprofen. These reduce the swelling in the brain that occurs with damage. They inhibit the formation of the cell signalling molecules that are involved in the inflammation response, and can block pain and reduce fever.

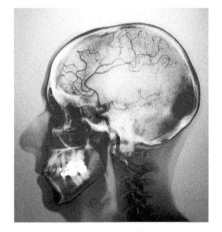

▲ Figure 2 Angiogram of the brain in a patient suffering a massive stroke due to blockage of the cerebral artery

Parkinson's disease results from loss of the brain cells that produce **dopamine**. This neurotransmitter helps to control muscle movement and, as less is produced, muscles become tense, resulting in symptoms such as tremor, joint rigidity and slow movements. Drugs used to treat Parkinson's include therapeutic drugs such as levodopa (which is metabolised to dopamine), dopamine agonists (which mimic dopamine) and monoamine oxidase inhibitors (MAO-B), which inhibit the enzyme that breaks down dopamine in neurones.

Recreational drugs

Others drugs that can modify brain activity are so called 'recreational' drugs. These frequently have damaging effects.

Alcohol affects coordination and motor skills by inhibiting neurotransmission across the synapses of some neurones. Alcohol binds to a number of receptors on the postsynaptic membrane, including GABA receptors. GABA is an inhibitory neurotransmitter which opens chloride ion channels, causing a negative change in the membrane potential inhibiting further transmission for a short time. The binding of alcohol to GABA receptors results in the chloride ion channel opening for longer, allowing more chloride ions to enter and therefore hyperpolarising the postsynaptic neurone. This prolongs the inhibition – hence alcohol acts as a depressant.

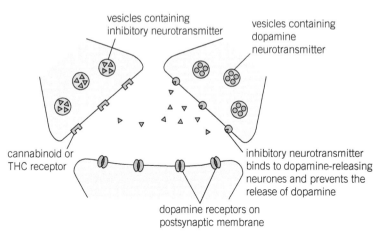

a *How an inhibitory neurotransmitter normally controls the release of dopamine*

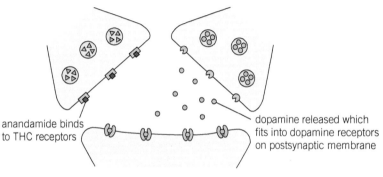

b *How anandamide causes the release of dopamine*

▲ Figure 3 *How THC affects the synapses*

Heroin is an example of a group of drugs called *opioids*. It is converted to morphine in the brain and binds to the opioid receptors. Heroin causes a number of effects, including clouded cognitive function, slowed heart function and reduced breathing rate, which may be life threatening.

Marijuana contains a chemical, THC, which is similar in shape to anandamide, a natural cannabinoid which, when produced by the body stimulates a temporary release of dopamine. However, THC does not break down as rapidly as anandamide, so dopamine is present in the body in larger amounts and for longer periods. THC blocks the action of the inhibitory neurotransmitter GABA, which normally prevents dopamine release.

Methamphetamines are a group of drugs that affect the excitatory synapses in neurones, by increasing levels of dopamine released at synapses. The drug is slow to be broken down, so the stimulatory affect of dopamine is prolonged.

The role of dopamine and the biological basis of dependency

The limbic system is the part of the brain that is associated with long-term memory, behaviour, and emotions. It influences both the endocrine system and the autonomic nervous system, acting as the brain's reward centre, including the 'highs' experienced in recreational drug taking. Dopamine is released by the hypothalamus, which is part of the limbic system. Dopamine plays an important role in reward-motivated behaviour or activities, such as exercise, which increase dopamine levels in the brain, and create an overwhelming desire or need to continue the activity. This is also the basis of dependency with some addictive drugs. In chronic drug abuse the ongoing dopamine release activates the limbic brain areas and seems to result in critical brain alterations that cause the change from occasional use to addiction.

Drug dependency

Whilst addiction may be described as uncontrolled, compulsive behaviour, drug dependency can sometimes be classed as psychological – where the person's body may not physically require the drug but there is a psychological need, which may stem from habit or social pressures. Full physical dependency is where the drug has become required in order for the body to function normally. In this case, if the drug is stopped abruptly withdrawal symptoms will occur, which may even result in death. With addiction, a tolerance to the drug builds up such that increasingly large amounts are needed to produce the same effect. This may be due to increasing concentrations of enzymes that break down the drugs, or a reduction in the sensitivity or distribution of receptors at synapses.

There are many possible reasons for dependency, which range from social or peer pressure, to emotional or anxiety issues, or depression, in which the pleasure reward system is malfunctioning. However, the consequences of dependency to the individual are enormous as the need to take increasingly large amounts of the drug can become overwhelmingly dominant, to the exclusion of all other aspects of life, including even basic functions such as eating, sleeping, personal hygiene and care for other family individuals. As these drugs are not legal, drug taking can lead to criminal behaviour to fund the habit and obtain the drugs.

The cost to society of dependency is high and includes:

- direct healthcare costs
- financial costs as a result of theft by abusers to support their drug habits
- policing and customs-related costs associated with law enforcement and detecting drug trafficking.
- loss of potential contributions that the drug abuser could make to society in terms of work and tax contributions, as well as reduction in benefits
- social breakdown as behaviour deteriorates.

Summary questions

1 Evaluate the use of different techniques in the assessment of brain damage and TBI.
(6 marks)

2 Using the information in this section and in other chapters explain how brain trauma may disrupt the ability of the body to control
a the water balance and
b the menstrual cycle.
(4 marks)

3 Suggest why anandamide is broken down rapidly by the body but THC is not broken down quickly. (2 marks)

Practice questions

1 Which of the following statements is/are true of the functions of the sympathetic nervous system?

   1 Increases heart rate.

   2 Increases the speed at which food moves through the gut.

   3 Decreases sweat production

   A 1, 2 and 3

   B Only 1 and 2

   C Only 2 and 3

   D Only 1   (1 mark)

2 What is the function of the cerebellum?

   A Coordination of balance and muscular movement.

   B Controls conscious thought processes.

   C Controls body temperature.

   D Regulates heart rate.   (1 mark)

3 One neuro-imaging technique is computed tomography. Which of the following is used during computed tomography?

   A Magnetic resonance

   B X-rays

   C Gamma rays

   D Positrons   (1 mark)

4 Which of the following is/are present in the membrane of an acetylcholinergic presynaptic neurone?

   1 Calcium ion channel

   2 Acetylcholine receptor

   3 Acetylcholinesterase

   A 1, 2 and 3

   B Only 1 and 2

   C Only 2 and 3

   D Only 1   (1 mark)

5 Outline the differences between the sympathetic nervous system and the parasympathetic nervous system. (6 marks)

6 The graph shows the changes that occur in the membrane potential of a neurone during the generation of an action potential.

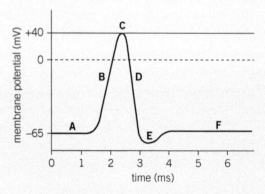

a Using the letters A to F, indicate the point or points on the trace that correspond to the following:

   (i) hyperpolarisation

   (ii) resting potential

   (iii) the membrane most permeable to potassium ions

   (iv) depolarisation   (4 marks)

b Puffer fish produce a poison called tetradotoxin, and some species store it in high concentrations in their bodies. Unless these fish are correctly prepared, eating them can be fatal.

Tetradotoxin is poisonous to humans because it blocks gated-sodium ion channels in cell membranes, preventing action potentials. This does not happen in the fish themselves.

   (i) Identify, using the appropriate letter, the part of the action potential trace, above, that will be affected by tetradotoxin.   (1 mark)

   (ii) Suggest why tetradotoxin is not toxic to the puffer fish.   (1 mark)

c Multiple sclerosis (MS) is an auto-immune condition in which the nervous system is damaged. This damage leads to loss of sensation. One form of damage is shown here.

   (i) Suggest why MS is described as an auto-immune condition.   (2 marks)

   (ii) Explain why this damage leads to a loss of sensation.   (2 marks)

OCR F214 2010

7 a The diagram represents a sensory neurone connected to its associated receptor cells.

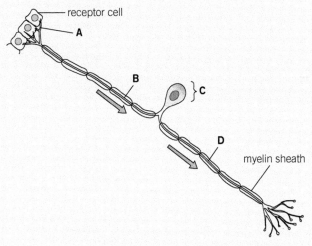

receptor cell

A

B

}C

D

myelin sheath

(i) Identify the parts of the neurone labelled A to D. *(4 marks)*

(ii) What is represented by the arrows on the figure? *(1 mark)*

b Describe, and explain, how the resting potential is established, and how it is maintained in a sensory neurone. *(4 marks)*

c The graphs show changes in the membrane potential of a sensory neurone when the receptor cells are stimulated and the strength of stimuli that correspond to these changes.

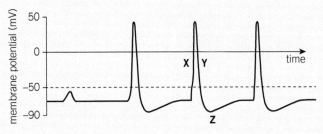

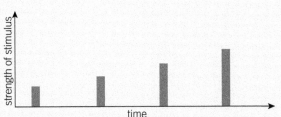

(i) State the terms used to describe what is happening at X, Y, and Z. *(3 marks)*

(ii) What term is used to refer to the value of −50 mV on the first graph? *(1 mark)*

(iii) Comment on the relationship between the strength of a stimulus and the resulting acion potential. *(2 marks)*

OCR F214 2013

8 a The diagram represents part of the axon of a neurone.

A

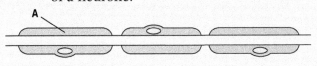

Describe the structure of the feature labelled A. *(2 marks)*

The table shows details of the diameter and the speed of conduction of impulse along the neurones of different animal taxa.

Type of neurone	Axon diameter (μm)	Speed of conduction (ms⁻¹)	Animal taxon
myelinated	4	25	mammal
myelinated	10	30	amphibian
myelinated	14	35	amphibian
unmyelinated	15	3	mammal
unmyelinated	1000	30	mollusc

b Using only data in the table, describe the effect of each of the following on the speed of conduction:

(i) Myelination *(2 marks)*

(ii) Axon diameter *(2 marks)*

c (i) Suggest why an increase in temperature results in an increase in the speed of conduction. *(1 mark)*

(ii) As the temperature continues to increase, it reaches a point at which the conduction of the impulse ceases. Suggest why. *(1 mark)*

d Outline the events following the arrival of an action potential at the synaptic knob until the acetylcholine has been released into the synapse. *(4 marks)*

OCR F214 2010

27 MONITORING VISUAL FUNCTION

27.1 The structure of the eye

Specification reference: 5.2.2

Structure of the eye

In the previous chapter you learnt about the nervous system. Receptors are a key component of any nervous system and in this chapter you will learn about the human light receptor – the eye.

Throughout the animal kingdom there are many types of eyes. For example, the compound eye of a fly can detect fast movements. The human eye starts to form two weeks after conception and is fully formed at birth, staying the same size throughout a person's life.

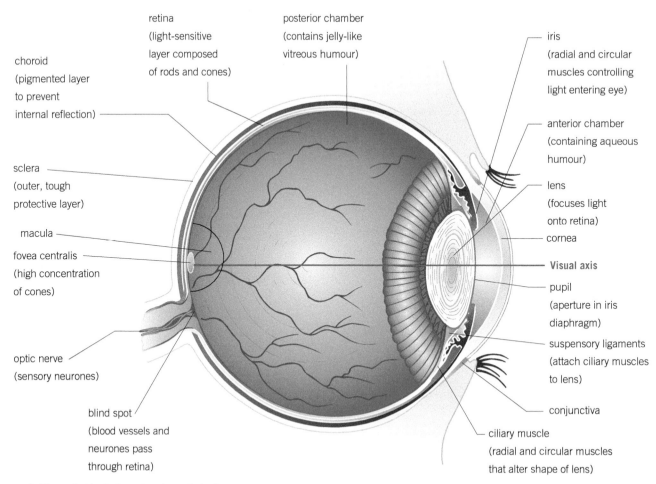

▲ **Figure 1** *Vertical section through the human eye*

Each eye is a spherical structure located in the bony socket of the skull called the **orbit**. Each eye can be rotated in the socket by two pairs of **rectus muscles** and one pair of **oblique muscles**. All of these muscles are attached to the tough, outer layer of the eye, the **sclera**. The sclera contains many collagen fibres to enable it to protect the eye and maintain its shape under the pressure of its fluid contents. Below the sclera is the **choroid** layer, which contains blood vessels and a layer of highly pigmented cells that prevent internal reflection. The inner layer is the retina, which contains ~107 million light sensitive cells (~100 million rod cells and ~7 million cone cells). At the front of the eye the sclera is transparent and forms the **cornea** – this *refracts* the light entering the eye. The **conjunctiva** is a thin, transparent membrane, which covers the surface of the cornea and is continuous with the eyelids. It is lubricated by tears produced from the **lachrymal gland**.

At the front of the eye the choroid is modified to form the **iris**. This is a heavily pigmented diaphragm of involuntary smooth radial muscles and circular muscles, which surround the pupil (a central hole). These antagonistic muscles contract and relax to change the size of the pupil and regulate the amount of light entering the eye.

Behind the pupil is the biconvex lens, which is made out of transparent protein that is contained in a capsule held in place by the suspensory ligaments attached to a ring of ciliary muscle. The ciliary muscles and ciliary processes (inward foldings of the choroid layer) form the ciliary body. The lens is flexible and elastic and has its shape altered by the ciliary muscles to change and fine tune the focusing of the light on to the part of the retina known as the macula. The internal space within the eye is divided into two chambers – the one in front of the lens contains a clear, watery fluid called aqueous humour – the much larger chamber behind the lens contains a transparent, jelly-like fluid called the vitreous humour, which maintains the eyeball's shape.

Study tip

Remember, it is only the ciliary muscles that can contract and relax (not the ligaments, as these are inelastic). Ligaments become taught or slack according to the pulling force applied to them by the muscles attached to them.

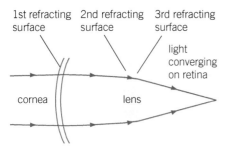

▲ **Figure 2** *The three refracting surfaces in the eye*

Focusing the light onto the retina

Accommodation is the process by which the elastic lens changes its shape to focus an object onto the retina and maintain a clear image even if the distance varies. As the light rays enter the eye they are refracted (bent) to focus them on to the retina. The distance of the object will determine how much refraction is necessary. Light rays from a nearby object need to be refracted more than those from objects further away.

When viewing nearby objects the biconvex lens bulges as the circular ciliary muscles contract to close the aperture around the lens. The suspensory ligaments slacken (as there is less tension on them) so the lens becomes more convex (spherical). This increases the degree of refraction and the eye is said to be *accommodated*.

When viewing distant objects the lens thins as the ring of ciliary muscle relaxes, widening the aperture around the lens. The suspensory ligaments are stretched taut, pulling the lens outwards and making it less convex (thinner). This position decreases the light refraction and the eye is said to be *unaccommodated*.

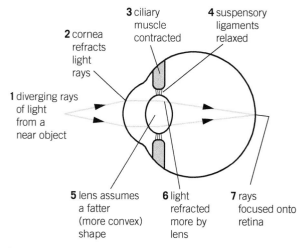

▲ **Figure 3** *Condition of the eye when focusing on a nearby object*

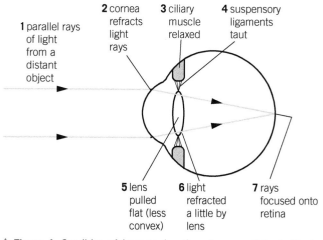

▲ **Figure 4** *Condition of the eye when focusing on a distant object*

1   Short-sightedness (myopia) is a common eye condition that causes distant objects to appear blurred, while close objects can be seen clearly. In childhood this is frequently due to an elongated eyeball. Suggest how this would affect the focusing function of the eye.   (1 mark)

2   Explain how laser surgery could be used to treat long- or short-sightedness.   (2 marks)

3   Suggest two further ways by which myopia could be treated. (3 marks)

Summary questions

1   Why is it essential that the choroid has a rich blood supply?   (2 marks)

2   Ciliary body melanomas are a rare form of tumour, but they are more commonly found in white Caucasians. Suggest why this is the case.   (2 marks)

3   As a result of diabetes sugars can build up in the lens, which can result in changes to vision. Suggest why a build up of sugar in the lens can result in eye defects.   (2 marks)

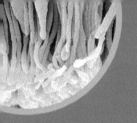

27.2 The structure and function of the retina

Specification reference: 5.2.2

Learning outcome

Demonstrate knowledge, understanding, and application of:

→ the structure and function of the specialised cells of the retina.

▲ **Figure 1** *Coloured scanning electron micrograph (SEM) of receptor cells in the human retina. Rod cells (white) perceive light of different intensities, whereas cone cells (yellow) allow colour vision. The cell bodies (red) of the receptor cells are located in a layer above the rods and cones, ×1500 magnification*

Study tip

Unlike most neurones, bipolar cells communicate via graded potentials, that is, changes in membrane potential that vary in size and duration. Graded potentials occur as a result of the summation of the individual actions of ligand-gated ion channel proteins and decrease over time.

The retina

Having looked at the overall structure of the eye in the previous topic, you will now learn about the actual light receptors in the eye – the cells in the retina. The retina covers approximately 65% of the eye's interior surface. It is ~0.5 mm thick and consists of specialised cells including photoreceptors. Light has to pass through several layers of neurones before it reaches the photoreceptor cells, which are embedded in the pigmented epithelium cells of the choroid. Light rays hitting the back of the retina produce an inverted image.

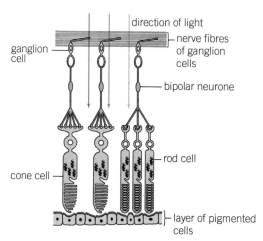

▲ **Figure 2** *Structure of the retina*

Specialised cells of the retina

Rod cells

Rod cells are sensitive to low light intensity, with a single photon of light being of sufficient energy to produce an action potential in a ganglion cell. Each rod cell consists of two segments:

Outer segment – this contains up to 1000 vesicles of rhodopsin (a photosensitive pigment), which is situated in the disc membranes. The rhodsopsin molecule consists of opsin (a protein) and retinal (a derivative of vitamin A). Retinal normally exists in its *cis*-isomer form but when light hits the retinal molecule it changes to its *trans*-isomer form. When this occurs the rhodopsin splits in a process called bleaching (Figure 3). The change in rhodopsin causes the rod cell to become impermeable to sodium ions.

Inner segment – this contains many mitochondria and the nucleus. The two segments are connected by a pair of cilia. Many rod cells form a synapse with a single bipolar (sensory) neurone, which means rod cells have low visual acuity (clarity of vision).

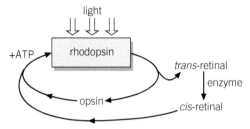

▲ Figure 3 *Breakdown and formation of rhodopsin*

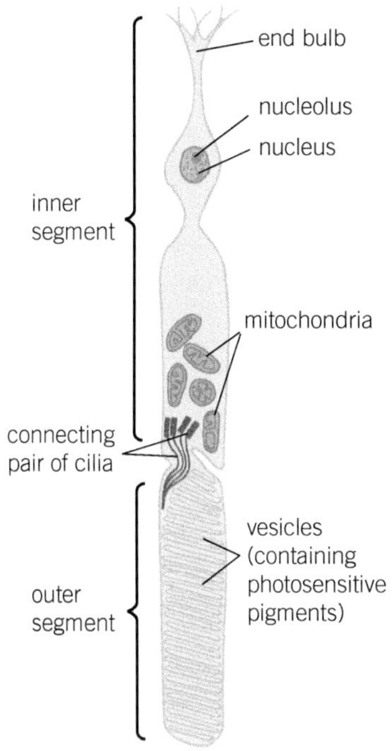

Cone cells

Cone cells have a similar structure to rod cells. They also contain vesicles that contain a photosensitive pigment, but in cone cells this is iodopsin. There are also three different types of cone cell, each of which possesses a different type of iodopsin, which absorbs either red, green, or blue light. Different cone cells differ in their sensitivity to photons of different wavelengths of light. Different degrees of stimulation of the different types of cone cell results in trichromatic vision, for example, yellow light stimulates an equal number of red and blue cones, which the brain interprets as yellow – orange light stimulates more red than green cones. The human eye has the ability to distinguish between eight million shades of colour, but cone cells cannot detect colour by themselves as colour vision requires comparison of the signal across different cone types.

Cone cells are not as sensitive to light so they only work in high light intensity, that is, they need more photons of light to generate an action potential. There are also fewer cone cells than there are rod cells in the retina. The highest concentration of cone cells is found in the macula. In the centre of the macula is the fovea centralis, which consists solely of cone cells. This is the part of the retina where the image is focused when you look directly at an object. Macular degeneration results in the loss of cone cells in the fovea.

Most cone cells connect to one specific sensory neurone, producing greater visual acuity than rod cells.

Bipolar cells

Bipolar cells – these are specialised neurones that have two processes (extensions) coming out of their central cell body (axon and dendrite). The processes closest to the photoreceptor cells are short and branch into many endings that form synapses with either a number of rod cells or a single cone cell. Impulses from a number of rod cells summate before triggering an impulse in the bipolar neurone, which causes low visual acuity. The other processes form synapses with a ganglion cell.

▲ Figure 4 *Structure of a single rod cell*

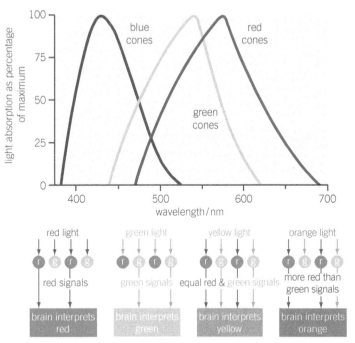

▲ Figure 5 *Top – Different wavelengths of light are absorbed by different cone cells. Bottom – These are compared and interpreted by the brain to see different colours*

The bipolar cell stimulates a generator potential in the sensory neurone. If the generator potential is large enough to overcome the threshold value it generates an action potential along the sensory neurone leading from the rod cell to the optic nerve. Before the rod cell can be stimulated again the rhodopsin is actively reformed.

If rod or cone cells are not stimulated by light they depolarise, causing them to release an inhibitory neurotransmitter onto the bipolar cell. This causes the bipolar cell to become hyperpolarised and prevents it from transmitting impulses to its related retinal ganglion cell.

Ganglion cells

Ganglion cells have cell bodies in the retina and possess many dendrites that form synapses with bipolar cells. Action potentials are first generated here in the retina. Ganglion cells are activated by depolarised bipolar cells. These cells have long axons to carry impulses to the brain – these impulses are passed at a low frequency when the ganglion cell is not stimulated and at an increased rate when it is stimulated. There is a place on the retina where all the neurones collect to form the optic nerve – which carries impulses from the eyes to the brain. Here there are no rod or cone cells and the area is known as the blind spot.

The function of the retina

Both types of photoreceptor cells, rod and cone cells, act as transducers – they convert light energy (visible electro-magnetic spectrum) into chemical energy in the form of action potentials. The action potentials are transmitted along the optic nerve to the visual cortex of the brain. This part of the cerebral cortex located in the occipital lobe is responsible for processing visual information.

At rest

The rod cell has a resting potential of $-40\,mV$ as a result of the action of the sodium–potassium pump and the presence of sodium channels and potassium channels in the cell surface membrane. The pump moves three sodium ions out of the cell for every two potassium ions it moves into the cell, hydrolysing ATP as it does so. A circulating electric current is generated by the constant flow of ions as channels in the outer segment allow sodium ions to diffuse through the cell surface membrane back into the cell, and open channels in the inner segment allow potassium ions to diffuse back out of the cell. The rod cell forms a synapse with an adjacent bipolar cell, and at rest, the synaptic bulb secretes glutamate into the synaptic cleft. Glutamate is an inhibitory neurotransmitter. This diffuses across the synaptic cleft and prevents generator potentials in the bipolar cell (which remains at its resting potential) so no impulse is sent to the brain.

When stimulated by light

Rod cells act as light receptors. When rhodopsin molecules in the disc membranes absorb light, the molecule changes shape (Figure 3) resulting in the closure of sodium ion channels. Sodium ions can no longer enter the rod cell, and the photoreceptor outer segment membrane becomes hyperpolarised (more negative on the inside). This change in the rod cell's membrane potential causes voltage-gated calcium ion channels to close and the intracellular calcium ion concentration falls. The fall in calcium ion concentration causes less glutamate to be released. This then removes inhibition of the bipolar cell and it becomes depolarised, creating a generator potential. This generator potential is transmitted across the synapse to a ganglion cell. From here an action potential is carried along the optic nerve to the brain.

▼ Table 1 *Comparison of rod cells in the dark and light*

Rod cell at rest	Rod cell when stimulated by light
no light present (or at too low level to stimulate the rod cell)	light is absorbed by rhodopsin
opsin and *cis*-retinal are joined to form rhodopsin	rhodopsin is broken down to form opsin and *trans*-retinal
sodium ion channels open	sodium ion channels close
sodium ions enter the outer segment of the rod cell	no current is formed
rod cell membrane is depolarised	rod cell membrane is hyperpolarised
neurotransmitter is released by exocytosis from the rod cell into the synaptic cleft	no neurotransmitter is released into the synaptic cleft
bipolar neurone is hyperpolarised	bipolar neurone becomes depolarised
inhibition of the release of neurotransmitter between the bipolar cell and ganglion occurs	release of neurotransmitter occurs into the synapse between the bipolar cell and ganglion
no generator potential	generator potentials summate to form an action potential
no action potential generated	action potential travels along the ganglion neurone

Study tip

Remember, sodium–potassium pumps work by pumping $3Na^+$ out and $2K^+$ in. Remember this by:

Na^+ is 3 characters, which matches 3 letters in 'out'.

K^+ is 2 characters, which matches 2 letters in 'in'.

Synoptic link

You will need to remember how enzymes work. You learnt about this in Topic 3.4, The structure of enzymes.

Summary questions

1   a   Explain the role of the pigmented cells underneath the rod and cone cells.   (*1 mark*)
    b   Explain why the inner segment of the rod cell contains many mitochondria.   (*2 marks*)
    c   Explain why rod cells are said to produce 'monochromatic vision'.   (*2 marks*)

2   a   The conversion of *trans*-retinal to *cis*-retinal requires an enzyme and occurs at a very slow rate. Explain how this accounts for some people claiming that 'they have gone blind' when they walk from a brightly lit room to a dark room.   (*4 marks*)
    b   Rod cells have a diameter of approximately 2 μm. Calculate the magnification of the diagram shown in Figure 1.   (*2 marks*)

3   Suggest why pilots wear red goggles in conditions with strong light.   (*4 marks*)

Learning outcome

Demonstrate knowledge, understanding, and application of:

→ how routine eye tests can be used to assess receptor activity:

- visual acuity
- colour vision
- optical coherence tomography.

Eye tests

In previous topics you learnt about the structure and function of the eye. In this topic you will learn about how eye function is assessed. Ophthalmologists (medical doctors), optometrists, and orthoptists all carry out routine eye examinations, including the assessment of a person's vision and ability to focus on objects. Healthcare professionals often recommend that all people should have thorough eye examinations every two years as many eye diseases are asymptomatic (patient has the disease but shows no symptoms). Eye examinations may also detect potentially treatable blinding eye diseases, signs of tumours, or other conditions such as diabetes.

Visual acuity test

Visual acuity refers to the clearness of vision, that is, the eye's ability to detect fine details and see a sharp image. The test is a quantitative measure of the eye's ability to see an in-focus image at a certain distance. It is usually assessed by a Snellen chart – this has a series of letters with the largest at the top and the smallest at the bottom. The lower down the chart a person can read, the better their visual acuity. The standard definition of normal visual acuity (20/20) indicates the person can read letters just under 1 cm in height at a distance of 20 ft. Depending on the assessment for any one person they will be advised what, if any, lenses they need to correct their vision.

Colour vision test

The Ishihara colour test, which consists of a series of pictures of coloured spots, is the test most often used to diagnose red–green colour deficiencies. A number (usually of more than one digit) is embedded in the picture as a number of spots in a slightly different colour, which can be seen with normal colour vision. People with colour-blindness do not see either (or both) numbers. There is a large variety of tests, which alter the colour combinations to allow the eye health professional to diagnose which particular visual defect is present.

▲ **Figure 1** *A Snellen chart is used to test a person's visual acuity*

▲ **Figure 2** *The Ishihara test, which uses pseudoisochromatic plates, is the most widely used colour vision test. Each card depicts a number formed from red dots on a background of green dots, or vice-versa. People with red–green colour blindness cannot distinguish the numbers*

Another test used by clinicians is the Farnsworth–Munsell 100 hue test. This involves asking the patient to arrange a set of coloured caps in sequence to form a gradual transition of colour between two points (Figure 3). The test is also used in industry to grade employee's colour discrimination. The caps are sorted into hue order in the four boxes. The boxes keep reds, greens, blues, and yellows separate.

Optical coherence tomography (OCT scan)

Optical coherence tomography, or OCT scans can capture image reflections from within tissue of the eye to provide cross-sectional images at very high resolution. These scans can be used to diagnosis age-related macular degeneration, diabetic retinopathy, macular holes, and macular oedema. An optical beam is directed at the tissue, normally the retina, and a small portion of this light that reflects from the retina is collected. However, most light is not reflected but, rather, scatters off at large angles. Usually this diffusely scattered light causes background noise and obscures an image. However, in OCT, a technique called **interferometry** can be used to build up clear 3D images of thicker samples of tissue by rejecting the background signal.

The technique is limited to imaging 1 to 2 mm below the surface of any biological tissue, but this is sufficient when studying the retina, which is ~0.5 mm thick. The laser output from the instruments is low, as eye-safe, near-infra-red light is used, so damage to the eye is unlikely. The test may require eye drops to be put into the eye(s) 30 minutes before the test, which causes vision to become blurred (an effect that can last 6–8 hours) but the scans are totally painless and take around 10–15 minutes or less.

▲ **Figure 3** *Caps of different colours, representing the hue circle of the visual spectrum, which are used to check for colour vision abnormalities. This is the Farnsworth–Munsell 100 hue test (F–M 100 hue), which can detect all colour vision defects from the mildest red-green colour blindness to total achromatopsia (absence of normal cone vision)*

▲ **Figure 4** *Patient having his eyes examined by an OCT (optical coherence tomography) scanner. This equipment produces a 3D image that reveals any common defects and disorders, and assesses the health of the eye's internal structures such as the retina. Here, the examination is being undertaken to check the patient's suitability for laser eye surgery*

Summary questions

1 a Describe what common features of colour vision tests make them effective in diagnosis. *(3 marks)*

   b Give an example of a situation where it may not be appropriate to use a colour vision test to determine if a person has colour-blindness. *(1 mark)*

   c Suggest how a colour vision test could be adapted to assess this group of people. *(1 mark)*

2 Suggest in what situations an OCT scan may be used to assess the effectiveness of treatment rather than the diagnosis of an eye defect. *(2 marks)*

3 Eye tests can detect a detached retina. Explain fully why the detachment of the retina from the underlying layer leads to faulty vision. *(2 marks)*

Practice questions

1 Which of the following statements is/are true of rod cells?

Statement 1: Sensitive to low light levels.

Statement 2: Contain rhodopsin.

Statement 3: Can pass signals to bipolar cells across synapses.

A   1, 2 and 3

B   Only 1 and 2

C   Only 2 and 3

D   Only 1 *(1 mark)*

2 What is the role of the choroid?

A   Circular muscle that controls how much light enters the eye.

B   A protective layer.

C   A layer composed of rods and cones.

D   A pigmented layer that prevents internal reflection. *(1 mark)*

3 Which of the following statements is/are true of rod cells when they are stimulated by light?

Statement 1: Rod cell membrane is depolarised.

Statement 2: Sodium ion channels close.

Statement 3: Rhodopsin is broken down into opsin and *trans*-retinal.

A   1, 2 and 3

B   Only 1 and 2

C   Only 2 and 3

D   Only 1 *(1 mark)*

4 The Ishihara test is used to diagnose which condition?

A   Colour vision deficiencies

B   Visual acuity deficiencies

C   Myopia

D   Long-sightedness *(1 mark)*

5 Cornea transplants account for approximately 40% of transplant operations each year. During a cornea transplant operation, the damaged area of the existing cornea is removed and a section of healthy cornea from a donor is grafted into place.

a   The cells that make up the cornea are kept transparent by the activity of a layer of endothelial cells, which form a layer between the cornea and the fluid in the eye. The endothelial cells actively remove water from the corneal cells.

The diagram shows the relative position of the tissues.

(i)   Name the fluid present in region M. *(1 mark)*

(ii) Explain why water moves into the corneal cells. *(2 marks)*

b   There are no blood vessels in the cornea. Tissue matching is not essential for successful cornea transplants.

Explain why the lack of blood vessels in the cornea means that tissue matching is not essential for cornea transplants. *(2 marks)*

c   Damage to the cornea may lead to a loss of visual acuity. A Snellen chart, which is used to assess visual acuity, is shown here.

Suggest two variables that must be controlled during a visual acuity test in order to achieve valid results. *(2 marks)*

OCR F225 2010

6 The diagram shows a computer-generated picture of cells in the human retina. The cell labelled A represents a rod cell.

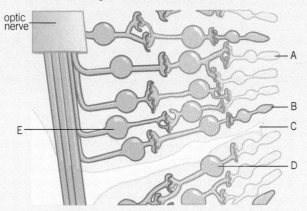

a **(i)** Which letters represent a ganglion cell, and a bipolar cell? *(2 marks)*

(ii) Describe how energy striking rod cells is converted into a nerve impulse in the optic nerve. Your answer should make clear the sequence of events. *(6 marks)*

OCR F225 2012

7 The diagram shows a section through the human eye. The arrows indicate the flow of aqueous humour, from where it is formed in the ciliary body to where it drains through the outflow channel.

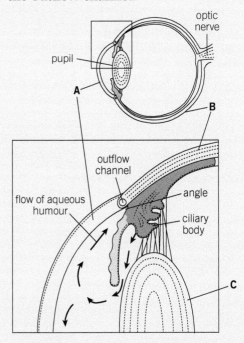

a Name the structures labelled A, B, and C. *(3 marks)*

b Glaucoma occurs when the aqueous humour fails to drain properly. This fluid builds up, raising the pressure in the eye. This may lead to blindness. There are two forms of glaucoma – open angle glaucoma and closed angle (acute) glaucoma.

(i) Suggest what is meant by the term 'acute' in the context of acute glaucoma. *(1 mark)*

(ii) Outline the effects on vision caused by glaucoma, which, if left untreated, may cause blindness. *(2 marks)*

c The table shows some of the causes of blindness in different regions of the world.

Economic status	Region	Cause of blindness (percentage of total number of blind people per region)			
		cataract	glaucoma	retinal damage due to diabetes	age-related macular degeneration (AMD)
Developed	United States of America	5	18	17	50
Developing	South America	40	15	7	5
Developed	Northern Europe	5	18	17	50
Developing	Africa	55	15	0	0

(i) Using the information, describe how the causes of blindness differ between the developed and the developing regions of the world. *(4 marks)*

(ii) Suggest why there are marked differences in different regions of the world in the percentage of blindness due to cataracts. *(2 marks)*

(iii) Discuss whether the data in the table supports the theory that type 2 diabetes is the main cause of retinal damage in diabetics. *(2 marks)*

(iv) The table indicates that there are fewer recorded cases of age-related macular degeneration in Africa than in other regions. Suggest why. *(2 marks)*

OCR F225 2012

28 THE EFFECT OF AGEING ON THE NERVOUS SYSTEM
28.1 Alzheimer's disease
Specification reference: 5.2.3

Learning outcomes

Demonstrate knowledge, understanding, and application of:

→ the causes of Alzheimer's disease

→ the signs and symptoms of Alzheimer's disease.

Alzheimer's disease

In the UK ~500 000 people are affected by Alzheimer's disease (AD), the most common type of dementia. The term 'dementia' describes a loss of cognitive ability associated with gradual death of brain cells. AD is a **degenerative** disease of the nervous system, which particularly affects the temporal and frontal lobes of the brain. The risk of the disease, as with other types of dementia, increases with age, and affects ~1 in 6 people over the age of 80 years. However, around 1 in every 20 cases is in people between 40 and 65 years of age.

Causes

Whilst the exact cause of Alzheimer's disease is as yet unknown it is considered a multifactorial disease. These factors include:

- increasing age – the risk doubles every five years over the age of 65
- a family history of the condition – the risk is higher if a parent, sister, or brother has the disease.

Some alleles of particular genes are known to cause AD and inheriting these genes is almost certainly likely to lead to the early onset of the disease (sometimes as early as 30–40 years of age). However, these alleles have only been found in a few hundred families worldwide. In these cases AD is called familial AD. Approximately 0.1% of cases are familial forms of autosomal dominant inheritance. The majority of autosomal dominant familial cases are caused by mutations in one of three alleles – those encoding amyloid precursor protein (APP) and presenilins 1 and 2.

Most cases of AD are not due to autosomal-dominant inheritance – these cases are called sporadic AD. In sporadic AD risk factors include:

- other genetic differences, for example, some variants of the apolipoprotein E gene (ApoE) may act as risk factors
- severe or repeated head injury
- stroke
- low level of education and intellectual activity

Dementia can also occur due to factors associated with cardiovascular disease, such as high blood pressure, smoking, and high blood cholesterol. This is usually referred to as vascular dementia as other symptoms of AD are not present.

Symptoms of AD

Initially, when the disease starts to destroy the brain cells there are no outward symptoms but eventually memory lapses may occur.

For example, forgetting about recent conversations or events, and forgetting the names of places and objects. Over a period of months and sometimes years, this memory loss becomes progressively more frequent and pronounced.

The symptoms of AD can be divided into three stages:

1 Early stages (mild) – memory loss and other cognitive deficits are compensated for by the person so they continue to function independently. They may start to forget family members' names, places, objects, and events, and find it difficult to express what they want to say. They may repeat themselves, their ability to make decisions becomes reduced and they show poor judgement. There may also be some early signs of mood changes, for example, increased anxiety, agitation, and confusion.

2 Mid-stage (moderate) – personality changes and physical problems develop and cognitive ability continues to decline. The person becomes more dependent on others for care and support for daily activities. Most patients at this stage are unaware that they are forgetting things and can't communicate coherently. They will show increasing confusion and disorientation, obsessive and repetitive/impulsive behaviour, aphasia (problems with speech and language), disturbed sleep, and frequent mood swings. By this stage patients may also have increasing difficulty performing spatial tasks, such as judging distance.

3 Late-stage (severe) – the symptoms become increasingly severe and distressing for the patient, and for their family, friends, and carers. The patient will likely have lost all control of their bodily functions – and so have developed urinary and bowel incontinence – and there may have been a devastating change in their personality. They will depend completely on the care and support of others for even simple daily activities. Confusion and disorientation become more frequent such that patients will require constant supervision for their own safety. Language is severely affected and can be lost completely. The person may become violent, demanding, and suspicious of people around them. In some cases the patient loses the ability to swallow (dysphagia), and the resulting considerable weight loss can lead to death.

Most people with AD live for ~8 to 10 years after they start to develop symptoms, but as with most diseases this can vary considerably from person to person. Often AD is not the actual cause of death but is a contributing factor. Pneumonia is a common cause of death as it may remain untreated because people with AD are not always able to recognise or communicate that they are ill.

Pathology

The destruction of brain cells in AD and the associated symptoms are caused by the presence of Tau protein tangles. Tau proteins are proteins that stabilise microtubules. They are abundant in neurones of the central nervous system. When Tau proteins become defective, neurone death results. Neurofibrillary tangles are twisted aggregates of Tau protein that have become hyper-phosphorylated and build up inside the neurones themselves. When this occurs, the microtubules

▲ **Figure 1** *Woman with Alzheimer's disease being spoon fed*

Study tip

The term 'dementia' is not synonymous with Alzheimer's disease. AD is one cause of dementia. In vascular dementia, memory and cognitive impairment occurs due to reduced blood flow to some parts of the brain – often in the form of 'mini strokes'. The underlying pathology associated with AD is absent.

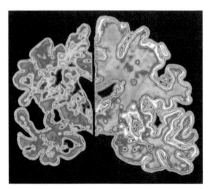

▲ **Figure 2** *Computer graphic of a vertical slice through the brain of an Alzheimer's disease patient (left) compared with a normal brain (right). The Alzheimer's disease brain shows a decrease in volume due to the degeneration and death of nerve cells. The surface of the brain is often more deeply folded than that of a healthy brain*

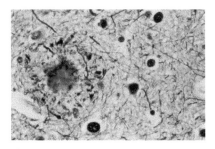

▲ **Figure 3** *Light micrograph of human brain tissue in Alzheimer's disease, showing a plaque (circular lesion at left), a characteristic histological feature of the disease. The other main feature of Alzheimer's disease is the formation of neurofibrillary tangles – masses of thickened filaments in the cytoplasm of neurones, approximately ×250 magnification*

disintegrate, causing the neurone's transport system to collapse, which in turn causes biochemical communication between neurones to malfunction, and later results in neuronal death. Proteins are found in the myelin sheath around neurones. Abnormal breakdown of these proteins produces β-amyloid proteins. These accumulate and form plaques outside and around axons and dendrites.

The loss of neurones and synapses in the cerebral cortex and certain subcortical regions causes gross atrophy of the temporal lobe, parietal lobe, and parts of the frontal cortex. Evidence from studies using MRI and PET have shown reductions in the size of specific brain regions in AD patients as their condition developed from mild cognitive impairment to Alzheimer's disease, when compared with similar images from healthy older adults.

Molecular causes of AD

There are many different theories for the molecular causes of AD, including:

- Herpes simplex virus type 1 has been suggested to play a causative role in people carrying the susceptible versions of the apoE gene.
- Age-related myelin breakdown in the brain. Iron released during myelin breakdown is thought to cause further damage.
- Inflammatory processes associated with ageing – for example, the formation of toxic proteins on the surface of brain neurones. This is thought to occur due to a derivative of β-amyloid called β-amyloid-derived diffusible ligands (ADDLs). β-amyloid fibrils kill a range of neurones in the brain but ADDLs appear to only affect those neurones involved in AD by binding to a surface receptor on the neurones and changing the structure of the synapse. This causes disruption to neuronal communication.

There is no cure for Alzheimer's disease, as yet, but medication is available that can help improve some of the symptoms and slow down the development of the condition in some people. Studies are ongoing into the effects of antioxidants such as vitamin E on the progression of Alzheimer's and early results are promising.

 Treating AD

There are not yet any drug treatments available which provide a cure for AD. Medicines, however, have been developed that can slow down progression and improve symptoms in some patients.

Patients with AD have been shown to have decreased acetylcholine transferase activity in their brains. This enzyme synthesises acetylcholine by transferring the acetyl group from acetyl CoA to choline:

Acetyl CoA + choline → acetyl choline

One of the most common forms of treatment is cholinesterase inhibitors. Research has shown that the brains of people with AD show a loss of nerve cells that use the neurotransmitter called acetylcholine.

Donepezil, rivastigmine, and galantamine prevent the enzyme acetylcholinesterase from breaking down acetylcholine in the brain. Increased concentrations of acetylcholine result in increased communication between the neurones, which may in turn temporarily improve or stabilise the symptoms of AD. Whilst all types

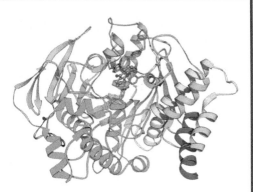

▲ **Figure 4** *This enzyme breaks down the neurotransmitter acetylcholine, stopping signal transmission*

of these cholinesterase inhibitors work in a similar way, one may suit an individual better than another, particularly in terms of side-effects experienced. However, these drugs only provide a short-term beneficial effect on memory. They do not stop the degenerative progression of the disease in the long term.

1 What type of reaction would be carried out by acetylcholinesterase to breakdown the acetylcholine?
2 Explain how the acetylcholinesterase inhibitor could result in the increase in acetlycholine.
3 Suggest how stem cell technology could be used to provide a treatment for AD.

Synoptic link

You may find it helpful to refer to the structure of the brain, and neurone structure and function in Topic 26.1, Nervous control.

Summary questions

1 a Suggest three different daily activities that a carer would have to assist an AD patient with. *(3 marks)*
  b Suggest what issues and problems carers may experience when looking after a member of the family with AD. *(3 marks)*

2 a Using the information in Figure 4 give two pieces of evidence that the molecule illustrated is a protein. *(2 marks)*
  b Explain why acetyl cholinesterase is said to have a quaternary structure. *(2 marks)*

3 Research studies have shown that approximately 5% of the neurones in the hippocampus are damaged and destroyed for every decade a person ages past the age of 50 years, and 20% over the age of 80. For every 100 neurones present in the hippocampus for a man who is 50 years old, calculate how many neurones he will have by the age of 90. Show your working. *(3 marks)*

28.2 Effects of ageing on the nervous system

Specification references: 5.2.3

Hearing impairment

Most people begin to lose a small amount of their hearing when they are 30 to 40 years old but deterioration has been found to start very early, from about the age of 18 years. Presbycusis is a progressive disorder and by the age of 80 years most people will have significant hearing problems. Age-related hearing loss affects high frequencies more than low frequencies, and men more frequently than women. As time passes the detection of high-pitched sounds becomes more difficult, speech perception is affected, and both ears are usually affected.

Age-related hearing loss occurs when the sensitive hair cells inside the cochlea (the inner ear) and the neurones in the auditory nerve gradually become damaged or die.

Study tip

Remember that one effect of ageing on the nervous system could be an increase in reaction time. You should review the practical described in Topic 26.2, Establishment and transmission of nerve impulses, and be able to describe how you would modify the procedure and statistical analysis to test the effect of ageing. (Hint – would you plan to use a paired or an unpaired *t*-test?)

Treatment of hearing impairment can involve the use of devices such as hearing aids and cochlear implants. A study in 2010 found that the water-soluble formulation of coenzyme Q10 (CoQ10) caused a significant improvement in hearing at certain thresholds. Drugs such as Tanakan have also been found to improve speech and hearing in elderly patients.

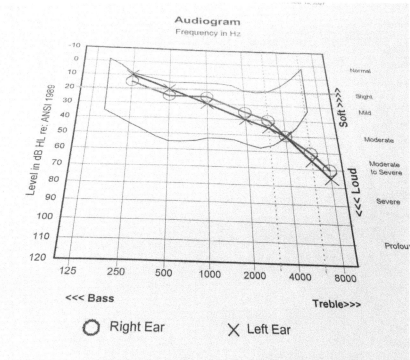

Right Ear: Best Word Rec. Score = 88% @ 60 dB HL

Left Ear: Best Word Rec. Score = 92% @ 65 dB HL

▲ **Figure 1** *An audiogram shows the softest sounds a person can hear at different frequencies*

Visual impairment

As people age inevitably there are disorders that affect vision. Poor vision can result from specific disorders, but the lens also becomes less elastic making accommodation more and more difficult.

Age-related macular degeneration (ARMD)

ARMD is a medical condition that usually affects older adults. It can result in a loss of vision in the centre of the visual field (the macula) due to damage to the retina. Two types of ARMD exist:

a   dry ARMD – cellular debris (called drusen) accumulates between the retina and the choroid causing the retina to become detached. This is the most common form of ARMD, accounting for ~90% of cases. Vision is lost slowly over a period of years, however 1 in 10 people will go on to develop wet ARMD.

b   wet ARMD – this is the more severe form, as blood vessels grow up from the choroid behind the retina. The blood vessels can leak blood and fluid into the retina, which can cause scarring and damage to the macula, resulting in visual distortion and blind spots. The retina can also become detached. Wet ARMD can be treated with laser coagulation and in some cases medication can stop (and sometimes reverse) the growth of blood vessels. Vision can deteriorate in days if left untreated.

ARMD is a major cause of blindness and visual impairment in adults over the age of 50 years. Macular degeneration can make it difficult or impossible to read or recognise faces, although enough peripheral vision remains to allow other activities of daily life. Although ARMD usually affects both eyes, because peripheral vision remains it will not usually cause complete blindness.

Cataracts

In the UK, about 1 in 3 people over the age of 65 have cataracts. They are the main cause of impaired vision worldwide. Cataracts may be partial or complete, stationary or progressive, hard or soft. An immature cataract has some transparent protein but with a mature cataract, all the lens protein is opaque.

There are three main types of cataract:

● Nuclear cataracts. These may make it difficult to recognise the intensity of colours but reading vision won't normally change.

● Cortical cataracts. These can cause problems with glare when driving. There may also be difficulty in reading and the low Sun in winter causes discomfort as it shines into the eyes.

● Subcapsular cataracts. These can make vision poor during the daytime, making it difficult for a person to drive and they are also likely to have problems reading.

As well as the natural process of ageing there are other factors which can increase the risk of cataracts:

● diabetes

● smoking

● regularly drinking excessive amounts of alcohol

● lifelong exposure of the eyes to UV sunlight (particularly UV-B)

▲ **Figure 2** *Cochlear implant transmitter on the scalp. Sound from a sound processor behind the ear is passed to this device and then sent to electrodes implanted in the cochlea of the inner ear. The electrical impulses from the cochlea are passed to the brain along the auditory nerve, allowing the person to hear*

▲ **Figure 3** *A scene as it would appear to a patient with macular degeneration. This eye disorder, which is usually age-related, affects the macula, an area near the centre of the retina. Degeneration of the macula leads to loss of vision at the centre of the field of view. It rarely causes complete loss of vision, but can lead to a person being officially registered as blind*

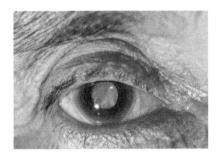

▲ **Figure 4** *A senile cataract seen during an eye examination*

- long term or high dose use of corticosteroid medication
- previous eye surgery or injury
- family history of cataracts.

Treating age-related cataracts

Cataracts can be treated using stronger glasses and brighter reading lights for a period of time. However, as cataracts worsen, surgery will be recommended when the loss of vision has a significant effect on the person's daily activities, such as driving or reading. Surgery involves removing the cloudy lens through a small incision in the eye and replacing it with a clear plastic one. Surgery is usually carried out under local anaesthetic as a day case and, almost everyone, experiences an improvement in their vision, although it can sometimes take a few days or weeks for their vision to settle.

Glaucoma

Glaucoma is a term that describes a group of eye conditions where, as a result of increased intraocular pressure caused by the failure of the aqueous humour to drain properly, the optic nerve is damaged. If left untreated glaucoma can lead to permanent blindness. In some cases one eye may develop glaucoma quicker than the other

There are four main types of glaucoma:

- Chronic open-angle glaucoma – this is the most common type and it develops slowly. It is caused by the blockage of drainage channels that normally allow the aqueous humour to drain away. The increased intraocular pressure is assessed in all people over the age of 40 years as part of a routine eye examination. Left untreated glaucoma will result in tunnel vision prior to blindness. It can be treated easily using eye drops, which reduce the amount of fluid produced by the eye or, in more severe cases, by opening up the drainage channels by laser surgery.

- Primary angle-closure glaucoma (also referred to as acute glaucoma) – this form is rare and can be chronic (occurring slowly) or acute (developing rapidly), with a sudden, painful build-up of pressure in the eye. The edge of the iris and cornea come into contact with each other, causing the pressure to rise suddenly as the aqueous fluid cannot reach the drainage channels between them. The eye becomes very red and painful causing the person to have severe headaches, blurred vision, nausea, and vomiting. This form of glaucoma needs immediate treatment to reduce the pressure in the eye, followed by laser treatment or surgery.

- Secondary glaucoma – this can occur as a result of an eye injury or another eye condition, such as uveitis (inflammation of the iris or inflammation of the iris and the ciliary body).

- **Developmental glaucoma (congenital glaucoma)** – this rare form of glaucoma can be very serious. It is usually present at birth or develops shortly after birth.

Memory loss

Periodical lapses in memory are a normal part of the ageing process and do not always indicate the onset of dementia or mental deterioration. Examples of these lapses include:

- Forgetting where regularly used items have been left, for example, glasses or keys.
- Forgetting the names of acquaintances or blocking one memory with another, such as calling their grandson by their son's name.
- Occasionally forgetting an appointment.
- Having trouble remembering what they have just read, or the details of a conversation.
- Walking into a room and forgetting why they went there.
- Becoming easily distracted.
- Not being able to retrieve information that is 'on the tip of their tongue'.

The main difference between age-related memory loss and dementia is that the age-related memory loss isn't disabling. The memory lapses have little impact on a person's performance and ability to complete normal daily activities. However, when memory loss becomes so severe that it disrupts the person's work, hobbies, social activities, and family relationships, it is possible that the person may be experiencing the warning signs of Alzheimer's disease, or another disorder that causes dementia.

Age-related memory loss can be reduced by making some or all of the following adjustments:

1 Regular exercise boosts brain growth factors and encourages the development of new brain cells.

2 Social interaction helps brain function by involving activities that challenge the mind. It also helps reduce stress and depression.

3 Eating plenty of fruit and vegetables and drinks such as green tea. These contain abundant antioxidants. Foods rich in omega-3 fats (such as salmon, tuna, trout, walnuts, and flaxseed) have been found to improve memory.

4 Managing stress as cortisol, the stress hormone, damages the brain over time and can lead to memory problems.

5 Sufficient sleep is necessary for memory consolidation (the process of forming and storing new memories). A lack of sleep also reduces the growth of new neurones in the hippocampus, and can cause problems with memory, concentration, and decision-making.

6 Not smoking decreases the risk of vascular disorders that can cause stroke and constrict arteries that deliver oxygen to the brain.

Summary questions

1 a Explain what changes other than to lens proteins, may occur in the lens which will affect vision as a person ages. *(2 marks)*
  b Suggest why a person may still need to wear glasses after surgery to treat cataracts. *(2 marks)*

2 a Suggest why ARMD does not normally result in complete blindness. *(2 marks)*
  b Suggest when ARMD may result in complete blindness. *(1 mark)*

3 a What term describes what is happening to the proteins in the lens that causes the lens to become opaque? *(1 mark)*
  b Suggest why the changes in proteins cause the proteins to clump together. *(2 marks)*

Practice questions

1 Which of the following statements is/are true of the pathology of Alzheimer's disease?

Statement 1: Alpha protein tangles are responsible for neurone death.

Statement 2: Beta-amyloid proteins can form plaques.

Statement 3: Herpes simplex virus type 1 has been linked with the onset of the disease.

  A 1, 2 and 3

  B Only 1 and 2

  C Only 2 and 3

  D Only 1   *(1 mark)*

2 What is the most accurate definition of a cataract?

  A Clouding of the lens, causing opacity.

  B Loss of vision due to retinal damage.

  C Optic nerve damage that results from an increase in intraocular pressure.

  D Inflammation of the iris.   *(1 mark)*

3 Which of the following activities is/are believed to reduce age-related memory loss?

  1 Eating foods containing omega-3 fats.

  2 Regular exercise.

  3 Engaging in activities that will increase cortisol concentration in the brain.

  A 1, 2 and 3

  B Only 1 and 2

  C Only 2 and 3

  D Only 1   *(1 mark)*

4 a Describe the techniques that can be used to improve memory in patients who have suffered a stroke.   *(4 marks)*

  b Memory loss is also a symptom of Alzheimer's disease. An investigation was carried out to distinguish between memory loss due to Alzheimer's disease and memory loss due to strokes.

    The activity of the brain was monitored during memory tests using positron emission tomography (PET).

    Describe the method used in carrying out a PET scan of the brain when investigating memory loss.   *(4 marks)*

  c Describe the changes in brain tissues that are associated with Alzheimer's disease.
                                        (3 marks)
OCR F225 2011

5 Two changes in the eye due to ageing that can lead to a reduction in colour vision are macular degeneration and cataracts.

    Suggest why macular degeneration and cataracts can lead to loss or reduction of colour vision.   *(3 marks)*
OCR F225 2012

6 One of the senses in the body which changes with increasing age is that of hearing. A hearing test is carried out as follows.

  • A set of headphones is placed over the patient's ears.

  • The audiologist selects the frequency of the noise to be generated (frequency is how high-pitched or low-pitched the sound is).

  • The sound is played into one ear, starting with a very quiet level (a low decibel level) and gradually increasing the level until the patient indicates they can hear the sound.

  • The test is repeated over a range of frequencies.

  • The test is repeated on the other ear.

  • The results are plotted on an audiogram.

    The graph shows the audiogram from a hearing test carried out on a 65 year-old male.

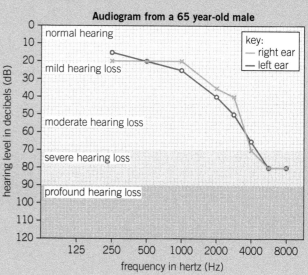
Audiogram from a 65 year-old male

a   Using the information in the figure, describe the pattern of hearing shown by this 65 year-old male. *(4 marks)*

b   Outline two possible social consequences of hearing loss for an elderly person.
(2 marks)

OCR F225 2013

7   The following passage is taken from the 2009 World Alzheimer Report:

Dementia is a syndrome due to disease of the brain, usually chronic, characterised by a progressive deterioration in intellect including memory, learning, orientation, language, comprehension, and judgement.

a   Dementia syndrome is linked to a large number of underlying brain pathologies, including Alzheimer's disease and vascular dementia.

The table shows some changes associated with Alzheimer's disease and vascular dementia.

(i)   Identify the two proteins, X and Y.
(2 marks)

(ii) Suggest two risk factors that might increase the chances of developing vascular dementia. *(2 marks)*

(iii) Using the information in the table, suggest why dementia due to Alzheimer's disease has a gradual onset, whereas dementia due to vascular disease can develop in sudden steps. *(2 marks)*

b   State two ways in which damage to the brain tissue can be assessed by health professionals. *(2 marks)*

OCR F225 2013

8   Outline the link between age and the deterioration of vision. Include examples of specific age-related conditions in your answer. *(6 marks)*

Type of dementia	Early characteristic symptoms	Changes in brain tissue	Proportion of dementia cases (%)
Alzheimer's disease (AD)	Impaired memory, apathy and depression with gradual onset	Build up of protein X outside neurones forming plaques, and build up of protein Y inside neurones forming neurofibrillary tangles	50–75
Vascular dementia (VD)	Similar to AD, but memory less affected, mood fluctuations more prominent, symptoms appear in sudden steps	Cerebrovascular disease due to deterioration of the blood vessels supplying the brain	20–30

29 THE PRINCIPLES AND IMPORTANCE OF HOMEOSTASIS
29.1 Homeostasis – the key principles
Specification reference: 5.3.1

Study tip

Homeostasis does not maintain *constant* values. Continuous fluctuations in physiological conditions occur, but homeostasis maintains these fluctuations within a narrow range.

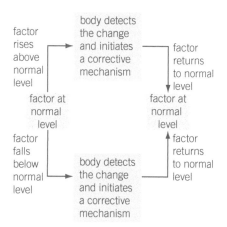

▲ Figure 1 *Negative feedback*

You have already learnt about the structure and function of the nervous system. One of the functions of the nervous system, in conjunction with hormones, is to regulate the internal environment of an organism. The maintenance and control of an organism's physiological conditions within narrow limits is known as **homeostasis**. Here you will learn about the key principles of homeostasis. In subsequent topics these general principles will be applied to specific examples, such as body temperature in Topic 29.2, and heart rate in Topic 29.3.

Why is homeostasis important?

Homeostasis enables factors such as temperature, pH, blood glucose concentration, and blood pressure to stay within restricted limits (i.e. within their normal ranges). The maintenance of a stable internal environment is important for several reasons:

* Optimum conditions for enzymes and other proteins within the body are maintained.

* The water potential of cells, blood, and tissue fluid is also kept within the correct range. This prevents cells shrinking due to water loss or expanding due to water entering by osmosis.

* Organisms are not as dependent on the temperature of their external environment. This enables mammals and birds to inhabit a wider range of ecosystems, from arid deserts to cold polar environments.

Negative feedback

Homeostasis relies on control systems that detect changes in internal conditions and produce the required responses to reverse these changes. This enables the internal environment to be returned to the desired state. These control systems involve hormones and nerve impulses. The corrective mechanism that allows only small fluctuations around a set point is known as **negative feedback**. The continuous monitoring of physiological factors enables continuous adjustments to be made. A control system must have these features:

* *A set point* – this represents the desired value that the negative feedback mechanism operates around. Physiological factors tend to vary over a small range either side of the set point. This represents the normal range of that factor.

- Receptors – they detect stimuli and deviations from the set point.
- Controller (communication pathway) – this coordinates the information from the receptors and sends instructions to the effectors. The nervous and hormonal systems act as controllers.
- Effectors – they produce the changes required to return the system to the set point.
- Feedback loop – the return to the set point creates a feedback loop that informs the receptors of the changes to the system.

In Topics 29.2 (body temperature) and 29.3 (heart rate), you will look at specific examples of these control systems.

▼ **Table 1** *Examples of ranges around a set point in human blood plasma*

Factor	Typical range	Units
Glucose	3.6–7.8	mmol dm^{-3}
Na$^+$	135–145	mmol dm^{-3}
pH (arterial)	7.35–7.45	No units
Osmolality (the balance between electrolytes and water)	275–295	mOsm kg^{-1}

Positive feedback

Negative feedback reverses a change in conditions – **positive feedback** increases the original change. Negative feedback inhibits the original stimulus, whereas positive feedback enhances the original stimulus.

One example of positive feedback is voltage-gated sodium ion channels and the action potential. However, generally speaking, positive feedback systems in organisms are destabilising and harmful. In Topic 29.2 you will learn about hypothermia and hyperthermia, which are examples of positive feedback damaging an organism.

Positive feedback is sometimes necessary in biological processes (for example, the attraction of platelets to a site of blood clotting).

Another example of a positive feedback system occurs during child birth. During labour, a hormone called *oxytocin* is released into the mother's bloodstream. This speeds up and intensifies uterine contractions. The increase in the frequency and strength of contractions causes more oxytocin to be secreted. This positive feedback mechanism continues until the baby is born. The birth stops the release of oxytocin and ends the positive feedback mechanism.

Study tip

Homeostasis is largely concerned with the control of factors in blood plasma. Internal cell environments, which you learnt about in Topic 2.2 (Mammalian and plant biofluids) are ultimately determined by the contents of blood plasma.

Synoptic link

You learnt about action potentials in Topic 26.1, Nervous control. You learnt about blood clotting in Topic 3.6, Blood clotting and enzymes.

▲ **Figure 2** *A woman in labour. The stimulation of uterine contractions by oxytocin relies on a positive feedback mechanism*

Synoptic link

The principles of homeostasis apply to some of the topics you have learnt about earlier in the course, such as blood composition (Chapter 1, Cells and microscopy) and water potential (Topic 2.6, Osmosis in cells).

Summary questions

1 In the context of a control system, state what is meant by the terms
   a set point   b normal range.   *(2 marks)*

2 Describe the positive feedback mechanism that controls uterine contractions during labour and suggest how positive feedback is beneficial during this process.   *(4 marks)*

3 Explain why homeostasis relies on negative feedback and not positive feedback.   *(3 marks)*

29.2 Regulation of body temperature and metabolic rate

Specification reference: 5.3.1

▲ **Figure 1** *Endothermic species such as penguins (top) and polar bears are able to maintain a high, stable body temperature, despite living in cold environments*

In the previous topic you learnt about the general principles of homeostasis. Here you will look at one example of homeostasis – the regulation of body temperature. You will learn about the methods used to measure a person's core body temperature. You will learn how metabolic rate can be fine-tuned in response to environmental temperature. In addition, you will learn about two situations in which the control of body temperature is lost – hypothermia and hyperthermia.

Control of body temperature

Mammals and birds are known as endotherms. Endothermic species maintain a core body temperature in the range 35–44 °C, which is favourable for metabolic reactions. This value represents a compromise between maintaining a temperature that is optimum for enzyme activity, and minimising the energy intake required to maintain such high temperatures. In humans, the optimum temperature is 37.4 °C. The ability to regulate body temperature is called thermoregulation.

The maintenance of a stable internal temperature, irrespective of changes in the surrounding environmental temperature, has enabled endotherms to occupy habitats with a wide range of ambient temperatures. In addition to behavioural mechanisms, endotherms also use physiological mechanisms to control their internal temperature. This is a form of homeostasis.

Which parts of the body are involved in thermoregulation?

Homeostatic control of core body temperature relies on temperature receptors that detect deviations away from the set point temperature (37.4 °C in humans). The core body temperature is defined as the temperature of structures deep in the body, such as the liver and heart.

Temperature-sensitive neurones (thermoreceptors) are located in the hypothalamus, which is a region of the brain responsible for autonomic responses. This part of the hypothalamus is called the thermoregulatory centre. The thermoreceptors detect changes in the temperature of the blood flowing through the brain.

Peripheral temperature receptors are located in the skin and other organs. These thermoreceptors pass information to the hypothalamus. This provides an early warning system – if the temperature of the extremities is decreasing, core body temperature is likely to decrease should conditions remain the same.

When a temperature change is detected by thermoreceptors, the hypothalamus sends nerve impulses to a variety of effectors – these include hair erector muscles, sweat glands, and blood vessels in the skin. These effectors will reverse the detected change, which could be an increase or a decrease in temperature.

What happens when the body is too hot?

When core temperature rises above the set point (e.g., 37.4 °C in humans), the heat loss centre of the hypothalamus initiates the responses shown in Figure 2, which lower the temperature through negative feedback.

Vasodilation (dilation of arterioles near the skin surface) allows blood to flow close to the surface of the skin. Heat is lost to the air, by conduction and radiation, which reduces body temperature.

Sweat glands secrete sweat, which evaporates from the skin. Heat from the body provides the energy required for the change of state from liquid to gas. Again, this lowers body temperature back towards the set point.

Hairs lie flat, thereby providing little insulation. More heat is lost by convection and radiation.

What happens when the body is too cold?

When core temperature drops below the set point, the heat gain centre of the hypothalamus initiates the responses shown in Figure 2, which raise the temperature through negative feedback.

Vasoconstriction (constriction of arterioles supplying blood to the skin) shuts off blood supply to capillaries near the skin surface. Smooth muscle in the arteriole wall contracts, which narrows the arteriole lumen. Less heat is lost from the skin's surface, which raises body temperature.

Hairs are pulled to a vertical position by the contraction of erector muscles at the base of hair follicles. This creates air pockets under each hair, which helps to insulate the body. Muscles can act as effectors in thermoregulation. When muscles receive nervous input from the hypothalamus, they contract, which generates heat from an increased rate of respiration. This is known as shivering. Respiration is an exothermic reaction that releases heat.

Metabolic rate and the long term control of temperature

Metabolic rate is the rate at which biochemical reactions occur in cells. Some reactions within cells (e.g., respiration) release energy as heat, which increases body temperature. Therefore adjustments to metabolic rate can fine-tune body temperature.

Thyroxine is the hormone that governs metabolic rate. It is secreted from the thyroid gland, which is situated in the neck. Thyroxine increases metabolic rate.

When a temperature decrease is detected, the hypothalamus secretes thyrotropin-releasing hormone (TRH). TRH stimulates the production of thyrotropin (thyroid-stimulating hormone, TSH) from the anterior

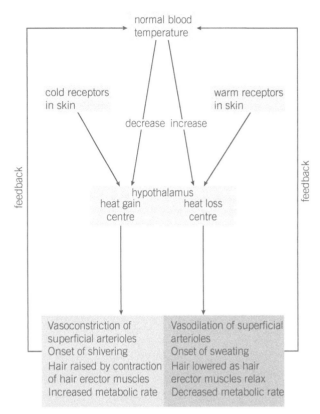

▲ Figure 2 *Thermoregulation coordinated by the hypothalamus through negative feedback*

cold conditons

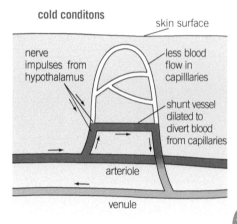

▲ **Figure 3** *Vasoconstriction*

warm conditons

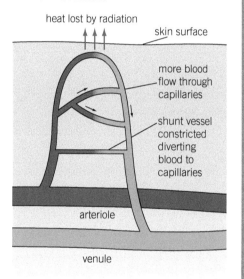

▲ **Figure 4** *Vasodilation*

pituitary gland. TSH is responsible for stimulating the secretion of thyroxine.

Thyroxine passes into the nucleus of cells. It increases the rate of transcription of several specific genes, such as those for mitochondria production and respiratory enzymes. The rate of respiration is raised, which increases heat production.

This form of temperature control is a slow response due to the time required for gene transcription. Extra thyroxine tends to be secreted when organisms are exposed to cold conditions for several days. This enables them to become acclimatised.

Measuring core body temperature

Temperature is measured using a thermometer. Modern thermometers tend to be digital – they contain infrared sensors that respond rapidly.

Core body temperature can be measured using oral (mouth), tympanic (ear), axillary (arm pit), and rectal methods.

Tympanic measurements accurately reflect core body temperature, because the eardrum shares its blood supply with the thermoregulatory centre in the hypothalamus. Oral and axillary measurements tend to give readings that are lower than the core body temperature.

Readings outside the normal range in adults and children indicate the possibility of hypothermia or hyperthermia. Raised temperature may be due to a fever. In such cases, the set point of the thermoregulatory centre is temporarily increased to a higher value.

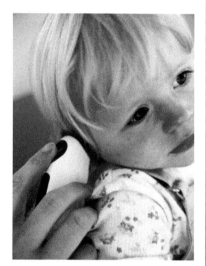

▲ **Figure 5** *Measuring the core body temperature of a child using the tympanic method*

Strip-type thermometers can be used as a cheap alternative to digital thermometers. A strip-type thermometer is held against a person's forehead. Explain why this has less accuracy than the methods described above.

The loss of control of body temperature

Hypothermia

Hypothermia is the lowering of the body's core temperature outside its normal range (in humans, below 35 °C). When this happens, metabolic reactions slow because molecules, including enzymes, have less kinetic energy. This creates a positive feedback effect, which you learnt about in the previous topic. Less metabolic heat is released, and this further reduces body temperature.

Causes

Exposure to cold weather or immersion in cold water can cause hypothermia.

Fuel poverty has increased the incidence of hypothermia. Households are considered by the UK government to be in fuel poverty when they would need to spend more than 10% of their household income on fuel to keep their home at a satisfactory temperature. The rising cost of energy has plunged more households into fuel poverty in recent years.

Symptoms

- shivering
- inability to pay attention
- weak pulse
- shallow breathing

Treatment

A person with hypothermia should be moved inside or insulated from cold ground. Wet clothing should be replaced with warm, dry blankets. Warm compresses should be applied to the neck, groin, and chest. Heating a person's arms and legs should be avoided because this forces cold blood back towards the major organs. If a person with hypothermia stops breathing, expired air resuscitation is given.

Hyperthermia

Hyperthermia is an increase in core body temperature above the normal range.

Causes

Exposure to high environmental temperatures can lead to hyperthermia. Other factors that increase the risk of hyperthermia include – illness, some medications, over-exertion, and being overweight.

Climate change is liable to increase the incidence of hyperthermia in some regions of the world and increase the risk of hypothermia in others.

Symptoms

- headaches
- mental confusion
- muscle cramps
- vomiting

Treatment

A person with hyperthermia should be taken to cool conditions, if possible, and rehydrated.

Synoptic link

You will have read about the structure of the autonomic nervous system, which governs thermoregulation and other aspects of homeostasis, in Topic 26.1. You learnt about the idea of metabolic rate back in Topic 16.1, Glycolysis.

Study tip

Remember, heat can be transferred via radiation, conduction, or convection. In each case, heat moves down a temperature gradient (i.e. from an organism to the cooler temperature in the surrounding air).

Summary questions

1 Name four structures located in the skin that are used by endotherms in thermoregulation. (*4 marks*)

2 Describe how negative feedback enables thermoregulation when an organism's body temperature rises above its set point. (*6 marks*)

3 Suggest why the food intake of an endothermic species is likely to increase during the winter compared to the summer months. (*3 marks*)

29.3 Regulation of heart rate

Specification reference: 5.3.1

Learning outcomes

Demonstrate knowledge, understanding, and application of:

→ the nervous and hormonal control of heart rate

→ the roles of the sympathetic and parasympathetic nervous system and adrenaline in the control of heart rate.

Synoptic link

You learnt about the initiation and coordination of the cardiac cycle in Topic 5.2, The cardiac cycle, and the effect of exercise on heart rate in Topic 5.3, Monitoring the heart. You learnt about the structure of the autonomic nervous system, including its division into the sympathetic and parasympathetic branches, in Topic 26.1, Nervous control.

Study tip

The cardiac muscle is *myogenic*, which means its contractions are initiated within the heart tissue rather than by an external stimulus. The accelerator and vagus nerves do not make the heart muscle contract, they regulate the rate at which the SAN fires.

Study tip

A *medulla* is the inner region of a tissue or organ. Other organs, such as the kidney, contain a medulla. Therefore you should refer to the medulla oblongata by its full name to avoid confusion.

You learnt about the general principles of homeostasis in Topic 29.1 and applied these principles to specific examples in 29.2. In this topic you will learn about another physiological factor that is under tight regulation – heart rate. Here you will learn about the hormonal and nervous control of heart rate.

The control of heart rate

Heart rate shows great variation across species. Blue whales' hearts beat six times per minute on average, whereas mice have resting heart rates of 670 beats per minute. The typical resting heart rate of an adult human is 70 beats per minute. Under resting conditions, this is controlled by the sinoatrial node (SAN). However, the ability to alter this resting rate to meet changing oxygen demands is essential. The factors influencing heart rate need to be regulated to ensure the heart beats at an appropriate rate.

The role of the medulla oblongata

The medulla oblongata is part of the autonomic nervous system. The autonomic nervous system coordinates homeostasis within organisms. Two branches of the autonomic nervous system exist – the sympathetic and the parasympathetic. Nerve impulses from these two branches have opposite effects on the body.

Changes in heart rate are controlled by the cardiovascular centre in the medulla oblongata. This structure in the brain has two regions that produce antagonistic responses:

- An increase in heart rate via the sympathetic nervous system. This involves nerve impulses being transmitted to the SAN of the heart via the accelerator nerve.

- A decrease in heart rate via the parasympathetic nervous system. This involves nerve impulses being transmitted to the SAN via the vagus nerve.

Two different types of receptor pass information to the medulla oblongata – chemoreceptors and pressure receptors.

Chemoreceptors

Chemoreceptors are located in the walls of the carotid arteries, which carry blood to the brain. These receptors detect changes in blood pH that result from changes in carbon dioxide concentration. As metabolic activity increases, the rate of carbon dioxide production from respiration is increased. This lowers blood pH. Chemoreceptors are sensitive to this change and send information to the sympathetic region of the medulla oblongata. The medulla oblongata generates nervous impulses along the accelerator nerve, which increases heart rate.

When carbon dioxide concentration in the blood decreases, chemoreceptors stimulate the medulla oblongata to lower heart rate.

Pressure receptors

Pressure receptors are found in both the carotid arteries and the aorta. When blood pressure is higher than normal, the receptors transmit nervous impulses to the medulla oblongata in order to decrease heart rate via the parasympathetic nervous system.

The opposite response is initiated when blood pressure drops below the normal range. The pressure receptors stimulate the medulla oblongata to raise heart rate via the sympathetic nervous system.

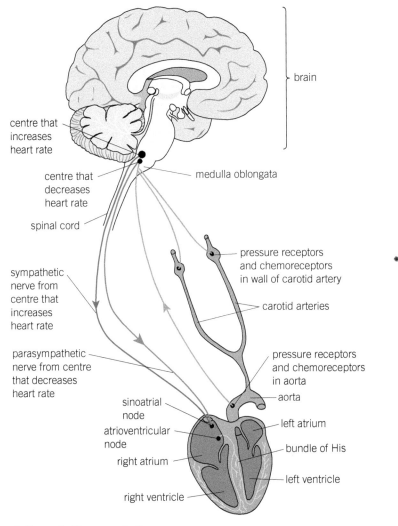

▲ **Figure 2** *The control of heart rate*

Adrenaline

Adrenaline is secreted when a person experiences anticipation, excitement, shock, or stress. It increases heart rate, helping to prepare the body for activity by binding to receptors on cells in the SAN. The binding of adrenaline initiates a secondary messenger system within cardiac cells. A G protein is activated, which leads to a rise in cAMP formation. This results in Ca^{2+} channels becoming more permeable, which enables pacemaker cells to depolarise more quickly by lowering the threshold for action potentials.

increased muscular/metabolic activity

↓

more carbon dioxide produced by tissues from increased respiration

↓

blood pH is lowered

↓

chemical receptors in the carotid arteries increase frequency of impulses to the medulla oblongata

↓

centre in medulla oblongata that speeds heart rate, increases frequency of impulses to SA node via the sympathetic nervous system

↓

SA node increases heart rate

↓

increased blood flow removes carbon dioxide faster

↓

carbon dioxide level returns to normal

▲ **Figure 1** *Effects of increased metabolic activity on heart rate*

Summary questions

1 Describe the role of receptors in regulating heart rate.
   (3 marks)

2 Explain why heart rate must be altered in response to increased physical activity.
   (3 marks)

3 Imagine that the vagus nerve of an organism is cut. Suggest what would happen if the organism's blood pressure increases above its normal value.   *(3 marks)*

Practice questions

1 Which of the following statements is an example / are examples of positive feedback?

Statement 1: The mechanism by which platelets are attracted to a site of blood clotting.

Statement 2: The interaction between oxytocin and contraction strength during child birth.

Statement 3: The development of hypothermia.

A 1, 2 and 3

B Only 1 and 2

C Only 2 and 3

D Only 1 *(1 mark)*

2 Which of the following is a response to the body's temperature rising above its set point?

A Vasoconstriction.

B Hairs lie flat.

C Increased muscle contraction within some tissues.

D Decreased sweat production. *(1 mark)*

3 Which of the following measurements is an accurate reflection of core body temperature because the region of measurement shares its blood supply with the hypothalamus?

A Oral

B Axillary

C Rectal

D Tympanic *(1 mark)*

4 Which of the following is a symptom / are symptoms of hyperthermia rather than hypothermia, in general?

1 Shivering

2 Vomiting

3 Muscle cramps

A 1, 2 and 3

B Only 1 and 2

C Only 2 and 3

D Only 1 *(1 mark)*

5 Describe how heart rate is controlled by both nervous and hormonal mechanisms. *(6 marks)*

6 The maintenance of a stable body temperature is an important aspect of homeostasis in endotherms. This is known as thermoregulation.

a (i) State where the core body temperature is monitored. *(1 mark)*

(ii) Name the type of sensory cell in the skin that detects changes in environmental temperature. *(1 mark)*

(iii) Name the corrective homeostatic mechanism that works to restore any changes in body temperature to the normal range. *(1 mark)*

b Endotherms respond in different ways to changes in environmental temperature. Some of these responses are listed here.

J	Secretion of adrenaline
K	Sweating
L	Shivering
M	Contraction of erector pili muscles (attached to base of hairs)
N	Curling up
O	Finding shade
P	Vasoconstriction of arterioles near to skin surface

Use the letters, J to P, to identify:

(i) The responses that conserve heat. *(1 mark)*

(ii) The responses that cool the body. *(1 mark)*

(iii) A physiological response that generates heat. *(1 mark)*

(iv) A behavioural response to a decrease in environmental temperature. *(1 mark)*

c Different endotherms have evolved different physiological and behavioural adaptations to assist with temperature control.

Explain how each of the following adaptations help the animal to control its body temperature.

(i) Elephants have large, thin ears that move backwards and forwards when hot. *(2 marks)*

(ii) Penguins living in cold climates have shunt blood vessels. These shunt vessels link arterioles carrying blood towards their feet with small veins that carry blood away from their feet. *(1 mark)*

OCR F214 2013

7 The control of body temperature is an important part of homeostasis. The accepted value for human body temperature is 37 °C. Body temperature does not vary significantly despite extreme variations in environmental temperature in different parts of the world.

a Explain why body temperature must be kept close to 37 °C in humans. *(3 marks)*

b An experiment was carried out to investigate how quickly humans can adapt to high environmental temperatures. A group of fit volunteers was subjected to repeated sessions of exposure to high temperature over a period of 9 days.

- Body temperature was measured using a temperature probe inserted into the ear.

- Sweat loss was estimated by measuring the loss of body mass during the period of heat exposure.

- The time between the start of exposure to high temperature and the onset of sweating was also measured.

    (i) Explain why measuring body temperature in the ear rather than in the mouth or at the skin surface gives a more accurate value. *(2 marks)*

    (ii) Suggest why measuring loss of body mass will provide only an estimate of the loss of water due to sweating. *(2 marks)*

c The graphs show the results of the investigation over the nine-day period. The graphs show the mean mass lost due to sweating, the mean ear temperature, and the mean time to the onset of sweating.

Using the information in the graphs, suggest how the body responds when exposed to high temperatures over a nine-day period. *(4 marks)*

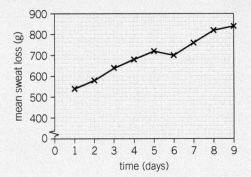

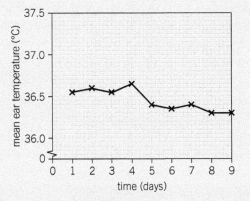

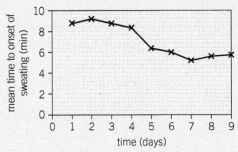

d Two of the volunteers in the original group withdrew from the experiment on the advice of the medical supervision team. This was because they had failed to adapt successfully to repeated exposure to high temperatures.

Suggest what signs and symptoms may have been recorded by the medical supervision team that resulted in the two volunteers being advised to withdraw from the experiment. *(3 marks)*

OCR F225 2011

Synoptic link

Insulin is initially produced as a polypeptide called *proinsulin*. Following translation, proinsulin is modified and processed in the Golgi apparatus. You learnt about the production and processing of proteins in Topic 4.7, Protein synthesis – translation.

In Chapter 29, The principles and importance of homeostasis, you learnt about the general principles of homeostasis, including the idea of negative feedback. You also looked at some examples of homeostasis: the regulation of body temperature, heart rate, and metabolic rate. Here, you will look in detail at the regulation of blood glucose, which is an example of homeostasis controlled by an organ called the pancreas.

The structure of the pancreas

The pancreas is a small, pale organ situated behind the stomach. It is an unusual organ because it acts as both an endocrine and exocrine gland. *Endocrine glands* secrete hormones into the blood. *Exocrine glands* secrete other substances, such as enzymes, into ducts. The pancreas secretes the hormones **glucagon** and **insulin** into the bloodstream and secretes digestive enzymes into the pancreatic duct. This topic will focus on its endocrine function and the production of insulin and glucagon.

The majority of pancreatic cells are involved in enzyme production. Scattered through the pancreas, however, are clumps of cells known as the *islets of Langerhans*. These cells produce hormones. There are two types of endocrine cells in the islets:

- *Alpha (α) cells* – they produce and secrete glucagon – the cells are distributed around the edge of the islets.
- *Beta (β) cells* – they produce and secrete insulin – the cells tend to be distributed towards the centre of the islets.

Insulin and glucagon are secreted directly into the many blood capillaries that lie adjacent to α and β cells. These two hormones together regulate blood glucose concentration.

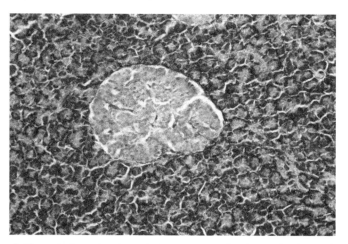

▲ **Figure 1** *A section through the pancreas showing endocrine and exocrine cells, which are shown as a red stain. Endocrine cells are shown in pale purple in the islet of Langerhans in the middle where the blue and white is the system of blood capillaries. Approximately ×128 magnification*

Regulation of blood glucose

Insulin and glucagon produce antagonistic effects. This enables blood glucose concentration to be controlled through negative feedback.

When blood glucose concentration rises:

- β cells act as receptors and detect the rising concentration of glucose
- β cells secrete insulin
- α cells stop secreting glucagon
- insulin binds to membrane-bound receptors on many cells, but principally liver and muscle cells
- these cells act as effectors by increasing their uptake of glucose
- more glucose is converted to fats or used in respiration in these cells
- liver cells convert some of the glucose to glycogen (glycogenesis), which is stored
- blood glucose concentration is reduced.

When blood glucose concentration drops:

- α cells act as receptors and detect the falling concentration of glucose
- β cells stop secreting insulin
- α cells secrete glucagon
- glucagon binds to membrane-bound receptors on liver cells
- less glucose is taken up by the liver cells (effector cells)
- the use of glucose for respiration in effector cells decreases
- more fatty acids are used in respiration as alternative substrates
- liver cells convert glycogen back into glucose (glycogenolysis), which is released into the bloodstream
- some amino acids and fats are converted into glucose (gluconeogenesis)
- blood glucose concentration is increased.

Blood glucose concentration, like other physiological factors that are controlled by homeostasis, fluctuates about a set point over a narrow range (usually 90 mg per 100 cm^{-3}). This tight control of glucose levels can go awry, however, if insulin production or effectiveness is reduced. This is what occurs in *diabetes*.

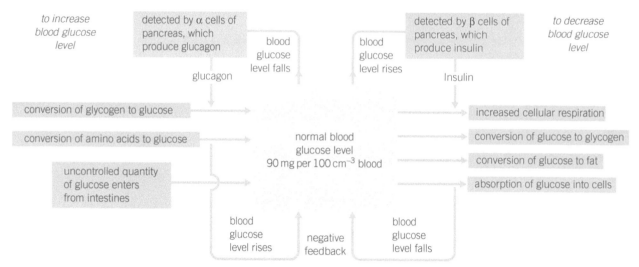

▲ **Figure 2** *The negative feedback mechanism used in blood glucose regulation*

Cellular changes produced by insulin and glucagon

Insulin

Insulin binds to a receptor in the plasma membrane of its target cell, which initiates a cascade of reactions inside the cell. The cascade of reactions causes more glucose transporter proteins to be inserted into the plasma membrane, which increases the uptake of glucose into the cell. The binding of glucose also activates enzymes that control glycogenesis (glycogen synthesis), and the conversion of glucose to fatty acids within the target cell.

Glucagon

Glucagon binds to a receptor in the plasma membrane of its target cell. This initiates a cascade of reactions similar to that produced by the action of insulin on its target cells. The intracellular cascade set in motion by glucagon causes enzymes involved in glycogenolysis and gluconeogenesis to be activated.

Summary questions

1 Explain the difference between exocrine and endocrine glands, using the pancreas as an example. *(4 marks)*

2 The set point for blood glucose concentration in humans is approximately 90 mg per 100 cm^{-3}. How would this value be expressed in g cm^{-3}? *(1 mark)*

3 Suggest why liver cells have specialised receptors for glucagon. *(3 marks)*

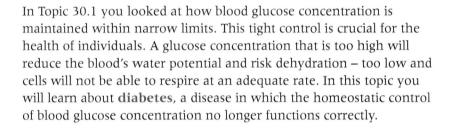

30.2 Diabetes

Specification reference: 5.3.2

In Topic 30.1 you looked at how blood glucose concentration is maintained within narrow limits. This tight control is crucial for the health of individuals. A glucose concentration that is too high will reduce the blood's water potential and risk dehydration – too low and cells will not be able to respire at an adequate rate. In this topic you will learn about **diabetes**, a disease in which the homeostatic control of blood glucose concentration no longer functions correctly.

The two types of diabetes mellitus

The blood glucose concentration of people with diabetes mellitus often deviates outside the normal range of values. Two forms of the disease exist – type 1 and type 2. In both cases, blood glucose concentration is liable to become too high without correct treatment and management. Initial symptoms can include thirst and increased urine production. Untreated diabetes results in glucose being present in a person's urine. Furthermore, elevated blood glucose concentrations can cause tissue damage, dehydration and, at its most serious, unconsciousness and death.

Type 1 diabetes

Type 1 diabetes is sometimes referred to as insulin-dependent diabetes. It is characterised by little or no insulin production. If untreated, a person with type 1 diabetes may lose consciousness or fall into a coma.

Risk factors and causes

Most cases of type 1 diabetes are caused by an autoimmune response. This is where a person's immune system attacks its own cells. In the case of type 1 diabetes, antibodies are produced that destroy β cells in the islets of Langerhans. This results in reduced production of insulin and, eventually, no insulin secretion.

Type 1 diabetes is heritable. Several gene variants have been implicated in the risk of developing the condition, including some HLA antigen genotypes. Among susceptible people, the onset of type 1 diabetes can be triggered by environmental factors, such as viral infections.

Type 2 diabetes

This form of diabetes is more common than type 1. Whereas type 1 diabetes is characterised by a lack of insulin production, type 2 is largely a result of insulin resistance. Although less insulin is often produced by people with type 2 diabetes, a reduced sensitivity to insulin in effector cells is the most significant problem. For this reason, type 2 diabetes is known as insulin-independent. Symptoms are generally less severe than those exhibited in type 1 diabetes.

Learning outcomes

Demonstrate knowledge, understanding, and application of:

→ the symptoms, causes, and risk factors of type 1 and type 2 diabetes

→ the techniques used to diagnose diabetes.

Synoptic link

You learnt about HLA antigens in Topic 23.3, Codominance.

Risk factors and causes

Type 2 diabetes has a number of risk factors:

- Obesity – insulin resistance is associated with a high body fat percentage and BMI (body mass index). More than 80% of people diagnosed with type 2 diabetes are overweight.

- Genetics – people with a family history of type 2 diabetes have an increased risk. The condition has a heritable component. A gene on chromosome 4 is associated with a higher risk of developing type 2 diabetes.

- Age – risk increases with age. The condition is most common in people aged over 40 years, but is becoming more prevalent among adolescents and young adults.

- High blood pressure.

- Low birth weight.

▼ **Table 1** *Summary of the differences between type 1 and type 2 diabetes*

Characteristic	Type 1 (insulin dependent)	Type 2 (insulin independent)
insulin production	little or none	often reduced
cause	usually autoimmune response	effector cells lose responsiveness to insulin
age at onset	childhood (juvenile-onset)	adulthood (late-onset)
speed of onset	quick	slow

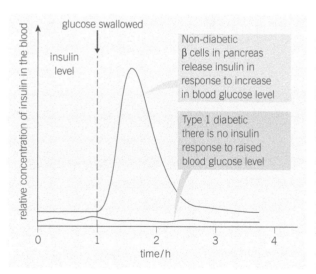
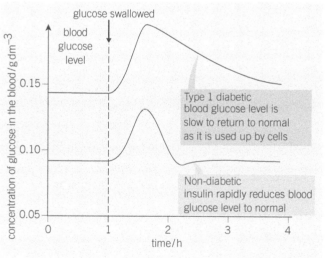

▲ **Figure 1** *The responses of a diabetic and a non-diabetic to the glucose tolerance test*

Maturity-onset diabetes of the young

Forms of diabetes mellitus exist other than type 1 and 2, although they are rare. Maturity-onset diabetes of the young (MODY) is a hereditary condition that constitutes 2% of diabetes mellitus cases. It shows autosomal dominant inheritance (you can revisit Topic 23.1, Monogenetic inheritance, for a reminder of inheritance patterns).

MODY tends to be diagnosed in people below 30 years of age. It is usually characterised by mild *hyperglycaemia* (high blood glucose concentration). MODY can be controlled in the early stages by planning meals, and may not require insulin injections. Patients are not insulin resistant – the problem lies with glucose metabolism or insulin secretion. Several different forms of MODY have been identified, each caused by a genetic variant at a single gene locus.

1 Suggest why MODY is sometimes referred to as 'monogenic diabetes'.
2 What is the range of probability that two parents with MODY will have a child with the condition?
3 MODY patients with heterozygous genotypes can often manage their condition through careful dietary planning. Suggest why MODY patients with homozygous genotypes are likely to require insulin injections.

Diagnosis of diabetes

A person suspected of having diabetes can be tested using either of two different methods: the fasting blood glucose test, or the glucose tolerance test.

Fasting blood glucose test

The person being tested should eat and drink nothing, other than water, for 8 to 12 hours beforehand. A blood sample is taken and blood glucose concentrations are measured.

Glucose tolerance test

The person is asked to eat as normal until the evening before the test. They fast from midnight until the test the next day. An initial blood sample is taken and glucose concentration measured. The person then drinks a glucose solution. Further blood samples are taken every 30 minutes for 3 hours after the consumption of the glucose. The results indicate the effectiveness of insulin at removing the excess glucose that has been absorbed.

▼ Table 2 *The interpretation of tests for diabetes*

Blood glucose concentration two hours after the start of the test ($mmol\,dm^{-3}$)	Fasting blood glucose test	Glucose tolerance test	Interpretation
	3.6–6.0	less than 7.8	normal glucose tolerance
	6.1–6.9	7.9–11.0	impaired glucose tolerance
	7.0 and above	11.1 and above	probable diabetes

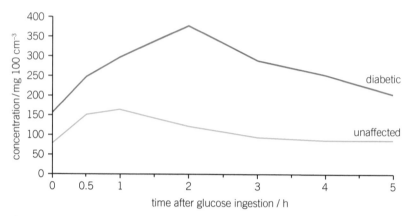

▲ Figure 2

Summary questions

1 Why is type 1 diabetes sometimes called insulin-dependent and type 2 called insulin-independent?   (*2 marks*)

2 Study Figure 2. Describe and explain the differences in the responses of diabetics and non-diabetics following the consumption of glucose.   (*3 marks*)

3 Compare the relative influences of genetics and the environment on the risk of developing type 1 and type 2 diabetes.   (*5 marks*)

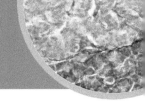

30.3 Treating and managing diabetes

Specification reference: 5.3.2

You learnt about the causes and symptoms of type 1 and type 2 diabetes in the previous topic. The two forms of the disease must be managed and treated using specific approaches because of their different causes and effects. In this topic you learn about the techniques that can be employed to treat type 1 and type 2 diabetes.

Monitoring blood glucose

Successful management of diabetes mellitus involves maintaining blood glucose concentration at a relatively constant level. The diabetic person must regularly monitor their glucose level using a biosensor. The measurement of blood glucose concentration involves the following steps:

- The person uses a sterile lancet to prick a finger and produce a small drop of blood, which is squeezed onto the test strip inserted into the biosensor.
- The biosensor contains the enzyme glucose oxidase.
- Glucose oxidase converts any glucose in the blood sample to gluconolactone.
- A small electric current is produced by this reaction.
- The current reaches an electrode, and a digital reading of glucose concentration is produced.

Glycosylated haemoglobin

Glucose molecules can attach to haemoglobin in the blood to form glycosylated haemoglobin. As plasma glucose concentration increases, the concentration of glycosylated haemoglobin also rises. A patient's blood concentration of glycosylated haemoglobin can be measured to monitor how well their blood glucose levels are being controlled. This test is performed by a doctor. It enables a patient's average blood glucose concentration over several weeks to be monitored.

Treating and managing diabetes

A person with diabetes should ensure that their blood glucose concentration stays within narrow limits (approximately 3.6–$7.8 \, \text{mmol dm}^{-3}$).

Type 1 diabetes

The principal treatment for type 1 diabetics is the use of insulin injections. Injections are administered either twice or four times per day.

The insulin is injected using a syringe. Fine control of the amount being injected is achieved through the use of an insulin pen. The amount of insulin given can be adjusted by turning a dial at the bottom of the pen. The tip of the pen holds the needle. Alternatively, an insulin pump can be used. These are battery operated and provide a continuous delivery of insulin, 24 hours a day. The rate of delivery can be adjusted.

Learning outcomes

Demonstrate knowledge, understanding, and application of:

→ the techniques used to monitor blood glucose concentrations

→ the treatment and management of type 1 and type 2 diabetes

→ the roles of health professionals in the management of diabetes

→ the future impact of diabetes on the human population.

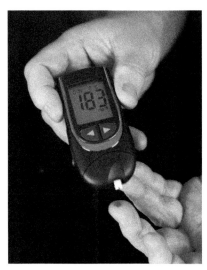

▲ Figure 1 A biosensor for testing blood glucose concentration

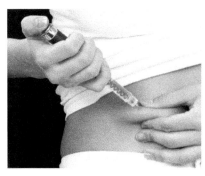

▲ Figure 2 A diabetic person injecting insulin

219

Type 2 diabetes

This form of diabetes often can be managed by carefully controlling diet and by reducing the intake of refined sugars. Meals should ideally be small but at regular intervals to avoid flooding the blood with excess glucose. Weight loss and an increase in the amount of exercise undertaken by a patient may help to alleviate symptoms.

Some drugs are available for patients. Metformin tends to be the first drug that is prescribed to people with type 2 diabetes. It has the advantage of not causing weight gain. Metformin works on three fronts: reducing the amount of glucose produced by the liver, lowering the concentration of glucose in the blood, and reducing the amount of glucose that is absorbed from food.

The role of health professionals

Diabetes nurses provide education, advice, and support to people with the disease. This can include advice for newly diagnosed patients with *hypoglycaemia* (low blood glucose levels) and support during a change of treatment.

Healthcare professionals monitor aspects of diabetic patients' health other than blood glucose concentrations. For example, diabetics are twice as likely as the general population to experience coronary heart disease – this necessitates regular checks on their blood pressure. Poor control of diabetes can damage blood vessels in the retina, resulting in blurred vision. Annual eye examinations are therefore important for diabetics. Bladder and kidney infections are more common among diabetics than non-diabetics, which means that kidney function tests are essential for patients.

Podiatry (foot care) is one of the most overlooked aspects of diabetes management. Diabetes can reduce the blood supply to a patient's feet, which may cause a loss of feeling, and means that foot injuries take longer to heal. Podiatrists screen patients' feet for signs of poor blood flow and offer advice, such as the type of footwear a patient should choose.

Diabetes treatment and management relies on evidence-based practice, which is the use of clinical data and research to inform decisions about the care of patients. Evidence-based practice has helped to move diabetes care from anecdotal experience and opinion to a stronger scientific foundation. For example, the nutritional advice now offered by medical professionals to diabetes patients is based on research studies. In 2014, the National Institute for Health and Care Excellence (NICE) recommended that the number of gastric bypass surgeries carried out in the UK should be increased. Clinical evidence indicates that the prevalence of type 2 diabetes would be reduced by encouraging this form of weight loss surgery.

The future impact of diabetes

Type 2 diabetes has traditionally been viewed as an adult disease. However, the prevalence of the condition among children is increasing. American children are now equally likely to be diagnosed with type 2 or type 1 diabetes. Rising levels of obesity among the young have been blamed.

Increased wealth in some countries has led to changes in lifestyle that increase the prevalence of risk factors for diabetes – more food, particularly processed food, is available, more people have cars and, overall, less exercise is taken. The average age of the global population is also increasing, which has further raised the prevalence of type 2 diabetes.

The increased incidence of diabetes is likely to have a significant financial impact on healthcare services, such as the NHS in the UK.

Summary questions

1 Why does a person with type 1 diabetes need to inject insulin rather than take it orally?   (2 marks)

2 Suggest why the concentration of glycosylated haemoglobin in red blood cells is used to monitor the long-term effectiveness of diabetes treatment.   (5 marks)

3 Suggest why the prevalence of type 2 diabetes is likely to have a greater impact on future populations than the prevalence of type 1 diabetes.   (5 marks)

Practice questions

1 Which of the following processes is a result / are results of the secretion of glucagon from the pancreas?

   1 Glycogenolysis

   2 Gluconeogenesis

   3 Glycogenesis

  A 1, 2 and 3

  B Only 1 and 2

  C Only 2 and 3

  D Only 1   *(1 mark)*

2 Which of the following characteristics is generally true of type 1 diabetes?

  A Slow speed of onset.

  B Late onset.

  C Caused by an autoimmune response.

  D Insulin-independent.   *(1 mark)*

3 The set point for blood glucose concentration in humans is approximately 9×10^{-4} g cm^{-3}. How would this value be expressed in mg 100 cm^{-3}?

  A 0.9

  B 9

  C 90

  D 900   *(1 mark)*

4 In the photomicrograph the white areas show damage to the endocrine tissue in a person with type 1 diabetes mellitus.

The dark areas contain cells which release digestive enzymes. These cells are unaffected by type 1 diabetes mellitus.

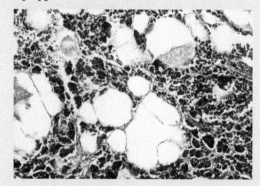

  a Name the organ from which the tissue sample was taken, and name the cells which would normally be found in the white areas.   *(3 marks)*

  b People with type 1 diabetes control their blood glucose concentration by injecting the hormone insulin. Insulin is transported to its target tissues by blood plasma.

Describe the effect of insulin on its target tissues.   *(5 marks)*

  c Human placental lactogen (HPL) is a hormone released during pregnancy.

HPL decreases the sensitivity of maternal tissues to insulin, and results in an increase in maternal blood glucose. HPL stimulates the breakdown of fats and the release of fatty acids.

A study was carried out to investigate the effect of increasing blood glucose concentration on insulin secretion in a large sample of pregnant and non-pregnant women.

The graph summarises the results of this study.

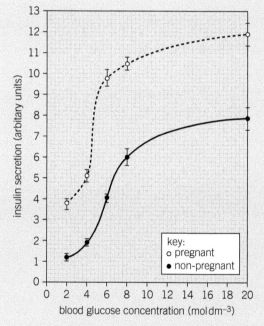

    (i) State what each data point represents and suggest what the error bars indicate about the data.   *(2 marks)*

    (ii) Using the figure, compare the response to rising levels of blood glucose in pregnant and non-pregnant women.   *(3 marks)*

    (iii) The effects of HPL on maternal glucose metabolism are advantageous to the fetus.

Suggest an advantage to the fetus and give a reason for your suggestion.

(*2 marks*)

d HPL can result in pregnant women displaying some symptoms of diabetes. This condition is referred to as gestational diabetes mellitus.

Type 2 diabetes mellitus is becoming increasingly common in the UK.

Discuss the similarities and differences between type 2 diabetes mellitus and gestational diabetes mellitus. (*5 marks*)

OCR F225 2011

5 One feature of homeostasis in humans is the maintenance of blood glucose concentration within a relatively narrow range. The range is normally between 82 and 110 mg dL^{-1}.

a Describe how blood glucose concentration is regulated in humans, and explain why glucose concentration needs to be maintained within a narrow range.

(*6 marks*)

b The UK's Foresight Programme aims to model future trends in science and health so the effective strategies can be developed.

One area that has been investigated is obesity. Some of the data used in the model to assess the impact of future trends in obesity are shown in the table and the graph. The data are for males aged between 21 and 60 years.

In the graph, the zone on either side of the plotted lines represents the confidence limits of the data.

BMI	Relative risk of developing type 2 diabetes
0–23	1.0
23–24	1.0
24–25	1.5
25–27	2.2
27–29	12.0
29–31	30.0
31–33	40.0
33–35	55.0
35+	90.0

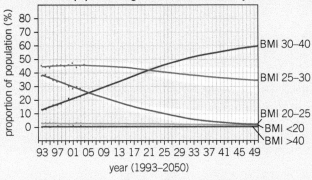

Predicted changes in proportions of BMI groups for the male population aged between 21 and 60 years

(i) In the period between 1993 and 2003, which BMI categories show the most reliable data?

Give a reason for your suggestion.

(*2 marks*)

(ii) Describe what happens to the confidence limits between 2003 and 2050 for the data in the BMI category 25–30.

Suggest a reason for the pattern you describe. (*2 marks*)

(iii) What evidence is there in the data to support the prediction that type 2 diabetes will become increasingly common? (*3 marks*)

(iv) Identify two problems with the presentation of the independent variable in the table and its use in the graph. (*2 marks*)

OCR F225 2013

31 KIDNEY FUNCTIONS AND MALFUNCTIONS

31.1 Kidney structure and excretory function

Specification reference: 5.3.3

Learning outcomes

Demonstrate knowledge, understanding, and application of:

→ the structure of the kidney and its nephrons

→ the role of the kidney in excretion

→ practical investigations into the biochemical composition of 'mock' urine, renal artery and vein plasma, and filtrate.

You looked at several examples of homeostasis in the previous two chapters. Here, you will learn about another organ with homeostatic functions – the kidney. This is the organ that is responsible for maintaining water potential within vertebrate organisms, and for removing unwanted and toxic substances. This topic will start by looking at the kidney's structure and how this enables it to excrete unwanted molecules from the body.

The gross structure of the kidney

Vertebrates usually have two kidneys, which are located just below the ribs, either side of the spine. Each kidney is supplied with blood from a renal artery, and a renal vein drains each organ. The **urine** produced by the kidneys passes through a tube called the ureter into the bladder – here it is stored before being excreted from the body.

The kidney has three distinct regions:

- The *cortex* (the outer region)
- The *medulla* (the inner region, consisting of renal pyramids, which are composed of sections of nephrons)
- The *renal pelvis* (the most central region, which leads into the ureter)

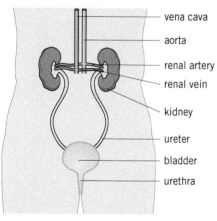

▲ Figure 1 *The organs of the excretory system*

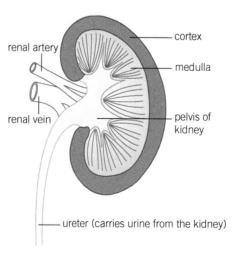

▲ Figure 2 *The structure of a kidney*

The kidney's role in excretion

Excretion is the removal of the waste products of metabolism. This includes the removal of carbon dioxide and water from the lungs, and the excretion of water and salts, in sweat, from the skin.

Another substance that must be excreted because of its toxicity is ammonia. Ammonia (NH_3) is formed when excess amino acids are **deaminated** – this occurs principally in liver cells. Ammonia is combined with carbon dioxide in a series of reactions called the *ornithine cycle* and converted to urea ($CO(NH_2)_2$), which is less toxic than ammonia. However, urea still needs to be excreted from the body, and this excretion is carried out by the kidneys.

Nephron ultrastructure and function

A large proportion of each kidney consists of **nephrons** – there are approximately one million in each kidney. Nephrons are the functional units of the kidney.

Each nephron begins in the kidney's cortex. A cup-shaped structure called the *Bowman's capsule* surrounds a dense network of capillaries known as the **glomerulus**. Blood pressure is very high in the glomerulus, which causes water, ions, and small molecules (e.g., urea, amino acids, and glucose) to be forced from the blood and filtered into the nephron. This process is called **ultrafiltration**.

The filtered fluid (filtrate) passes through the nephron, which is a heavily folded tubule that can be divided into four sections:

- proximal convoluted tubule
- loop of Henle
- distal convoluted tubule
- collecting duct

The composition of the filtrate changes as it moves through the nephron. Useful substances, such as glucose, some water, and certain ions, are reabsorbed back into the tissue fluid and blood capillaries through the process of **selective reabsorption**. Unwanted substances travel to the collecting duct. The fluid that reaches the end of the collecting duct is called urine. This moves through the renal pelvis, down the ureter, and into the bladder. Urine is excreted from the bladder via a tube called the urethra.

Ultrafiltration

Ultrafiltration is filtering at a molecular level. Large molecules remain in the blood, but smaller molecules (with a relative molecular mass below 65 000) pass into the Bowman's capsule.

The hydrostatic pressure in the glomerular capillaries is higher than the pressure in the Bowman's capsule. Blood is brought into the glomerulus by an **afferent** arteriole and leaves in the **efferent** arteriole. The smaller diameter of the efferent arteriole causes a build-up of hydrostatic pressure. Water and small molecules are therefore forced into the Bowman's capsule. Blood cells and most proteins remain in the blood.

The structure of the capillaries in the glomerulus and the cells in the Bowman's capsule provide a filter that allows only small molecules through:

● Endothelium cells in capillaries are separated by very narrow gaps, known as *fenestrations*. This prevents blood cells from escaping.

● The basement membrane of the capillaries is composed of a fine mesh of collagen and glycoprotein. This prevents large proteins from being filtered out of the plasma.

● The epithelial cells of the capsule have finger-shaped projections to create small gaps between cells. These cells are called *podocytes* and the gaps are referred to as *filtration slits*.

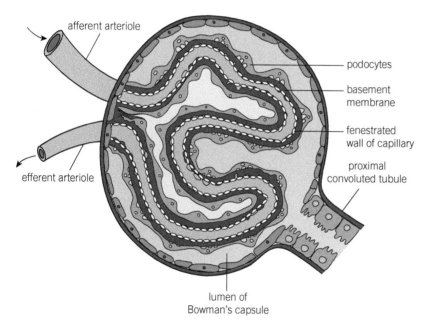

▲ Figure 3 *The structure of the Bowman's capsule, containing the glomerulus*

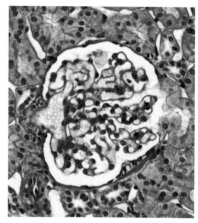

▲ Figure 4 *Light micrograph of a kidney glomerulus within a Bowman's capsule (white arc), ×600 magnification*

Selective reabsorption

Approximately 85% of the water in the filtrate is reabsorbed from the proximal convoluted tubule (PCT). All glucose is reabsorbed here, unless an individual has diabetes mellitus (see Topic 30.2). Amino acids and some ions are also reabsorbed from the PCT.

Cells lining the PCT are specialised for reabsorption in the following ways:

● *Microvilli* (folds in the surface membrane lining the lumen) increase the surface area for reabsorption.

● *Co-transporter proteins* are present in the membrane. These enable the facilitated diffusion of glucose or amino acids, in association with sodium ions, from the tubule fluid. Glucose and amino acids are then able to diffuse into blood capillaries.

- The membrane on the other side of PCT cells (the basal membrane), adjacent to capillaries, contains *sodium-potassium pumps*. These pump sodium ions out of the PCT cells, into tissue fluid. The sodium ions then diffuse into capillaries. The subsequent low sodium ion concentration in the PCT cells enables co-transport of glucose or amino acids to be carried out.

- The high concentration of *mitochondria* in PCT cells provides energy for active transport.

Toxic compounds (e.g., urea), excess water, and some ions remain in the nephron filtrate to be excreted as urine. You will look at the reabsorption of water from the nephron and the kidneys' role in **osmoregulation** in Topic 31.2, Kidney functions – osmoregulation and endocrine roles.

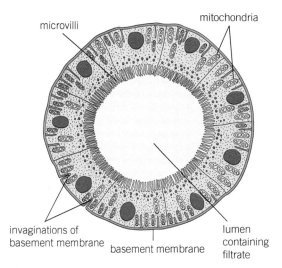

▲ **Figure 5** *Transverse section of a proximal convoluted tubule*

Investigating the composition of body fluids

The different compositions of urine, glomerular filtrate, renal artery blood, and renal vein blood can be investigated using 'mock' samples. Solutions can be produced that mimic the concentrations of solutes found naturally in the various body fluids.

Glucose concentration will be virtually identical in renal arteries, renal veins, and the initial glomerular filtrate. No glucose should be present in healthy urine. However, as you saw in Chapter 30, the urine of an untreated diabetic person will contain glucose. Mock urine samples can be tested for the presence of glucose using Clinistix strips or the Benedict's test (Topic 2.4, Detecting and measuring carbohydrates).

The presence of protein in the mock samples can be assessed using the biuret test (Topic 2.5, Detecting and measuring proteins). Filtrate and urine should have very low concentrations of protein in comparison to blood plasma.

The concentration of urea in the renal vein should be much lower than that found in the renal artery, filtrate, and urine.

The pH of the mock samples can also be tested and compared.

> Apply your knowledge from Chapter 2, The importance of water, to describe how the concentration of glucose in a urine sample can be measured using colorimetry.

Summary questions

1   Outline the likely differences in composition of blood in the renal artery and blood in the renal vein.   *(2 marks)*

2   Describe how glucose is reabsorbed into the blood from the proximal convoluted tubule.   *(5 marks)*

3   The cells lining the proximal convoluted tubule use endocytosis and exocytosis, in addition to active transport and facilitated diffusion, in order to move molecules across their membranes. Suggest why endocytosis and exocytosis are used.   *(3 marks)*

Synoptic link

You learnt about the general principles of homeostasis in Topic 29.1, Homeostasis – the key principles. It is also important to understand the concepts of water potential (Topic 2.6, Osmosis in cells), active transport (Topic 1.11, Active moment across cell membranes), and facilitated diffusion (Topic 1.10 Passive movement across cell membranes), when considering how the kidney functions.

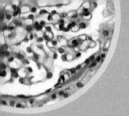

In the previous topic you examined the role of the kidney in the excretion of unwanted and toxic substances. Here, you will look at another function of the kidney – osmoregulation. You will also learn about the hormones that are secreted by the kidney in its role within the endocrine system.

Osmoregulation

Osmoregulation is the regulation of the water potential of body fluids. The kidneys govern this aspect of homeostasis. The kidneys control how much water is expelled from the body and how much is reabsorbed back into the blood.

Approximately $125\,cm^3$ of water enters nephron tubules in the kidneys each minute. The majority of this is reabsorbed back into the blood from the proximal convoluted tubule, as you learnt in Topic 31.1, Kidney structure and excretory function. The next section of a nephron, the loop of Henle, creates a very low water potential in the kidney medulla outside the nephron. This enables more water to be reabsorbed from the collecting duct (the final section of the nephron tubule) if necessary.

The loop of Henle

The loop of Henle has a descending limb, which dips down into the kidney medulla, and an ascending limb, which rises back into the kidney cortex. It is arranged as a hairpin **countercurrent multiplier**. What this means is that the descending and ascending limbs pass close to each other, with their fluids flowing in opposite directions. This enables exchange of content between the two limbs.

Ions and water are transferred through the following processes:

● Water diffuses out of the descending limb by osmosis and is reabsorbed by capillaries.

● At the bottom of the loop, ion concentration is very high and water potential is very low.

● Sodium and chloride ions passively diffuse out of the nephron at the base of the ascending limb and into the tissue fluid in the medulla.

● Water is unable to leave the ascending limb because its wall is impermeable to water.

● The filtrate that reaches the distal convoluted tubule has a high water potential (and low solute concentration).

● A low water potential has been created in the medulla, through which the collecting ducts pass.

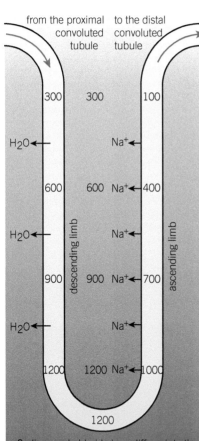

• Sodium and chloride ions diffuse into the descending limb down a concentration gradient.
• Further up the ascending limb, sodium and chloride ions are actively transported into the tissue fluid.

▲ Figure 1 Solute concentrations in the loop of Henle

The collecting duct

The fluid in the nephron flows from the loop of Henle, via the distal convoluted tubule (DCT), to the collecting duct. Active transport in the DCT adjusts the concentrations of ions. The fluid entering the collecting duct has a high water potential and the surrounding kidney medulla tissue has a low water potential. Water will naturally pass by osmosis from the collecting duct into the medulla tissue and then into blood capillaries. However, the permeability of the collecting duct can be altered to control how much water is reabsorbed into the blood.

Homeostatic control of water reabsorption

The permeability of the collecting duct walls is changed by the level of antidiuretic hormone (ADH) in the blood. When the body needs to reabsorb extra water, more ADH is produced and released. This increases the permeability of the walls of the collecting duct, resulting in an increase in urine concentration and a reduction in volume. Lower levels of ADH reduce permeability, meaning that less water is reabsorbed and the urine that is produced is dilute. This is another example of negative feedback. The water potential of blood is maintained close to a set point. The negative feedback mechanism works in the following way.

Decrease in blood water potential below the set point:

- *Osmoreceptors* in the *hypothalamus* detect the low water potential.
- Neurosecretory cells (specialised nerve cells) in the hypothalamus are stimulated to release *ADH* from the *posterior pituitary gland*.
- ADH travels through the blood to the kidneys.
- ADH binds to receptors on the cell surface membrane of cells in the collecting duct wall.
- The concentration of cyclic AMP (*cAMP*) in these cells is increased.
- cAMP acts as a second messenger and causes *aquaporins* to be inserted into the membrane of cells in the collecting duct wall.
- Aquaporins are membrane-spanning channel proteins that increase the permeability of the collecting duct wall. They allow water to diffuse through but prevent the passage of ions.
- More water is reabsorbed from the collecting duct by osmosis.
- Urine with a high solute concentration is produced.
- The water potential of the blood is increased.

Increase in blood water potential above the set point:

The same system responds if blood water potential rises above the set point. Less water is reabsorbed into the blood from the collecting duct. A large volume of dilute urine is produced.

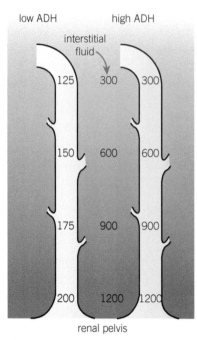

low ADH   high ADH

interstitial fluid

▲ Figure 3 *Solute concentration in the collecting duct in response to changing ADH*

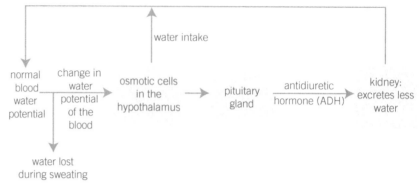

▲ Figure 2 *A summary of osmoregulation as a homeostatic process*

Endocrine function

The kidneys take part in homeostatic mechanisms other than osmoregulation by producing the hormone erythropoietin (EPO) and the enzyme renin (angiotensinogenase). EPO is secreted when oxygen levels in the blood are low. It stimulates the bone marrow to produce more red blood cells (erythrocytes). Renin is secreted by the kidneys in response to low blood volume and catalyses the conversion of angiotensinogen to the hormone angiotensin. Angiotensin increases blood pressure by constricting blood vessels, stimulating more ADH to be produced, and activating the thirst reflex in the hypothalamus.

Summary questions

1 Identify a receptor, a controller, and an effector in Figure 2.
   (3 marks)

2 Describe how renin raises blood pressure.   *(3 marks)*

3 Study the graph below, which shows the relationship between thirst and blood plasma solute concentration.
   a Describe and explain the relationship.
   b State and explain how ADH concentration would change with plasma solute concentration.   *(5 marks)*

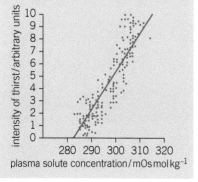

Diuretics – treating, cheating, and energy-depleting?

A *diuretic* is a substance that increases the volume of urine produced by an individual. There are many examples, which produce their effects through a variety of different mechanisms.

Diuretics have many medical uses. High blood pressure (hypertension) can be treated with thiazide diuretics. Thiazides target cells in the distal convoluted tubule wall. They inhibit sodium–chloride ion co-transport, which leads to more water being retained in the urine.

Some athletes use diuretics to cheat by invalidating their drug tests. By raising their urine volume, athletes can dilute performance-enhancing drugs and their metabolites. The diuretics act as masking agents.

Alcohol acts as a diuretic by inhibiting ADH production. Hangovers are not only a result of alcohol toxicity – they are also caused by the effect on brain cells of blood with a high solute concentration.

1 Suggest why thiazide diuretics result in a higher volume of urine being excreted.
2 Explain how drinking plenty of water after heavy alcohol consumption can prevent a hangover the next morning.

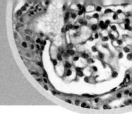

You looked at how a healthy kidney works in the previous two topics. Here, you will consider why a kidney might stop functioning correctly, and you will learn about the treatments that are available to remedy kidney failure.

Kidney failure

Causes

Kidney failure is either acute or chronic and can occur for several reasons:

Acute kidney failure
- bacterial infection, for example, with *E. coli* O157
- kidney stones (or another problem that results in a blockage within the organ)
- medication side effects.

Chronic kidney failure
- uncontrolled diabetes mellitus, either type 1 and 2
- hypertension (high blood pressure)
- certain inherited diseases, for example, polycystic kidney disease.

Diagnosis

A reduction in urine volume is symptomatic of kidney failure. A person's urine may be bloody or cloudy. Another warning sign is *oedema*, which is the accumulation of fluid in tissues. The feet, hands, and areas around the eyes appear puffy or swollen as a result.

Kidney failure is diagnosed by analysing the composition of blood or urine samples in the laboratory. For example, *erythrocytes* (red blood cells), *leucocytes* (white blood cells), and proteins may be present in urine samples. A diabetic patient or a person with high blood pressure will have protein levels in their urine checked at least once every year.

Blood samples with a high concentration of creatinine are indicative of kidney failure. Creatinine is a waste substance that would normally be filtered from the blood by a healthy kidney. Creatinine levels increase as kidney disease progresses.

Structural abnormalities or blockages in the kidneys can be identified through the use of ultrasound or CT scans.

Consequences

Kidney failure prevents the body from removing excess water and waste products, such as urea and excess salts, from the blood. Toxins begin to accumulate in tissues. This leads to death if untreated.

Successful treatment of acute renal failure will enable kidney function to return to normal over time. Chronic kidney damage, if not diagnosed at an early stage, is often irreversible.

Study tip

Remember that an *acute* condition has a quick onset, but often a fast recovery time. A *chronic* condition tends to develop more slowly, but usually lasts a long time.

Synoptic link

You looked at the differences between acute and chronic diseases in Topic 15.1, Pollutants and lung disease. In Topic 30.2, Diabetes, you learnt about diabetes mellitus.

A diseased kidney is likely to produce more of the enzyme renin (see Topic 31.2) than is a healthy kidney. Renin is produced in response to low blood pressure. Low blood pressure in the glomerulus of a poorly functioning kidney is detected. The kidney responds by secreting more renin into the blood plasma. This can lead to *hypertension* (high blood pressure) throughout the body.

Erythropoietin (EPO) production may be increased in a kidney with cysts or cancer. However, the normal response in a diseased kidney is for less EPO production, which leads to anaemia. Patients can be given genetically engineered EPO (RhEPO) to offset the natural reduction caused by kidney disease.

Research has shown that cardiovascular disease is linked with the development of kidney disease. Conversely, kidney disease can also trigger cardiovascular disease. This occurs because blood pressure rises as a result of the increase in renin production associated with kidney disease, which causes hypertension and damage to the endothelium in arteries.

> **Synoptic link**
>
> You learnt about the illicit use of RhEPO to enhance athletic performance in Topic 17.4, Enhancing athletic performance.

Treatment of kidney failure

Two treatments are available for kidney failure: **dialysis** or transplant surgery.

Dialysis

Dialysis is a process that artificially filters a patient's blood. It relies on a partially permeable membrane that separates the patient's blood from a dialysis fluid, which is designed to match the composition of body fluids. The dialysis membrane allows the exchange of substances between the blood and dialysis fluid. Any molecules that are in excess in the blood diffuse across into the dialysis fluid. Any substances that the blood lacks diffuse across the membrane in the other direction, from the fluid. As is the case with a real kidney, blood cells and almost all proteins are too large to pass out of the blood across the membrane.

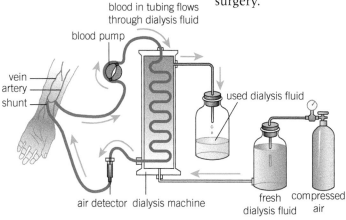

▲ Figure 1 *A diagrammatic representation of a haemodialysis machine*

In *haemodialysis*, blood is passed from the patient's vein into a dialysis machine, where exchange occurs across a partially permeable membrane. Heparin is added to prevent blood clotting. This treatment is usually performed three times per week at a clinic, although some patients are able to carry it out at home. Each session takes several hours.

In *peritoneal dialysis*, the patient's own abdominal membrane (peritoneum) acts as the dialysis membrane. A surgeon implants a tube in the patient's abdomen. This enables the dialysis fluid to be pumped into the patient's body cavity, and exchange can happen across their peritoneal membrane. The procedure must be performed several times per day. The patient is able to walk around and work while the dialysis happens, which is an advantage of this form of treatment. However, the dialysis needs to be carried out daily rather than a few times per week.

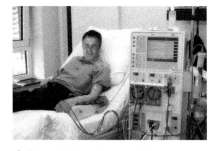

▲ Figure 2 *A patient receiving dialysis treatment*

Transplant surgery

Dialysis helps patients to feel better and live longer, but it does not provide a cure for kidney failure. Kidney transplant surgery can remove the need for dialysis. A kidney can be donated to a patient from a living relative or from someone who has died. The suitability and compatibility of a donor is assessed based on which antigens they have on their red blood cells and in their haplotype. The antigens should be a close match to those of the patient. The patient's original kidneys are left in place unless they are cancerous or likely to cause infection.

The surgery required to implant a new kidney is complex. The new organ must be attached to both the blood supply and to the bladder. Although ideally the donor organ will be a close genetic match to the recipient, the patient's immune system will still recognise foreign antigens on the organ and begin an immune response. Patients are therefore given immunosuppressant drugs to prevent organ rejection.

The future of transplant surgery

In the future, it may be possible to overcome the rejection of transplanted kidneys by using kidneys grown from a patient's own stem cells. This could be achieved through the use of induced pluripotent stem cells (iPSCs). iPSCs are stem cells that have been reset from differentiated adult stem cells.

Embryonic stem cells could also be used to clone organs. In 2013, scientists were able to grow a small amount of human kidney tissue from embryonic stem cells.

The potential development of transplantable organs from stem cells is an example of therapeutic (non-reproductive) cloning. This is different to reproductive cloning, in which whole organisms are cloned from stem cells. Research into human reproductive cloning is illegal in most countries.

▼ Table 1 *A comparison of the two forms of dialysis*

	Haemodialysis	Peritoneal dialysis
Advantages	• trained professionals are present for hospital dialysis • daily dialysis not required	• can be done at home • diet can contain more fluids and salts than the diet recommended for haemodialysis
Disadvantages	• if hospital dialysis is given, patients must travel	• self-administered dialysis requires attention to hygiene during exchanges • infection is a risk • requires daily dialysis

Synoptic link

You learnt about haplotypes in Topic 23.2, Gene mutations.

Synoptic link

You have also already learnt about two important aspects of transplant surgery – erythrocyte and haplotype antigens (Topic 23.3, Codominance) and the use of stem cells for therapeutic cloning (Topic 8.4, Stem cells).

Summary questions

1 Describe four symptoms of kidney failure. (4 marks)

2 Describe how therapeutic cloning may be used to improve kidney transplantation in the future. (2 marks)

3 Evaluate the costs and benefits of dialysis and transplant surgery as treatments for kidney failure. (6 marks)

Practice questions

1 Which of the following is an adaptation / are adaptations in a nephron to enable ultrafiltration?

   Statement 1: Cells in capillary walls are separated by narrow gaps.

   Statement 2: The endothelium cells of the capillaries have finger-like projections called podocytes.

   Statement 3: Basement membranes allow large proteins to pass through into the nephron filtrate.

   A 1, 2 and 3

   B Only 1 and 2

   C Only 2 and 3

   D Only 1   *(1 mark)*

2 In which region of a nephron does the majority of selective reabsorption occur?

   A Proximal convoluted tubule.

   B Loop of Henle.

   C Distal convoluted tubule.

   D Collecting duct.   *(1 mark)*

3 Which of the following is / are true of antidiuretic hormone?

   1 It is released from the posterior pituitary gland.

   2 It binds to cells in the collecting duct wall.

   3 It results in aquaporins being inserted into the membrane of cells in the collecting duct wall.

   A 1, 2 and 3   B Only 1 and 2

   C Only 2 and 3   D Only 1   *(1 mark)*

4 Which of the following statements is true of peritoneal dialysis?

   A It is usually performed three times per week.

   B A patient's blood is passed through a dialysis machine and returned to their body.

   C It must be performed every day.

   D It must be carried out by medical professionals.   *(1 mark)*

5 The kidney is responsible for the removal of nitrogenous waste from blood plasma. The diagram shows a section through the cortex of the kidney.

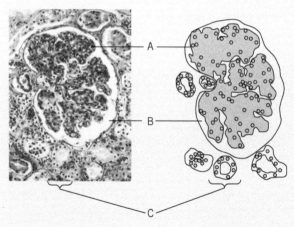

a Identify the structures labelled A to C.
                                              (3 marks)

b Describe how the process of ultrafiltration takes place in the kidney and describe how the structures in the kidney are adapted to allow ultrafiltration to occur.   *(6 marks)*

c One method of assessing kidney function is to determine the glomerular filtration rate (GFR). The GFR is the volume of fluid that is filtered out of the blood per minute. Inulin is a polysaccharide that is used to determine the GFR.

   Inulin is not removed completely from blood plasma by ultrafiltration, and it is not reabsorbed or secreted into the kidney tubules. Inulin is given intravenously. The blood plasma concentration of inulin is then measured and urine samples are collected. The GFR can be determined from the concentration of inulin in a known volume of urine. An inulin molecule is composed of a chain of 32 monosaccharide units.

   Inulin's structure is shown in the diagram.

(i) Name the bond joining the monosaccharide units in a molecule of inulin. (*1 mark*)

(ii) Suggest two properties of the inulin molecule that make it suitable for use in determining the GFR. (*2 marks*)

d Using inulin to determine the GFR is expensive. As an alternative, the GFR can be estimated from the blood plasma concentration of a substance called creatine.

Creatine is a breakdown product of creatine phosphate, a substance found in muscle cells.

The graph shows the relationship between plasma creatine concentration and the GFR.

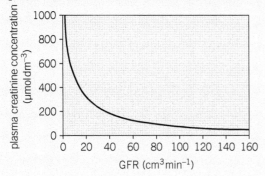

(i) In normal kidney function, the GFR range is between 90 and 150 cm³ min⁻¹. Estimate the normal range of plasma creatine concentration using the graph. (*2 marks*)

(ii) Suggest why measuring the plasma creatine concentration can only give an estimate of the GFR. (*2 marks*)

e When kidney function is severely impaired, renal dialysis may be necessary.

Outline the advantages and disadvantages of haemodialysis in the treatment of kidney disease. (*4 marks*)

OCR F225 2010

6 The diagram shown here is of a nephron.

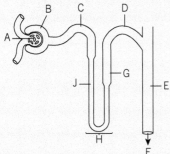

Copy and complete the table by inserting the correct letter for each description.

Description of region of nephron	Letter
region where numbers of aquaporins vary in tubule wall cells	
region where tubule wall cells are modified to provide filtration slits	
region where aquaporins and glucose transport proteins are present in tubule cell walls	
region where aquaporins are always present in the tubule wall cells but no other transport proteins are present	
region where no aquaporins are present in the tubule cell walls	

(*5 marks*)

OCR F225 2011

7 Blood pressure is regulated in several ways. One mechanism involves renin (angiotensinogenase). The role of renin in regulating blood pressure is outlined in the flow diagram.

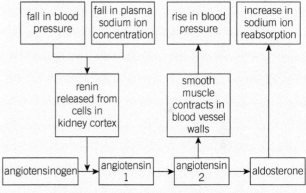

a Using the information in the figure, explain why control of blood pressure is an example of homeostasis. (*6 marks*)

b Aldosterone is a steroid hormone that is produced by the adrenal cortex. Release of aldosterone stimulates an increase in the reabsorption of sodium ions in the kidney.

(i) Suggest the precise location of the receptors for aldosterone in the kidney. (*2 marks*)

(ii) Explain why an increase in the reabsorption of sodium ions causes an increase in blood pressure. (*2 marks*)

OCR F225 2012

The diagram shows a generalised plant cell with the organelles labelled P – U. The organelles are not drawn to scale.

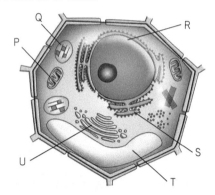

1   Which letter correctly identifies an organelle which is surrounded by the tonoplast?

A   P

B   Q

C   T

D   U   (*1 mark*)

2   Which of the following in the diagram identifies some of the organelles that would be present in a companion cell in phloem tissue?

A   P, Q, R, S

B   P, R, S, T

C   P and S only

D   R and T only   (*1 mark*)

3   Which organelles in Fig.1 would be visible using a light microscope?

A   P, Q, R, and T

B   R and T only

C   Q and T only

D   Q, R, and T only.   (*1 mark*)

4   Which of the following statements could be applied to a phase 1 clinical trial?

A   A placebo is used

B   The group is randomised as to whether they get the drug or the placebo

C   The trial could involve 20 people

D   The trial compares the new drug with existing treatments.   (*1 mark*)

5   Aspirin has been found to have an effect on the rate of transpiration. Which of the following plants would not be suitable to investigate the effect of aspirin on transpiration?

A   *Salix cinerea* (grey willow)

B   *Pelargonium* sp. (geranium)

C   *Ligustrum ovalifolium* (privet)

D   *Solanum tuberosum* (potato)   (*1 mark*)

6   The graph shows the changes in volume achieved during one inhalation and one forced exhalation.

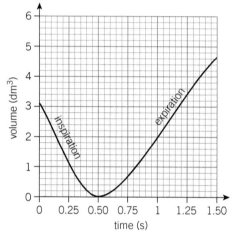

Which of the following statements is a correct interpretation of the graph?

A   The tidal volume is 3.5 dm^3

B   The vital capacity is 3.5 dm^3

C   The FEV_1 is 4.75 dm^3s^{-1}

D   The FEV_1 is 4.75 dm^3 (*1 mark*)

7   Measurements of lung volumes can be used in the diagnosis of lung disease. An obstructive lung disease is diagnosed when FEV_1 is less than 70% of the forced vital capacity (FVC).

The table gives data on FEV_1 and FVC for four patients, W, X, Y, and Z.

Which values are likely to be correct for a patient with emphysema?

Patient	FEV_1 (a.u)	FVC (a.u)
W	3.3	4.7
X	4.7	3.2
Y	4.2	4.7
Z	3.2	4.7

A   W   C   Y

B   X   D   Z   (*1 mark*)

8 The image here shows a breast produced from a routine screening test for breast cancer.

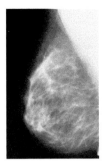

Which of the following methods was used to produce this image?

A X-ray

B MRI

C Biopsy

D Ultrasound   (*1 mark*)

9 The diagram shown here is of the HIV virus. HIV virus particles escape from infected cells by budding off the cell surface membrane.

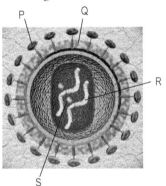

Donated blood is screened for the presence of HIV antibodies. Which letter corresponds to the part of the virus responsible for triggering the production of antibodies?

A P

B Q

C R

D S   (*1 mark*)

10 Using the diagram, which components of the HIV virus consist of molecules which contain phosphorus?

A P and Q

B Q and R

C R and S

D S and P   (*1 mark*)

11 The table shows the sequence of amino acids in three different animal species, X, Y, and Z. Each letter represents a different amino acid in part of the haemoglobin molecule.

Species	Amino acid sequence
X	A E E K B B V T A L W A K V N V E.....D S...S
Y	A E E K S B V T B L W A K V N V D.....D S...S
Z	B E E K S B V T B L W B K V N V E.....E B....T

Which of the following statements is/are true?

Statement 1: species X is most closely related to species Y.

Statement 2: X, Y, and Z are in the same domain.

Statement 3: the primary sequence of amino acids in species X and Y is the same.

A 1,2 and 3

B Only 1 and 2

C Only 2 and 3

D Only 1   (*1 mark*)

12 The diagram shown here represents the mitotic cell cycle.

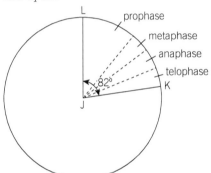

Which of the following statements is/are true?

Statement 1: the cell is a eukaryotic cell.

Statement 2: the cell could be a palisade mesophyll cell.

Statement 3: the DNA content will double in prophase.

A 1,2 and 3

B Only 1 and 2

C Only 2 and 3

D Only 1   (*1 mark*)

13 The time taken each for cell cycle was 48 hours. Using the information given in the figure, the number of hours spent in interphase was:

A 18.5

B 11

C 37

D 17   (*1 mark*)

14 The image shows a killer T lymphocyte (cell T) attacking a cancer cell (cell V).

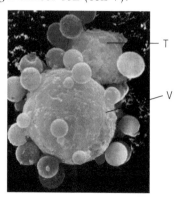

Which of the following is a correct description of the event taking place in the image.

A   Phagocytosis triggered by a specific immune response.

B   Phagocytosis triggered by a non-specific immune response.

C   Apoptosis triggered by a specific immune response.

D   Apoptosis triggered by a non-specific immune response.   *(1 mark)*

15 One of the factors which affects the rate of entry of substances into cells is the surface area.

a   (i)   Assuming that a cell is spherical in shape, calculate the surface area of a cell which has a diameter of 7 μm. Give your answer to 1 significant figure.   *(2 marks)*

    (ii) Human erythrocytes have a diameter of approximately 7 μm. How will the surface area of an erythrocyte compare to the value you calculated in part (a)(i)? Explain your answer.   *(1 mark)*

b   In blood samples, potassium ions are found both inside erythrocytes and in the blood plasma. The concentration of potassium ions is 20 times greater in the cell cytoplasm than in the plasma.

    Suggest how potassium ions enter erythrocytes. Give a reason for your suggestion.   *(2 marks)*

c   Samples of blood taken for the measurement of potassium ion concentration must be stored correctly.

Suggest why each of the following could result in a **false high** reading of plasma potassium ion concentration:

(i)   prolonged storage of blood at low temperatures   *(2 marks)*

(ii) the presence of water in the blood collection tube   *(2 marks)*

16 All cells require folic acid in order to synthesise nucleic acids and amino acids such as methionine.

a   Copy and complete the following table on the similarities and differences between amino acids and nucleic acids by inserting either a tick (✔) or a cross (✗) in the appropriate column.

Feature	Nucleic acid	Amino acid
Polymer		
Contains nitrogen		
Contains phosphorus		

(3 marks)

b   In bacterial cells, folic acid is synthesised using an enzyme called DHPS.

• The substrate for this enzyme is PABA.

• Antibiotic drugs known as sulfonamides act as inhibitors of DHPS.

The graph shows the effect of increasing the concentration of PABA on the activity of DHPS.

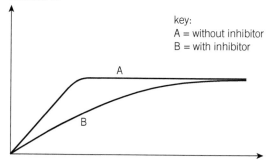

key:
A = without inhibitor
B = with inhibitor

(i)   Using the information in the graph, explain how sulfonamides act as inhibitors of DHPS   *(3 marks)*

(ii) Suggest why sulfonamides have no effect on human cells   *(1 mark)*

c   Sulfonamide antibiotics have a bacteriostatic effect on bacterial cultures.

(i)   what is meant by the term 'bacteriostatic'   *(1 mark)*

(ii) Suggest why sulfonamides are bacteriostatic in their action.   *(2 marks)*

17 a The diagram shows a suggested evolutionary relationship between bears, raccoons and two species of panda, the giant panda, *Ailuropoda melanoleuca*, and the red panda *Ailurus fulgens*.

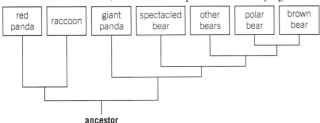

ancestor

(i) Using the diagram, name the two species that share the most recent common ancestor *(1 mark)*

(ii) State whether pandas form a distinct taxonomic group. Use information from the diagram to justify your answer. *(1 mark)*

b The evolutionary relationship of the giant panda and red panda has been a matter of scientific debate for many years. It was hoped that molecular evidence would provide a definite answer.

Some of the results of scientific studies are listed in the table.

Year	Protein sequenced	Conclusion
1985	albumen	Giant panda is more closely related to bears, and red panda is more closely related to raccoons, than pandas are to each other.
1986	haemoglobin	Giant and red panda are more closely related to each other than the giant panda is to bears or the red panda is to raccoons.
1993	cytochrome c	Giant panda is more closely related to bears, and red panda is more closely related to raccoons, than pandas are to each other.

Comment on what the results in the table show about the nature of scientific knowledge and the role of the scientific community in validating new knowledge. *(2 marks)*

Research on another protein from the giant panda was carried out in 2008. This protein, called crystallin, is found in the lens of the eye, and has a sequence that has been highly conserved in all mammals.

c The panda crystallin protein obtained was 175 amino acids long.

Explain why a protein that is 175 amino acids long is coded for by 528 base pairs of DNA *(3 marks)*

(OCR F215 2012)

18 A student was investigating the distribution of stomata on a leaf. The image shows the appearance of stomata on leaf as seen under a light microscope.

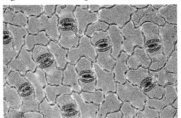

a Explain how a light microscope and graticule could be used to measure the number of stomata per mm^2 on the leaf (the stomatal density). *(3 marks)*

b The student sampled the upper and lower surface of a single leaf and obtained the following results:

Upper surface density (stomata mm^{-2})	Lower surface density (stomata mm^{-2})
43	42
56	7
3	4
56	57
5	51
5	46
45	3
Mean = 30.4	Mean = 30.0
s.d = 24.9	

(i) Calculate the standard deviation for the lower surface density. *(2 marks)*

(ii) Explain how the evidence supports the conclusions that the leaf was taken from a monocotyledon plant. *(1 mark)*

c Describe how changes in the water potential of guard cells bring about the opening of stomata. *(3 marks)*

AS Depth of knowledge questions

1 Dietary reference values are given for a number of nutrients. Some of these values change during pregnancy. The table shows how DRV change for different populations of women.

Nutrient	Women (19–50)	Pregnant women
Protein (g)	45.0	51.0
Calcium (mg)	700	700
Iron (mg)	14.8	14.8
Vitamin A(µg)	600	700

a (i) Calculate the percentage increase in the DRV value for protein in pregnancy. *(2 marks)*

(ii) Explain why the DRV value for protein changes in pregnancy. *(2 marks)*

b Both meat and fish are good sources of protein. The table compares some of the nutrients found in one type of fish and one type of meat.

Nutrient	Mass per 100g	
	Mackerel (fish)	Steak (meat)
Protein (g)	18.7	30.9
Calcium (mg)	11.0	15.0
Iron (mg)	0.8	3.0
Vitamin A(µg)	45.0	0.0

(i) Outline how the presence of protein could be detected in a sample of meat or fish. *(3 marks)*

(ii) A pregnant woman decides to include fish rather than meat in her diet as a source of protein.

*Using the information in the Table, outline the roles of vitamins and minerals in pregnancy and evaluate the consequence of including fish rather than meat as a protein source. *(6 marks)*

2 In 2009, approximately 1.7 million people worldwide died from tuberculosis (TB). This figure includes 380 000 people with HIV infection.

a Describe how TB is transmitted from one person to another. *(2 marks)*

b The graph shows the estimated mean percentage change in the incidence of TB for different groups of countries, from 1997 to 2006.

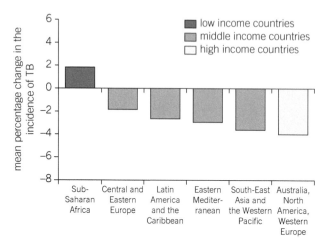

(i) Suggest why the mean percentage change in the incidence of TB is given as an estimate and not as a direct measurement. *(1 mark)*

(ii) What conclusions can be drawn about the changes in incidence of TB from the information in the graph? *(3 marks)*

c The photo shows an agar plate on which colonies of the bacterium responsible for causing TB are growing

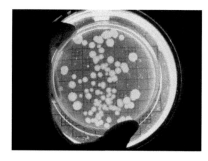

(i) State the genus to which the bacterium which causes TB belongs. *(1 mark)*

(ii) How could you assess that the agar plate shown in the photo contained a pure culture of the TB bacterium? *(2 marks)*

(iii) The agar plate shown in the photo has been produced from a serial dilution of a culture of the TB bacteria.

On a similar plate:

- the dilution used was 10^{-5}.
- 100 µl of the culture was spread onto the surface of the agar plate.
- 90 colonies were found to be growing on the agar plate.

Calculate how many cells were growing in the original culture. Show your working (give your answer in cells cm^{-3}). *(2 marks)*

3 A student used a potometer to investigate the effect of leaf area on the rate of transpiration. This apparatus is shown in Fig. 3.1.

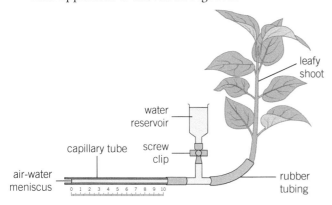

The student presented the results of their investigation in a table as shown.

Number of leaves present on shoot attached to potometer	Mean rate of bubble movement
0	7
2	28
4	49
6	73
8	92

a **(i)** State what information the student has not included in their table of results *(2 marks)*

(ii) Describe and explain the data shown by the students results. *(3 marks)*

b As part of the evaluation of the experiment, the student wrote the following statements:

1 One limitation is that the leaves were not all the same size

2 I assembled the potometer under water and the leaves got wet.

3 During my investigation the sun came out and the lab warmed up very quickly.

For each statement, explain why this may affect the results **and** suggest how the student could improve the investigation. *(6 marks)*

c The top photomicrograph shows a transverse section through a leaf from a hydrophyte. The bottom photomicrograph shows a transverse section through a leaf from a xerophyte.

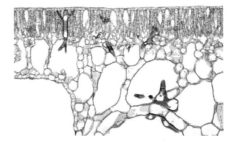

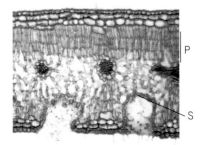

The region labelled P is the palisade mesophyll tissue. S indicates the location of stomata.

*Using the information in both photomicrographs, explain how features of hydrophytes and xerophytes adapt them to survive in extremes of water availability. *(6 marks)*

4 In the United Kingdom, all children are tested shortly after birth for conditions known as inborn error of metabolism (IEMs). IEMs occur due to non-functioning of an enzyme in a metabolic pathway. One of the most common IEMs is Phenylketonuria (PKU).

In PKU, the enzyme which converts the amino acid phenylalanine to the amino acid tyrosine does not function.

The diagram shows a simplification of this pathway.

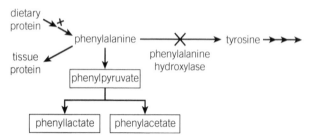

a (i)   Suggest how the structure of tyrosine differs from that of phenylalanine.
          (1 mark)

   (ii) State one source of tyrosine **other than** from the conversion of phenylalanine.
          (1 mark)

b Phenylalanine is incorporated into tissue proteins on the ribosomes. Explain the role of RNA in the assembly of amino acids such as phenylalanine into proteins on ribosomes. *(4 marks)*

c The presence of PAH (phenylalanine hydroxylase) in human tissue samples can be detected using a test called ELISA.

   The diagram outlines how an ELISA detects the presence of PAH.

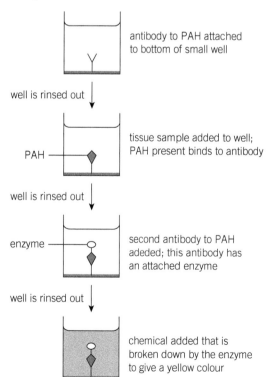

(i)   Describe the general structure of antibodies such as the attached antibody. *(3 marks)*

(ii) Explain why both the attached and second antibodies are specific to PAH. *(1 mark)*

d The photo shows the results from an ELISA test. Each circle (well) represents one sample that has been tested. The intensity of the colour can be measured using an optical reader similar to a colorimeter. There is a positive correlation between the colour intensity and the PAH concentration.

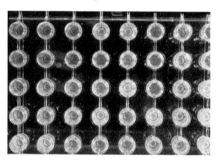

Outline how you would use the results from an ELISA to determine the exact concentration of enzyme present in a sample. *(3 marks)*

e Several different mutations can occur in the gene for PAH which result in a child being born with PKU. A mutation is a change in the sequence of bases in DNA. One mutation replaces the amino acid arginine with the amino acid tryptophan in the PAH enzyme.

   (i)   Explain why a change in the amino acid sequence of PAH results in an enzyme which is present but does not function. *(3 marks)*

   (ii) Suggest why an ELISA test on samples with the mutated enzyme could give a negative result. *(1 mark)*

5 The diagram shows a vertical section of the heart to show the position of certain structures.

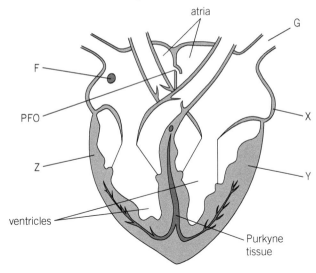

atria

G

F

PFO

X

Z

Y

ventricles

Purkyne tissue

a Name the structure F and G *(2 marks)*

b The statements below were made to a group of students. Explain why each statement is true.

(i) The difference in thickness of the walls of the chambers, as shown by the letters X, Y, and Z is related to the functions of the different chambers. *(3 marks)*

(ii) Without the Purkyne tissue, blood would not be pumped out of the heart efficiently. *(2 marks)*

c Recent research has shown that there may be a link between migraines (severe headaches) and the minor heart defect PFO (patent foramen ovale).

In PFO, the small flap shown in the diagram fails to close completely at birth. Suggest how PFO might lead to a migraine. *(3 marks)*

6 One group of pathogens that can cause diseases in humans are the fungi.

The diagram shows part of a fungal filament (hypha).

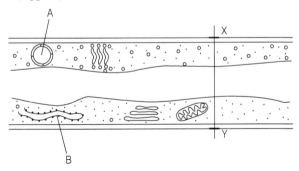

A

X

Y

B

a Name two features visible in the diagram which show that this organism is eukaryotic. Do not include those features labelled A and B. *(2 marks)*

b Assuming fungal filaments are cylindrical in shape, calculate the surface area of a section of a filament 4 mm in length if the distance XY is 12 μm. Show your working (giving your answer in μm²). *(2 marks)*

c The fungus *Trichophyton* is responsible for the condition known as athletes foot in humans. Protease enzymes are secreted by the filaments. The enzymes break down keratin which is a fibrous protein present in skin.

(i) Give **two** reasons why the protease enzymes synthesised by *Trichophyton* do not break down intracellular proteins in the fungal filament *(2 marks)*

(ii) Outline how the protease enzymes leave the filament and come into contact with human skin. *(2 marks)*

d The protease enzymes produced by *Trichophyton* have been shown to trigger an allergic response.

Complete the passage below which describes how an allergic response occurs.

In the first allergic immune response to the enzyme .... lymphocytes respond by producing .... antibodies. These antibodies attach to receptors on .... cells. This process is called sensitisation.

On a second encounter with the enzyme, the enzyme molecule cross links with the antibodies and stimulates the release of the chemical .... This chemical triggers the inflammation response causing blood vessels near the site to become more.................... leading to the area becoming .... in colour. *(6 marks)*

e Asthma is also an allergic response.

The diagram on the following page shows drawing made from cross sections of the upper bronchioles of a non-asthmatic (X) and an asthmatic (Y). The sections were drawn using observations made using a light microscope.

Upper bronchioles normally have an epithelium with a few scattered goblet cells.

X – a non-asthmatic

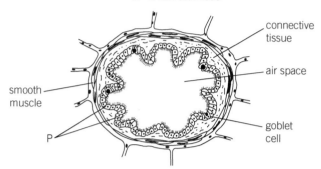

Labels: connective tissue, air space, goblet cell, smooth muscle, P

Y – an asthmatic

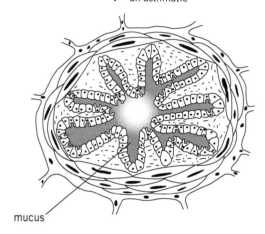

Label: mucus

(i) Describe the function of the cells labelled P on the top diagram in the gas exchange system. *(3 marks)*

(ii) Use the information given in the diagrams to explain the following observations made on the bronchioles of an asthmatic during an asthma attack:

the bronchioles fill with mucus ....

the cross sectional area of the air spaces in the bronchioles decreases...................
(3 marks)

(iii) Describe the symptoms of asthma and outline the changes that occur in the airways during an asthma attack *(4 marks)*

7   In an experiment to measure the rate of diffusion, a student placed tubes of agar jelly containing an indicator into dilute hydrochloric acid. The indicator changes from pink to colourless in acidic conditions.

The student used cubes of different sizes and recorded the time taken for the pink colour in each cube to disappear completely.

The students results are recorded in the table

Length of side of cube (mm)	Surface area of cube (mm²)	Volume of cube (mm³)	Surface area to volume ratio	Time taken for pink colour to disappear (s)	Rate of diffusion (mm s⁻¹)
2	24	8	3.0:1	50	0.020
5	150	125	1.2:1	120	0.021
10	600	1000		300	0.017
20	2400	8000	0.3:1	700	0.014
30	5400	27000	0.2:1	1200	0.013

a (i) Calculate the surface area to volume ration of the cube with 10mm sides.

Show your working. *(2 marks)*

(ii) **Using the data in the table**, describe the relationship between the rate of diffusion and the surface area to volume ratio. *(2 marks)*

(iii) Explain the significance of the relationship between the rate of diffusion and the surface area to volume ratio for large plants. *(2 marks)*

b   Another student used the same raw data obtained in the experiment but calculated a different rate of diffusion for each cube.

This student's results are shown in the Table

Length of side of cube (mm)	Time taken for pink colour to disappear (s)	Rate of diffusion (mm s⁻¹)
2	50	0.040
5	120	0.042
10	300	0.033
20	700	0.029
30	1200	0.025

In this students table, the calculation of the rate of diffusion is incorrect.

(i) Suggest the method used to calculate the rate of diffusion. *(1 mark)*

(ii) State why the method in **(b)(i)** is not correct. *(1 mark)*

A level Paper 1 practice questions

1 One role of ATP in cells is to act as a monomer in the synthesis of nucleic acids. Which of the following statements is correct about ATP and nucleic acid synthesis:

   A It is incorporated into DNA in the nucleus

   B It is incorporated into mRNA in the nucleus

   C It is incorporated into DNA at the ribosome

   D It is incorporated in mRNA at the ribosome

   (*1 mark*)

2 The diagram represents the cycling of nitrogen in an ecosystem.

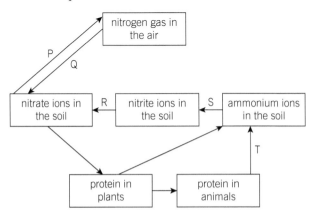

   Which of the following statements correctly identifies the bacteria represented by the letters in the diagram?

   A Q is *Rhizobium* and R is *Nitrobacter*

   B Q is *Rhizobium* and R is *Nitrosomonas*

   C R is *Nitrosomonas* and S is *Nitrobacter*

   D R is *Nitrobacter* and S is *Nitrosomonas*

   (*1 mark*)

3 Which of the following terms correctly describes what is happening during process T?

   A Hydrolysis of peptide bonds and denitrification.

   B Hydrolysis of ester bonds and denitrification.

   C Hydrolysis of peptide bonds and deamination.

   D Hydrolysis of ester bonds and deamination.

   (*1 mark*)

4 Which of the following statements are correct about mitochondria and chloroplasts?

   A They both contain NADP

   B They both contain decarboxylase enzymes

   C They both contain ATP synthase

   D They both contain ribulose bisphosphate

   (*1 mark*)

5 The diagram shows a section through the retina of a human eye.

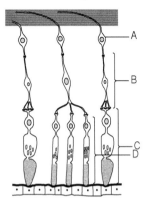

   Which letter is labelling a cell found only in the fovea? (*1 mark*)

6 The photo shows a cross section through a collecting duct in the mammalian kidney.

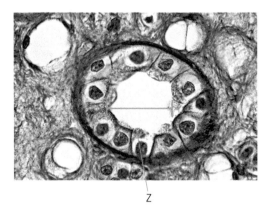

   ×250 magnification

   The diameter of the lumen is shown by the line X – Y. Which of the following is the actual cross sectional area of the lumen in μm^2 ?

   A 4072

   B 226

   C 254

   D 57

   (*1 mark*)

7 Letter Z in the photo indicates the location of the cell surface membrane. Which of the following correctly describes the changes which happen to cells in the wall of the collecting duct?

  A If the water potential in the blood is too low, more aquaporins are inserted into the membrane at Z.

  B If the water potential in the blood is too high, more aquaporins are inserted into the membrane at Z.

  C If the water potential in the blood is too low, ADH binds to receptors on the membrane at Z.

  D If the water potential in the blood is too high, ADH binds to receptors on the membrane at Z. *(1 mark)*

8 Which of the following correctly describes the action of thromboplastin – an enzyme involved in blood clotting?

  A The substrate for thromboplastin is fibrinogen.

  B The product of the reaction catalysed by thromboplastin is fibrin.

  C The substrate for thromboplastin is thrombin.

  D The product of the reaction catalysed by thromboplastin is thrombin. *(1 mark)*

9 The diagram shows the occurrence of the genetic disease haemophilia in one family. The disease is due to a mutation in the gene coding for clotting factor VIII.

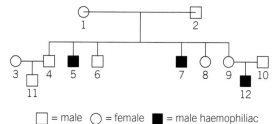

Which of the following statements are correct?

  A Individuals 1 and 2 are both heterozygous for haemophilia.

  B Individuals 1 and 9 are both heterozygous for haemophilia.

  C Individual 8 must be homozygous for the normal allele for factor VIII.

  D Individual 7 inherited a copy of the haemophilia allele from each parent.

                                            (1 mark)

10 The photomicrograph shows a blood smear.

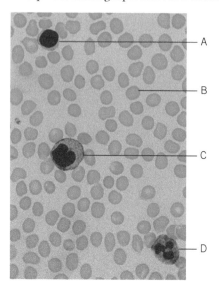

Which of the following statements is/are true?

Statement 1: The blood smear was stained with a differential stain.

Statement 2: Cells such as B transport oxygen and carbon dioxide.

Statement 3: Cells A, C and D are phagocytic cells.

  A 1,2 and 3   **B** Only 1 and 2

  C Only 2 and 3   **D** Only 1   *(1 mark)*

11 The image shown is an electron micrograph of cardiac muscle showing Z-lines and mitochondria **(M)**.

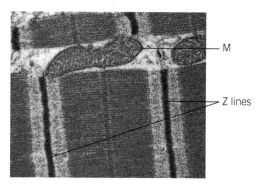

Which of the following statements is/are true?

Statement 1: The image was taken with a transmission electron microscope.

Statement 2: The distance between the z-lines shortens in ventricular systole.

Statement 3: Cardiac muscle is striated.

  A 1,2 and 3   **B** Only 1 and 2

  C Only 2 and 3   **D** Only 1   *(1 mark)*

12 The breakdown of a macromolecule in aerobic respiration is given by the following equation:

$$2C_{18}H_{34}O_2 + 50O_2 \rightarrow 36CO_2 + 34H_2O$$

Which of the following statements is/are true?

Statement 1: The macromolecule forms a polymer with glycerol.

Statement 2: The respiratory quotient (RQ) for respiration of this macromolecule is 0.7.

Statement 3: The macromolecule contains one carbon:carbon double bond.

A 1,2,and 3   B Only 1 and 2

C Only 2 and 3   D Only 1   (*1 mark*)

13 The diagram shows the structure of a glucose molecule.

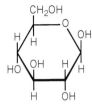

Which of the following statements is/are true?

Statement 1: The molecule polymerises by forming glycosidic bonds.

Statement 2: The molecule is a reducing sugar.

Statement 3: The molecule forms a polymer found in plant cell walls.

A 1,2,and 3   B Only 1 and 2

C Only 2 and 3   D Only 1   (*1 mark*)

14 The diagram shows some of the cells in the root of a dicotyledonous plant.

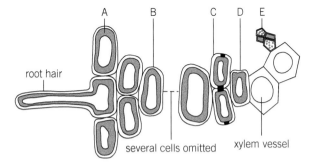

Which of the following statements is/are true?

Statement 1: Cell A is in the epidermis and cell C is in the endodermis.

Statement 2: Water moves from cell B to cell D through the apoplast pathway.

Statement 3: Cell E contains a nucleus and other organelles but not chloroplasts.

A 1,2, and 3   B Only 1 and 2

C Only 2 and 3   D Only 1   (*1 mark*)

15 Cereals such as wheat and rice are important staple crops. One feature of wheat and rice plants is the production of side shoots or 'tillers'. Each of these tillers can go on to branch further, flower, and produce grain.

a   What type of cell division occurs in the production of tillers?   (*1 mark*)

b   Allele 'T' codes for a protein which regulates transcription. Expression of allele T allows stimulation of cell division in the buds which become tillers.

Allele 't' has a 'stop' triplet within its DNA sequence as well as at its ends.

(i)   State what is meant by a stop triplet.   (*2 marks*)

(ii) Describe the effect of a 'stop' triplet within the DNA sequence of allele 't'.   (*3 marks*)

c   A copy of allele **'T'** was introduced into **tt** rice plants by genetic engineering.

*Describe how plants such as rice can be genetically modified to contain a copy of the 'T' allele.   (*6 marks*)

d   The number of tillers per plant and the number of times each tiller branched were recorded for wild type 'TT' plants and for 'tt' plants which had been given a copy of allele 'T' by genetic engineering.

The results are shown in the graph.

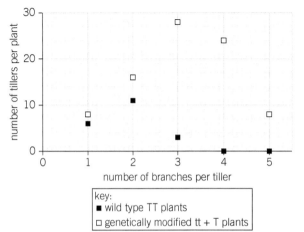

Use the information in the graph to evaluate and explain the effect of the two genotypes on tiller growth.   (*4 marks*) OCR 2805-02 2006

16 Each Winter the UK government recommends that vulnerable members of the public are vaccinated against the influenza (flu) virus.

a Suggest why the influenza vaccine has to be changed each year   (2 marks)

b The graph shows the concentration of antibodies in a patient's blood stream following an influenza vaccination and then infection with the influenza virus.

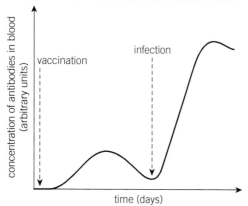

(i) Using the information in the graph, state two differences between the primary and secondary immune response   (2 marks)

(ii) Memory cells are produced when a patient is vaccinated against influenza. Describe the role of these memory cells when an influenza virus enters the body.   (3 marks)

c Tamiflu® is an antiviral drug that can be used to treat influenza patients.

Neuraminidase is an enzyme which is present on the protein coat of the influenza virus. This enzyme is used to break down the host cell membrane and allow the influenza virus to leave the infected cell. Tamiflu® is a neuraminidase inhibitor.

Suggest how Tamiflu® could inhibit neuraminidase.   (2 marks)

17 As blood passes through the kidney it is filtered and the urine formed in the nephron leaves the kidney through the ureter.

The first table shows the concentration of some of the components of blood, glomerular filtrate and urine. The second table shows the presence or absence of erythrocyes in blood, glomerular filtrate, and urine.

Component	Blood (g 100 cm^{-3})	Glomerular filtrate (g 100 cm^{-3})	Urine (g 100 cm^{-3})
glucose	0.10	0.10	0.00
urea	0.03	0.03	1.80
amino acids	0.05	0.05	0.00
large proteins	8.00	0.00	0.00
inorganic ions (total)	0.90	0.90	variable, up to 3.60

Component	Blood	Glomerular filtrate	Urine
erythrocytes	present	absent	absent

a *Explain the changes in fluid composition shown in the two tables.   (6 marks)

b Kidney function can be assessed by measuring the Glomerular Filtration Rate (GFR). GFR is a measure of the rate at which blood is filtered by the kidneys.

The GFR is estimated using the concentration of creatinine in the blood plasma. This compound is produced naturally by the body and is normally filtered from the blood by the kidneys and excreted. A formula is used to obtain a value for GFR that takes into account the various factors that contribute to the concentration of creatinine in the blood.

One factor is body surface area.

Body surface area can be estimated using a nomogram such as that shown in the diagram.

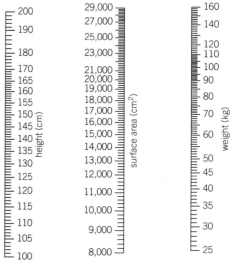

nomograph for estimating body surface area from height and weight

Height and weight are both measured. A straight line is drawn between these two points on the left and right scales. The point where the line crosses the middle scale corresponds to the surface area.

A typical person is assumed to have a body surface area of 1.73 m².

In order to obtain an estimate of GFR (eGFR) for individuals who are smaller or larger than a typical person, the following calculation is performed:

$$eGFR = GFR \times \frac{1.73}{individual's\ body\ surface\ area}$$

A man has a height of 180 cm and a weight of 70 kg.

(i) Use the diagram to calculate his surface are in m². Give your answer to 2 significant figures. (*2 marks*)

(ii) Assuming a GFR of 82 cm³min⁻¹, use your answer to part (i) to calculate the eGFR for this person. Give your answer to the nearest whole number.

(*2 marks*)

(iii) Chronic kidney disease is divided into 5 stages according to the eGFR value. These stages are listed in the table.

CKD stage	eGFR (cm³ min⁻¹)	Effect on kidney
1	greater than 90	little or no damage
2	60–90	some or no damage
3	30–59	moderate reduction in function
4	15–29	severe reduction in function
5	less than 15	kidney failure

Use the information in the table to:

• identify the CKD stage indicated by the eGFR that you calculated in (b) (ii)

• determine the effect on the kidney of this man. (*1 mark*) OCR F214 2013

18 A student investigated the movement of substances through the cell surface membrane yeast cells using an indicator.

• The student was supplied with a suspension of yeast cells in a slightly alkaline solution

• The indicator used is yellow in alkaline conditions but turns red in acidic conditions.

The student mixed the indicator with the yeast suspension and labelled the tube containing this suspension A. The suspension was red/pink in colour.

a (i) The student took a small sample from tube A and centrifuged this sample.

After centrifuging, the student observed that the liquid portion was colourless but the cells at the bottom were red/pink.

Suggest the mechanism by which the indicator enters the cell and suggest the component of the plasma membrane involved. (*2 marks*)

a (ii) The student took a small sample from suspension A and added alkaline ammonia solution. There was no colour change. What could the student conclude about the permeability of the yeast plasma membrane? (*1 mark*)

a (iii) The student then took another sample from suspension A and boiled it. When this boiled suspension was centrifuged the liquid portion was yellow and the cells at the bottom of the tube were red/pink.

The student suggested that the liquid in the suspension was yellow because boiling the yeast had damaged the plasma membrane, allowing the indicator out of the cells.

Describe the effect of high temperature on the structure of the yeast cell membranes.

(*3 marks*) OCR F211 2015

Pre-released Case Study

CRISPR: a tasty new DNA editing technique?

DNA editing techniques, such as gene therapy and gene knockouts, have existed for many years. However, a new procedure called CRISPR has been developed in recent years. Some scientists believe CRISPR will revolutionise the manipulation of genomes.

CRISPR is a natural defence mechanism used by bacteria to protect themselves from viruses. The basic CRISPR machinery comprises a DNA-cutting enzyme and a guide molecule (a short piece of RNA). When viral DNA is detected in its cytoplasm, a bacterium produces a short RNA fragment that matches the viral DNA sequence. The RNA binds to the viral DNA and forms a complex with an enzyme called Cas-9, which cuts and disables the viral DNA.

Scientists have adapted the bacterial mechanism to enable DNA to be edited either *in vitro* or *in vivo*. The guide RNA can be designed to match a particular target sequence and therefore be specific to a target gene.

CRISPR causes genes to be edited when a cell attempts to repair the cuts in DNA caused by Cas-9. Often the repair is inexact, which means the base sequence is altered (i.e., a mutation is introduced) and the gene no longer works. This is equivalent to a gene knockout and enables researchers to study the role of the disabled gene. By introducing a replacement DNA sequence once Cas-9 has cut the cell's DNA, faulty alleles can be replaced by functional versions. The introduction of new DNA has a low rate of success, but could represent a new method of gene therapy.

The CRISPR method could be used as somatic cell gene therapy to combat diseases in the future. Scientists are researching the possibility of correcting disease-causing alleles in monogenic conditions such as cystic fibrosis and haemophilia. One advantage of CRISPR is that it has the potential to target many genes at once, which may be useful when tackling complex polygenic diseases. Other diseases that could be treated with CRISPR include cancer (by targeting genes such as BRCA1) and blood diseases such as beta thalassemia. HIV therapy may be possible.

The gene coding for the CCR5 protein could be switched off in T lymphocytes. CCR5 is a protein used by HIV to enter T lymphocytes.

CRISPR has also been used in stem cells and in gametes of various species for research purposes. The CRISPR components can be injected into egg cells with sperm using Intracytoplasmic Sperm Injection (ICSI).

1   The following questions are based on the pre-released case study **CRISPR: a tasty new DNA editing technique?**

   a   Describe how the enzyme Cas-9 is able to catalyse the breakdown of DNA.   *(3 marks)*

   b   Suggest and explain an advantage of CRISPR over traditional gene therapy techniques.   *(2 marks)*

   c   Copy and complete the table using ticks (✓) to indicate the features of the three genetic techniques listed. The first row has been completed for you. Rows may contain more than one tick.

Technique	CRISPR DNA editing	Insertion of a gene into a plasmid	RNA interference using siRNA
Requires DNA ligase		✓	
Can produce transgenic DNA			
Involves the binding of RNA to DNA			
Can stop or reduce the expression of a gene			

(3 marks)

   d   The CRISPR method could be used in the future to target genes such as mutant BRCA1 alleles.

   Explain why using CRISPR to target mutant BRCA1 alleles reduces the risk of breast cancer but is not guaranteed to prevent breast cancer.   *(2 marks)*

e Suggest how switching off the CCR5 gene would reduce the spread of HIV infection within a patient's body. *(2 marks)*

f Why would blood diseases such as beta thalassemia be easier to treat with CRISPR than diseases of other tissues? *(1 mark)*

g ICSI can be used to introduce CRISPR components into an egg cell. State one other procedure for which ICSI can be used. *(1 mark)*

2 The process of vernalisation in some plant species ensures that they flower at the correct time of year.

In thale cress (*Arabidopsis thaliana*) the process of vernalisation is controlled by epigenetic gene silencing.

A period of cold temperature switches off the expression of the Flowering Locus C (FLC) gene. The FLC gene codes for a transcription repressor that stops the expression of the genes that control flowering in plants. Flowering can therefore begin once the FLC gene is switched off.

a The FLC gene is switched off through epigenetic chromatin remodelling.

Suggest what chromatin remodelling involves and how this stops the expression of FLC. *(2 marks)*

b A team of scientists investigated the effect of vernalisation on the timing of flowering in four varieties of daikon (*Raphanus sativus*). The four varieties are called Jumbo Scarlet, E40, Minowase, and Everest.

The scientists recorded the number of days it took each plant to flower after the end of treatment. The results of their experiments are shown in the table.

Treatment	Number of days to flowering (following the end of treatment)			
	Jumbo Scarlet	E40	Minowase	Everest
no vernalisation (18.5 °C)	no flowering	39	47	42
vernalisation (6.5 °C) for 15 days	34	29	29	27
vernalisation (6.5 °C) for 20 days	35	26	32	29

(i) State and explain which variety of daikon should be chosen when early flowering is a desirable trait. *(2 marks)*

(ii) State and explain which variety should be chosen as a model organism to study the effects of the environment on vernalisation and flowering. *(2 marks)*

(iii) State two additional factors that need to be considered to confirm the validity of this data. *(2 marks)*

c Pea plants (*Pisum sativum*) are able to flower without vernalisation. Their flowers can be purple or white.

Flower colour in *P. sativum* is controlled by a gene locus that has two gene variants.

P = a dominant allele coding for purple flowers

p = a recessive allele coding for white flowers.

(i) A purple-flowered pea plant is crossed with a white-flowered pea plant.

The genotype of the purple-flowered plant is unknown. Calculate the range of probabilities that an offspring of these two plants will have white flowers. *(2 marks)*

(ii) In one population of pea plants, 25% of the plants have white flowers.

Calculate the percentage of plants that are heterozygous for the flower colour genotype. *(3 marks)*

(ii) The photograph shows a flowering pea plant.

Suggest whether the flower is adapted for insect or wind pollination. Explain your answer using evidence from the photograph. *(3 marks)*

3 More than 95% of the species that have existed on Earth during the past 3.5 billion years are now extinct.

a In 2014 scientists calculated the relative threat posed to biodiversity by different human activities. 37% of the total threat was estimated to be from human exploitation of ecosystems (e.g., fishing and hunting).

State two other threats to biodiversity posed by human activity. *(2 marks)*

b The highest estimate for the extinction rate of species is 0.7% of existing species per year.

Assuming an extinction rate of 0.7% per year and a current global species count of 5 million species, how many species will exist after

(i) 1 year *(1 mark)*

(ii) 3 years *(1 mark)*

Give your answers to four significant figures.

c The International Union for Conservation of Nature (IUCN) decide which species are considered to be threatened with extinction. For example, 26% of mammals were placed on the IUCN's Red List of Threatened Species in 2014.

Describe how sampling techniques and data analysis can be used to monitor the population size of a mammalian species to decide whether it should be placed on the Red List of Threatened Species. *(6 marks)*

d Efficient cycling of nitrogen within ecosystems plays a key role in the maintenance of biodiversity.

Copy and complete the passage using the most appropriate terms.

Nitrogen gas is converted into nitrogen-containing compounds by nitrogen-fixing bacteria such as *Rhizobium* and free-living .... *Rhizobium* species have .... relationships with plants. Ammonium ions in the soil can be oxidised to .... and then nitrate ions, which are either absorbed by producers or converted back to nitrogen gas in a process called .... *(4 marks)*

A level Paper 3 practice questions

1   While studying the nitrogen cycle, a group of students attempted to grow nitrogen fixing bacteria on Agar plates. The students investigated two different nitrogen-fixing bacteria from two sources:

- *Azotobacter* sp in soil samples
- Rhizobium sp from the roots of clover plants

a   Why were clover plants selected as a potential source of *Rhizobium*?   (*1 mark*)

The students were provided with two different types of agar. The composition of the agar is shown in the table.

Agar for soil samples	Agar for root samples
$FeCl_3$	NaCl
K_2HPO_4	K_2HPO_4
$MgSO_4$	$MgSO_4$
glucose (0.2%)	$CaCl_2$
agar	agar
	mannitol (0.1%)
	year extract

b   All organisms require a source of nitrogen, such as yeast extract, for growth.

   (i)   Name an organic polymer found in bacteria for which the following minerals are necessary:
   - $K_2HPO_4^{*}$
   - $MgSO_4$ (*2 marks*)

   (ii)   Suggest why a nitrogen source was omitted from the agar used to grow *Azotobacter* but included in the agar to grow *Rhizobium*.   (*4 marks*)

   (iii)   Calculate how much glucose would be required to make up $500\,cm^3$ of agar to grow bacteria from soil samples. Show your working.   (*2 marks*)

c   To grow Azotobacter, the students were provided with agar plates for the soil samples.

- Small fragments of soil were placed on the surface of the agar.
- The lid of the plate was taped securely to the base.
- The plate was incubated at 25°C

To grow *Rhizobium*, the students were provided with a sample of surface sterilised roots from clover plants. The roots had been liquidised. Agar plates were inoculated using this material and incubated at 25°C.

The photo shows a plate with colonies of *Rhizobium growing*.

   (i)   *Describe how an agar plate would have been inoculated to obtain the pattern of growth shown in the photo.   (*6 marks*)

   (ii)   Describe the morphology of the colonies shown in the photo. (*2 marks*)

d   The plates obtained from soil samples were examined without removing the lids. A range of different types of colonies were observed. What do you conclude about the procedure used to isolate *Azotobacter* from soil samples? Justify your conclusion.   (*2 marks*)

2   An investigation was carried out on the effect of different carbohydrate solutions on samples of epidermal cells from red onion (*Allium cepa*). A piece of onion epidermis was placed in a beaker containing a solution.

The photo shows the appearance of the epidermal cells under a light microscope after being placed in a **sucrose** solution for 15 minutes.

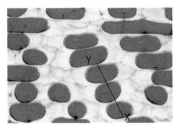

×200 magnification

a (i) Make a labelled drawing of the three adjacent cells under the line XY to indicate the position of the plasma membrane and cytoplasm. (*4 marks*)

(ii) The actual length of line XY is 0.1 mm.

Calculate the **true magnification** of the drawing you have produced. Show your working. Give your answer to the nearest whole number. (*2 marks*)

b Explain why most of the cells in the photo are plasmolysed. (*2 marks*)

c The investigation was repeated using the same onion and distilled water as a control with two different carbohydrate solutions.

- glucose   • starch

All solutions appeared clear and colourless.

(i) Describe the procedure you could use to identify the solutions. (*3 marks*)

(ii) Identify **two** variables which must be controlled in order to make a valid comparison of plasmolysis in the four solutions. (*2 marks*)

d Quantitative data was obtained by observing 10 cells from each of the samples in the three solutions. A student made the following prediction:

> All 10 cells in the glucose solution will be plasmolysed but very little plasmolysis will happen in samples that have been in the water or starch – maybe only in a couple of cells (2) in each.

(i) Use your knowledge of carbohydrates to justify the student's prediction. (*4 marks*)

(ii) The table shows the results obtained by the student for cells in water, glucose, and starch.

Solution	Number of cells plasmolysed
water	1
glucose	9
starch	4

Using the Chi-squared test, analyse the student's results. The formula for χ^2 is given here. You can find formulae for statistical tests and critical value tables in the appendix. (*3 marks*)

(iii) Does the evidence support the student's prediction? Explain your answer. (*3 marks*)

$$\chi^2 = \sum \frac{(O - E)^2}{E}$$

3 Germination in seeds requires the seeds to take in water (imbibe). This intake of water triggers a sequence of events which results in an increase in metabolic activity including respiration. Respiration requires substrates which are produced from the breakdown of storage molecules within the seed.

a Outline the sequence of events within imbibed seeds which results in the breakdown of storage molecules such as starch. (*3 marks*)

Aerobic respiration results in the production of hydrogen peroxide which is toxic. Hydrogen peroxide is broken down by the enzyme catalase:

$$2H_2O_2 \rightarrow 2H_2O + O_2$$

An investigation was carried out on the effect of germination age on catalase activity in mung bean seeds. Germination age was measured as time following first exposure to water and the following samples were provided:

12 hours, 2, 4, 6, and 10 days.

An outline of the method for measuring catalase activity is given here:

- 10 g of germinating seeds were crushed in a pestle and mortar and placed in a conical flask.

- A 100 cm³ measuring cylinder, **graduated every 5 cm³** was filled with water and inverted in a water bath. A delivery tube from a stopper was placed below the cylinder.

- 20 cm³ hydrogen peroxide was introduced into the flask which was then sealed with the stopper.

- The volume of oxygen collected after 30 seconds was recorded.

The diagram shows the apparatus involved.

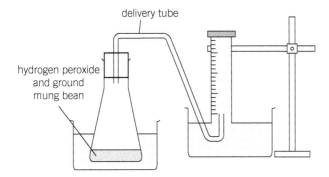

The results obtained from the experiment are shown in the table.

Germination age (days)	Volume of oxygen collected (cm³)
	10.0
	30.5
4	67.5
	31.0
	18.5

b (i) Copy and complete the table by inserting the remaining values for the independent variable in ascending order. *(1 mark)*

(ii) Using the space provided, plot a suitable graph. *(3 marks)*

(iii) *Evaluate the method and apparatus used to determine changes in catalase activity and suggest how improvements could be made. You should support your argument using calculated uncertainties where possible. *(6 marks)*

c Following an investigation such as that described in part (b), some students carried out a review of the literature on catalase activity and seed germination.

(i) The students found some data on a school website showing results from a student project on germinating peas.

- a respirometer was used to measure oxygen uptake
- catalase activity was measured by the method described in part (b).

This data obtained in the project is given in the table. The data had not been analysed.

Germination age (days)	Oxygen uptake in respiration (cm³)	Volume of oxygen released from hydrogen peroxide (cm³)
2	9	26
4	18	55
6	27	64
8	26	85
9	30	76

The students concluded that the data suggested that as the respiration rate of the peas increased so did catalase activity. Choose a statistical test. Analyse the data in the table using your chosen test. Formulae for statistical and critical value tables are given in the appendix. *(4 marks)*

(ii) Is the conclusion drawn by the students supported by the data? *(2 marks)*

d In a further search of the literature, the students found a paper on catalase and respiration in plants. The authors were S.Ranjan and A.K. Mallik. The paper was published in 1931 in volume 30 of a journal called New Phytologist. The paper was in the fifth issue of the journal in that year and appeared on page 335.

The students used information from this paper in writing up their practical work on catalase activity in mung beans. The reference was cited in the text and included in a list of references.

How should this paper:

- be cited in the text
- appear in the bibliography?

You do not need to include a title. *(2 marks)*

Glossary

action potential a change in the electrical membrane potential which causes the transmission of a nerve impulse.

activation energy the energy that needs to be put in to cause a reaction. The activation energy is lowered by the presence of an enzyme (biological catalyst).

active immunity a type of resistance developed in an organism through production of specific antibodies in response to an exposure to a pathogen (natural) or to a vaccine (artificial).

active site a group of usually 3–12 amino acid R-groups that makes up a region on the surface of the enzyme into which a complementary substrate temporarily bonds to forming an enzyme-substrate complex.

active transport movement of molecules from a region of low concentration to a region of high concentration against a concentration gradient requiring an input of ATP and involving transport proteins.

acute disease a disease that has a sudden onset and lasts a short time.

adaptation a trait that benefits an organism in its environment and increases its chances of survival and reproduction.

adenosine triphosphate (ATP) a phosphorylated nucleotide composed of a nitrogenous base (adenine), a pentose sugar, and three phosphate groups. The universal energy currency for cells.

adhesion the force of attraction between two different molecules e.g., water and molecules of lignin.

adrenaline a hormone secreted in response to stress.

afferent leading towards or incoming (e.g. an afferent nerve or blood vessel is one that leads into a tissue or organ).

agglutination the clumping together of antigen-bearing cells, microorganisms, or particles in the presence of specific antibodies.

allele a gene variant.

alveoli small air sacs in the lungs which allow for rapid gaseous exchange.

amino acid an organic compound that has a central carbon atom to which an amine group (NH_2) and carboxyl group (COOH) and variable residual group are attached. They are joined together by condensation reactions to form a polypeptide chain.

amniocentesis a procedure for sampling fetal cells from the amniotic fluid.

anaphase the stage of mitosis in which chromatids separate and are pulled towards opposite poles of the cell by the spindle.

aneurysm a localized bulge of an artery, vein, or the heart wall. The wall of the blood vessel or organ is weakened and may rupture.

antenatal care the care received by a pregnant woman.

anther the part of a plant that holds pollen.

antibiotics a substance produced by a living organism that kills or inhibits the growth of microorganisms, has no effect on viruses.

antibodies globular protein molecules (immunoglobulins) produced by plasma cells (B lymphocytes) in response to stimulation by an antigen.

anti-coagulant a substance that prevents blood from clotting e.g., Sodium citrate, heparin.

anticodon a sequence of three bases at the end of a tRNA molecule that allow complementary binding to a codon of the mRNA molecule being translated at the ribosome.

antigen a toxin or other foreign substance which induces an immune response in the body, especially the production of antibodies.

antiparallel a feature of the two strands in a DNA molecule. The 5′ (5 prime) end of one strand is directly opposite the 3′ (3 prime) end of the parallel strand. The two strands run in opposite directions.

aorta the major artery of the body, supplying oxygenated blood to the circulatory system.

apoplast pathway the transport route taken by water and dissolved substances through the cell walls and intercellular spaces of plants.

apoptosis programmed cell death.

artery a thick-walled vessel that carries blood away from the heart.

aseptic techniques any techniques/manipulations of equipment or materials that are designed to prevent contamination by microorganisms.

asthma a respiratory condition characterised by the inflammation and narrowing of the bronchi.

atrio-ventricular node (AV Node) a patch of tissue in the septum of the heart that conducts the electrical stimulus from the atria in the heart through to the Purkyne fibres.

atria the upper chambers of the heart which receive blood returning from the organs and vessels of the body.

autonomic nervous system the branch of the motor nervous system that controls the non-conscious actions of the body. The autonomic system controls the actions of involuntary muscles and glands. Consists of two branches – the sympathetic and parasympathetic nervous systems.

bacteriocidal describes a chemical substance that kills bacteria.

bacteriostatic describes a chemical substance that prevents the reproduction of bacteria.

Benedict's test a biochemical reaction to test for the presence of a reducing sugar, for example glucose. The test can be semi-quantitative or quantitative depending on the procedure used.

benign a tumour that stays in its original location and does not shed any cells into the blood plasma or lymph system.

biodiversity the variety of life, which can be measured on a genetic, species or ecosystem level.

biosensor a device which uses a living organism or biological molecules, especially enzymes or antibodies, to detect the presence of specific chemicals.

biuret test a biochemical reaction to test for the presence of proteins. It is a qualitative test.

blind trials a clinical trial in which participants are unaware whether they are receiving a placebo or a medicinal drug.

blood clot a structure formed from fibrin fibres which traps red blood cells and platelets in response the damage to a blood vessel.

blood group also known as blood type – the classification of blood depending on which antigens are present on the plasma membrane of the erythrocytes.

blood pressure the force exerted by the blood against the walls of the blood vessels.

B lymphocyte a specialised leucocyte that is produced in the bone marrow and matures in the bone marrow. It forms plasma cells after contact with a specific antigen and produces antibodies.

Bohr effect haemoglobin's oxygen binding affinity is inversely related to the concentration of carbon dioxide and hence the acidity of the blood – an increase in blood CO_2 concentration causes a decrease in blood pH which results in oxyhaemoglobin releasing oxygen – a decrease in carbon dioxide causes an increase in pH which results in haemoglobin picking up more oxygen.

boosters an additional dose of an immunizing agent, such as a vaccine, given at a time after the initial dose to sustain the immune response elicited by the previous dose of the same agent.

bradycardia a slowness of the heartbeat, usually under 60 beats per minute in adults.

broad-spectrum antibiotics antibiotics that are effective against a large variety of organisms.

bronchi the two main branches from the trachea that go into the lungs.

bronchioles one of the smaller subdivisions of the branched bronchial tree that connects the trachea to the alveoli.

buffer a chemical solution which has the ability to absorb or donate hydrogen ions (protons) to maintain the pH of the solution.

bundle of His specialised cardiac muscle fibres that run from the atrioventricular node to the base of the heart.

Calvin cycle the reactions of the light-independent stage of photosynthesis.

cancer a disease usually caused by a mutation that causes uncontrolled cell division and the subsequent formation of a tumour. Some of these (primary tumour) cells may break away and be transported in the plasma or lymph system to form a secondary tumour in a different location.

capillaries very small blood vessels where water, solutes and respiratory gases are exchanged with body tissues.

capsid the outer protein coat of a virus.

cardiac output the amount of blood the heart pumps through the circulatory system in a minute.

carcinogen a chemical or form of radiation that causes cancer.

cardiac monitor a device for the continuous observation of cardiac function.

carrier protein protein found within a cell membrane that carries a specific molecule or ion across the membrane by active transport.

casparian strip a band of impermeable *suberin* found in the walls of endodermal cells in plant roots.

catalyst a substance that speeds up the rate of the reaction without itself being altered or used up in the chemical reaction.

cell cycle the series of events that take place in a cell leading to its division to produce two daughter cells.

cell surface membrane the phospholipid bilayer that forms the membrane surrounding the outside of a cell – sometimes known as the plasma membrane.

cellulose a polysaccharide made from the condensation of many β-glucose molecules to form fibrils. It is used to form plant cell walls.

cell wall a freely permeable structure lying outside of the cell surface membrane of plant, fungal, and bacterial cells.

centrifugation the process of separating molecules and organelles on the basis of their density by spinning them at different speeds in a centrifuge.

centriole(s) two cylinders composed of microtubules which are involved in the process of mitosis and cell division in some eukaryotic cells.

centromere the region of a chromosome that joins two sister chromatids and attaches to spindle fibres during mitosis.

cerebellum a structure located at the back of the brain in vertebrates, which coordinates and regulates muscular activity.

cerebrum the largest part of the brain which is formed as two hemispheres.

channel protein a protein pore that spans a cell membrane to enable water soluble molecules and small ions to passively cross the membrane.

chemotherapy destroying cancerous cells using drugs that affect cancerous cells more than other cells in the body.

chiasmata the points at which crossing over occurs between homologous chromosomes.

chi-squared test a statistical test that enables an investigator to determine whether differences exist between two sets of data.

chloroplast the organelle in which photosynthesis occurs.

chordae tendinae also known as the heart strings, they are cord-like tendons that connect the papillary muscles to the tricuspid valve and the mitral valve in the heart.

chorionic villus sampling a procedure for sampling fetal cells from the placenta.

chromatid a DNA molecule; during prophase and metaphase of mitosis, a chromosome consists of two identical chromatids from the replication of DNA in S phase.

chromatography a technique used to separate substances in a mixture according to differences in their solubility.

chromosome a single DNA molecule, linear, and associated with proteins such as histones in eukaryotic cells.

chromosome mutation a random change to the structure or number of chromosomes due to translocation or non-disjunction.

chronic disease a disease that lasts a long time and has symptoms that worsen over time.

classification the organisation of organisms into groups based on similarities in biochemistry, anatomy, behaviour and embryology.

climax community the stable community of species that exists at the end of ecological succession.

clinical trials a series of controlled studies in which a new medicinal drug is tested.

codominant alleles at a gene locus that are both expressed and therefore both contribute to the phenotype.

coenzyme an organic non-protein molecule that binds temporarily with the substrate to the active site in order for an enzyme to function.

codon a sequence of three bases on the template strand of the DNA or the mRNA that codes for one amino acid.

cofactor a molecule or ion which aids the function of an enzyme – it can be an inorganic ion or a coenzyme.

cohesion the attraction between water molecules due to hydrogen bonding.

cohesion-tension theory a theory of intermolecular attraction that explains the process of water flow upwards (against the force of gravity) through the xylem of plants.

colony morphology the characteristics of a bacterial colony, in cultures, in terms of shape, colour, edge, and elevation.

colorimeter a device that measures the absorbance of particular wavelengths of light by a specific solution, most commonly used to determine the concentration of a known solute in a given solution.

communicable a disease or infection caused by microorganisms capable of being communicated or transmitted to another organism.

companion cells a cell in the phloem tissue involved in actively loading sucrose into sieve tube elements. The companion cell is closely associated with the phloem sieve element, to which it is linked by many plasmodesmata.

compartmentalisation the use of intracellular membranes to separate metabolic processes within the cell e.g., the nuclear envelope around the nucleus.

compensation point the light intensity at which the rate of photosynthesis matches the rate of respiration.

competitive inhibitor a molecule that has a similar shape to the natural substrate and which competes for the active site on the enzyme preventing the formation of enzyme substrate complexes and instead forming enzyme-inhibitor complexes.

complementary therapy treatments that involves procedures that are not part of mainstream medicine e.g., acupuncture, aromatherapy.

computed tomography (CT) a radiographic technique that produces an image of a detailed cross section of tissue.

condensation reaction a chemical process in which two molecules are combined to form a more complex molecule with the removal of a molecule of water to form a covalent bond.

correlation the relationship between two variables e.g. a linear relationship.

co-transporter proteins in the cell surface membrane that allows movement of one molecule when linked to the movement of another molecule by active transport.

cotyledon in seeds, the part of a plant embryo that becomes the first leaf.

countercurrent multiplier any mechanism that uses energy and countercurrent flow to establish concentration gradients.

cristae fold in inner membrane of mitochondria.

crossing over the process in which homologous chromosomes exchange alleles during prophase I of meiosis.

cytokines cell signalling molecules which are used for communication between cells, allowing some cells to regulate the activity of others.

cytokinesis the division of a cell to form two new cells.

cytoskeleton a network of microtubules and microfilaments that give the cell shape and maintain its structure. They can attach to organelles and move organelles within the cytosol.

deamination the removal of the amino group from an amino acid.

defibrillator an apparatus used to control heart fibrillation by application of an electric current to the chest wall or heart.

deflected succession the changes resulting from human activities (such as farming) that produce a stable community called a plagioclimax. A plagioclimax community and a climax community formed by natural, unimpeded succession have different compositions.

degenerate code situation where more than one codon codes for the same amino acid.

denaturation usually permanent change to the tertiary structure of a protein resulting in the loss of function. This can be caused by large changes in pH or high temperatures.

denitrification the production of nitrogen gas from nitrates in the soil.

deoxyribonucleic acid (DNA) the molecule responsible for the storage of genetic information.

depolarisation the loss of the difference in charge between the inside and outside of the plasma membrane of a muscle or nerve cell due to a change in permeability and migration of sodium ions to the interior.

diabetes mellitus a condition in which a person cannot control blood glucose levels.

diaphragm a sheet of muscular and fibrous tissue separating the chest cavity from the abdominal cavity. Plays an important role in ventilation.

diastole period of relaxation and repolarisation of the cardiac muscle when the chambers fill with blood.

dicotyledons a plant that produces flowers and has two cotyledons (seed leaf) inside the seed, which develops wide leaves with veins.

differential staining staining processes that use more than one chemical stain, for example, multiple stains can be used to distinguish between different microorganisms or structures/cellular components of a single organism.

differentiation the development of unspecialised cells to form specialised cells.

diffusion the movement of molecules down their concentration gradient as a result of random motion. In cells diffusion may occur passively directly across the plasma membrane (simple) or via membrane proteins (facilitated).

diploid cells that have two copies of each chromosome.

disaccharide a dimer made from the condensation of two monosaccharides joined by a glycosidic bond.

disulfide bond a S—S chemical bond between two sulfur atoms in the R groups of two cysteine amino acids.

dominant an allele that is expressed in the phenotype, the presence of which prevents a recessive allele from being expressed.

dopamine dopamine is a neurotransmitter (or chemical in the brain) that either increases or reduces the activity of neurons (nerve cells). A precursor to both adrenaline and noradrenaline.

double-blind trials a clinical trial in which neither participants nor scientists are aware whether a placebo or a medicinal drug has been issued.

double circulatory system a type of blood circulation system in which the blood flows through the heart twice for each full circuit of the body.

ecosystem the organisms and non-living components of a specific area, and their interactions.

efferent leading away from or outgoing (e.g., an efferent nerve or blood vessel is one that leads away from a tissue or organ).

electrocardiogram (ECG) a graph showing the electrical activity in the heart during the cardiac cycle (heartbeat).

electroencephalogram (EEG) technique where electrodes attached to the scalp detect small electrical impulses transmitted between brain cells. Impulses are recorded and monitored for abnormal readings.

electrolytes ions such as sodium, potassium, and chloride dissolved in water.

electron microscope a microscope that uses a beam of electrons to view a magnified image of an object giving it greater resolution than a light microscope. There are two main types – scanning and transmission electron microscopes.

electron transport chain a series of compounds that can transfer electrons from electron donors to electron acceptors via a series of redox reactions resulting in the transfer of protons (H+ ions) across a membrane and the generation of an electrochemical proton gradient. This gradient drives the formation of ATP.

electrophoresis a method used to separate molecules (e.g. DNA fragments) based on their relative size.

endemic describes a disease that is ever present in a population. May also mean a species that is found only in a particular area and nowhere else.

endocytosis the inward transport of large quantities of molecules through the cell surface membrane to form a vesicle within the cytosol. This requires an input of energy in the form of ATP.

endodermis a ring of cells between the cortex of the root and the area housing the xylem and phloem.

endoplasmic reticulum (ER) a series of membranes forming a system of tubes within the cytoplasm which is joined to the nuclear envelope and which may have ribosomes attached (Rough ER) or not (smooth ER).

endosperm the food source for the embryo within certain seeds.

enzyme a globular protein that acts as a biological catalyst.

epidemic a sudden increase in the incidence of a disease that spreads to many people quickly and affects a large proportion of the population.

epidemiology the study of patterns of disease and the factors that influence their spread.

epigenetics the study of DNA modifications that alter gene expression rather than altering the genetic code.

erythrocyte a red blood cell. Red blood cell production is Erythropoiesis and this is regulated by the hormone Erythropoietin.

ester bond a C—O—C chemical bond formed by a condensation reaction e.g., Between a fatty acid and a glycerol molecule.

eukaryotic cell a cell that has a true nucleus (contained by a nuclear envelope) and contains membrane-bound organelles in the cytosol e.g., Golgi apparatus.

excitatory postsynaptic potential (EPSP) a postsynaptic potential which increases the likelihood of an action potential occurring in the postsynaptic neurone.

exocytosis the outward transport of large quantities of molecules through the cell surface membrane. This requires an input of energy in the form of ATP.

exon a DNA base sequence within a gene that codes for an amino acid sequence.

extrinsic proteins proteins which are attached to a monolayer within a cell membrane.

eye piece graticule a square grid that appears imposed in the field of view when inserted into the eye piece of a microscope.

FAD flavin adenine dinucleotide is an enzyme cofactor which can be reduced by accepting two electrons and two protons (or oxidised by losing two protons and two electrons). It acts as a prosthetic group in several metabolic reactions.

false negative test result indicates that a condition failed while it actually was successful, or indicated that a substance was not present when it was.

false positives test result indicates that a condition was successful while it actually was not successful, or indicated that a substance was present when it actually wasn't.

fetus the unborn offspring of a mammal.

fibrillation a state in which the chamber walls of the heart contract out of rhythm.

fibrin a fibrous protein that forms long fibrils to form a mesh to trap red blood cells and platelets in clot formation.

fibrinogen a soluble protein found in blood plasma that can be converted to insoluble fibrin during the clotting process.

flagellum (plural Flagella) whip like structure which functions in the movement of both prokaryotic and eukaryotic cells. The structure of a flagellum varies between the two types of cells.

follicle stimulating hormone (FSH) a hormone secreted by the pituitary gland that triggers the maturation of germ cells.

food security the ability of a population to obtain sufficient amounts of safe and nutritious food.

founder effect a type of genetic bottleneck in which a group of organisms leave an original population to form a new, smaller population.

functional MRI (fMRI) a functional neuroimaging procedure using MRI technology that measures brain activity by detecting associated changes in blood flow.

gamete a reproductive (sex) cell that fuses with another gamete during fertilisation.

gametogenesis the series of nuclear and cellular divisions that lead to the production of gametes.

gene a length of DNA on a chromosome that codes for the production of a specific polypeptide chain.

gene therapy any therapeutic technique in which a functional allele is placed in cells that lack the allele.

genetic bottleneck a drastic reduction in population numbers, with an accompanying reduction in genetic diversity.

genetic engineering the insertion of a gene into the genome of a recipient organism, which is often a different species, to produce a transgenic organism.

genome all the DNA that makes up the organism.

genotype an individual's set of alleles. This can refer to the two alleles at one gene locus (e.g. Aa) or to combinations of alleles at more than one gene locus (e.g. Aa Bb).

germination the process by which a plant seedling begins to grow from a seed.

gibberellins plant hormones that have a variety of roles, including initiating germination, in plants.

globular protein proteins that are generally soluble in water and form spherical structures, often having a role in the metabolism of cells.

glomerulus the network of capillaries located inside the Bowman's capsule in a kidney nephron.

glucagon the hormone released by pancreatic alpha cells that stimulates the conversion of glycogen to glucose in liver cells.

glycogen a branched polysaccharide made from the condensation of many α-glucose molecules. It is used as an energy storage molecule in liver and muscle cells.

glycoprotein a protein molecule with a carbohydrate portion (glycocalyx) added to it.

glycosidic bond a type of C—O—C covalent bond that joins two saccharides together as a result of a condensation reaction.

Golgi apparatus organelle involved with modification of proteins and production of vesicles (small membrane bound sacs) for storage of enzymes in cells (lysosomes) or exporting molecules from cells (exocytosis).

grana stacks of thylakoid membranes in chloroplasts (sing. granum).

haemoglobin a globular protein made from four polypeptide chains and four prosthetic haem groups which can combine with four oxygen molecules to form oxyhaemoglobin.

haploid cells that have one copy of each chromosome.

heart rate the number of heart beats per minute. The heart rate is based on the number of contractions of the ventricles per minute. Frequently measured as pulse rate.

heterozygous possessing two different alleles (gene variants) for a particular gene.

herd immunity a form of immunity that occurs when the vaccination of a significant portion of a population (or herd) provides a measure of protection for individuals who have not developed immunity.

histone proteins that are associated with DNA within chromosomes that enable the DNA to condense and coil during nuclear division.

homeostasis the maintenance of a stable equilibrium in the conditions inside the body.

homologous chromosomes a pair of chromosomes, one maternal and one paternal, that have the same gene loci.

homozygous possessing two identical alleles for a particular gene.

hormone replacement therapy the replacement of hormones to relieve the symptoms of menopause in women.

hormones chemicals released by endocrine glands which act as cell signalling molecules. They circulate in the blood and regulate the activity of a range of tissues and organs.

Huntington's disease a neurodegenerative genetic disorder that affects muscle coordination and leads to mental decline and behavioural symptoms.

hydrogen bond a relatively weak chemical attraction formed between the slightly positive charge on a hydrogen atom and the slightly negative charge on another atom (usually oxygen) on an adjacent molecule e.g., between water molecules, found in secondary and tertiary strictures of proteins.

hydrolysis the breaking of a covalent bond within a large molecule to form two smaller molecules by the addition of a molecule of water.

hydrophilic a molecule able to associate with water as it possesses a charge which interacts with the dipoles on water molecules.

hydrophobic a molecule not able to associate with water as it is uncharged i.e., repels water.

hydrophobic interactions tendency of nonpolar substances to aggregate in aqueous solution and exclude water molecules – commonly seen in hydrocarbon R groups which repel water from the centre of the folded polypeptide chain.

hydrostatic pressure pressure created by a fluid pushing against the sides of a vessel.

hyperpolarisation the process where a membrane becomes more highly polarised, with the inside more negative than the usual resting state.

hypertension a condition in which the resting blood pressure (particularly the diastolic pressure) is raised for prolonged periods.

hypothalamus a portion of the brain that contains various receptors that monitor the blood. Also involved with controlling the autonomic nervous system.

immunotherapy treatment of disease by inducing, enhancing, or suppressing an immune response.

incidence rate the number of new cases of a disease in a given population, usually in a certain time period.

independent assortment the random assortment of chromosomes during anaphase I of meiosis and chromatids during anaphase II of meiosis.

inflammatory response a tissue reaction to injury or an antigen that may include release of histamine, pain, swelling, itching, redness, heat, and loss of function.

inhibitor a molecule that reduces the rate of an enzyme-controlled reaction by reducing the ability of the enzyme to bind to its substrate(s).

inhibitory post synaptic potential (IPSP) post synaptic potential which decreases the chances of an action potential occurring in the post synaptic neurone.

insulin the hormone released by pancreatic beta cells that results in blood glucose levels falling.

intercostal muscles muscles between the ribs, responsible for moving the rib cage during breathing.

interphase the phase of the cell cycle in which new DNA and organelles are synthesised.

introns portions of DNA within a gene that do not code for a sequence of amino acids within a polypeptide chain. These are removed from the mRNA after transcription.

ionic bond a strong chemical bond caused by the attraction between a negatively charged ion (anion) and a positively charged ion (cation).

karyotype the number and appearance of chromosomes in a cell.

leucocyte a white blood cell. There are various types e.g., Neutrophil, T-lymphocyte, phagocyte.

light-dependent stage (photosynthesis) the first stage of photosynthesis, which uses light energy to produce ATP and reduced NADP.

light-independent stage (photosynthesis) the second stage of photosynthesis, which uses carbon dioxide, ATP and reduced NADP to produce organic molecules.

limiting factor a variable that limits the rate of a particular process.

linkage The presence of gene loci on the same chromosome (i.e., two loci on the same chromosome are linked).

locus the position occupied by a gene on a chromosome.

logarithmic scale nonlinear scale used when there is a large range of quantities e.g., pH.

luteinising hormone (LH) a hormone secreted by the pituitary gland that stimulates ovulation, among other functions.

lymph a colourless fluid containing leucocytes that bathes the tissues and drains through the lymphatic system into the bloodstream.

lysosomes membrane bound vesicles made by pinching off from the Golgi body. They contain strong digestive enzyme to break down old cellular components.

magnetic resonance imaging (MRI) a technique that uses a magnetic field and radio waves to create detailed images of the organs and tissues.

magnification the number of times larger an image appears compared to the real specimen.

malignant a tumour that sheds cells that can spread through the body via the blood plasma or lymph system to initiate a secondary tumour in a new location(s).

mammography X-rays of the breast that are usually carried out to detect and screen for the early stages of cancer.

mantoux test a skin test for tuberculosis.

mass flow the movement of fluid in one direction, usually through tube like vessels.

mass transport the transport of molecules in bulk from one part of an organism to another.

medulla oblongata part of the brain which is responsible for regulating breathing rate and blood flow around the body.

meiosis a type of nuclear division in which the chromosome number is halved.

menopause the occurrence of the last menstrual cycle in a woman.

menstrual cycle the cycle in women that involves development of the uterus lining, ovulation, and shedding of the uterus lining. The cycle is regulated by several hormones.

mesosome an infolding of the cell surface membrane found in prokaryotic cells.

metaphase the stage of mitosis in which chromosomes line up at the equator of the cell.

metastasis process in which cancer cells break from the primary tumour and spread in the plasma or lymph system to initiate a secondary tumour formation at a different location(s).

microtubule protein polymers that form the mitotic spindle and are components of the cytoskeleton.

mitosis a form of nuclear division in which two genetically identical nuclei are formed; occurs in somatic cells.

model organism a non-human species that is used in research and gives insights into the biology of other species, including humans.

monocotyledons a flowering plant that produces seeds with only one cotyledon (seed leaf).

morbidity the incidence or prevalence of a disease or of all diseases in a population.

mortality the number of deaths in a population.

mRNA a single stranded polynucleotide formed as a result of the transcription of a gene from the template strand of the DNA.

mutagen a chemical or form of radiation that causes a change in the amount or structure of DNA

mutation a change to the structure of DNA. This can be either a change to the DNA base sequence (a gene mutation) or a chromosome mutation.

myelin sheath the insulating envelope of myelin that surrounds the core of an axon; it facilitates the transmission of nerve impulses.

myogenic describes muscle tissue (heart muscle) that generates its own contractions.

myosin a fibrous protein making up thick filaments in a myofibril – the protein has two head groups that can form cross-bridges with actin to bring about muscle contraction.

NAD Nicotinamide adenine dinucleotide is a coenzyme molecule found in all living cells which can be reduced by removing hydrogen atoms and accepting electrons from substrates. It is involved in many different metabolic processes including glycolysis, Krebs cycle, photosynthesis as well as being involved in cell signalling.

NADP a coenzyme that is reduced (gains hydrogen atoms) in the light-dependent stage of photosynthesis and is oxidised (loses hydrogen atoms) in the light-independent stage.

narrow-spectrum antibiotics antibiotics that are effective against a narrow range of organisms.

natural selection the mechanism for evolution. The best-adapted organisms will be more likely to survive and reproduce, thereby passing on favourable alleles to the next generation.

negative feedback a process in which a change in a parameter leads to the reversal of the change. This causes the parameter to remain relatively constant.

nephron tubules in the kidneys that produce urine.

net primary productivity the rate of production of new biomass in producers (primary productivity minus the energy losses from respiratory heat).

neurones a cell specialised for the conduction of nerve impulses.

neutrophils a type of granulated phagocytic white blood cell.

nitrification the conversion of ammonium ions to nitrites, and the subsequent conversion of nitrites to nitrates. These are examples of oxidation. The reactions occur in soils and are carried out by bacteria.

nitrogen fixation the conversion of nitrogen gas into ammonia by organisms.

node of Ranvier a gap in the myelin sheath of a nerve, that forms between adjacent Schwann cells.

non communicable disease a medical condition or disease which is non-infectious and non-transmissible among people.

non competitive inhibitor a molecule that can bind to a binding site separate from the active site on the enzyme. This changes the shape of the active site such that it is no longer complementary to the the substrate.

non-disjunction failure of a homologous pair of chromosomes to separate during nuclear division.

non-polar a substance that contains no permanent dipolar molecules.

non reducing sugar a monosaccharide or disaccharide that cannot donate electrons to other molecules and therefore cannot act as a reducing agent. Sucrose is the most common non-reducing sugar.

non-specific response a resistance manifested innately by a species, it protects you against all possible antigens.

nucleotide monomer made from a nitrogen-containing base, phosphate group and a pentose sugar. Nucleotides can be joined together to form polynucleotides.

objective lens the lens closest to the specimen. The magnification can usually be varied to give low, medium, and high power images.

oestrogen a hormone that stimulates the development of tissue in the uterus, among other functions.

oncogene allele formed when a proto-oncogene mutates which allows cells to divide uncontrollably resulting in cancer.

oncotic pressure also known as colloid osmotic pressure. The force due to the tendency of plasma proteins and other substances to lower the water potential.

opsonin an antibody which binds to a pathogen making it more susceptible to phagocytosis.

organ a collection of tissues that work together to perform a specific overall function or set of functions within a multicellular organism.

organelle a structurally distinct part of a cell that is specialised to carry out specific function(s). Usually they are surrounded by a membrane (true organelle).

osmoregulation the homeostatic control of the water potential of blood plasma in mammals.

osmosis the passive movement of water from an area of high water potential to an area of low water potential down a water potential gradient through a partially permeable membrane.

ovules structures that contain female gametes in plants.

pandemic describes an epidemic occurring worldwide, or over a very wide area, crossing international boundaries and usually affecting a large number of people.

parasympathetic the part of the autonomic nervous system which regulates physiological functions when the body is at rest.

Parkinson's disease a progressive disease of the nervous system marked by tremor, muscular rigidity, and slow, imprecise movement.

passive immunity immunity acquired by the transfer of antibodies. It may be natural (placental transfer of antibodies during pregnancy and via breast milk) or artificial (injection of antiserum).

pathogen a microorganism that causes disease.

Peptide bond a CO—NH covalent bond that is formed via a condensation reaction between the carboxyl group (COOH) of one amino acid and the amine group (NH_2) of the adjacent amino acid.

pH a measure of the number of protons (hydrogen ions, H^+) in a solution.

phagocyte a specialised leucocyte that destroys bacteria and other foreign material by phagocytosis e.g., neutrophil, macrophage.

phagosome a vacuole inside a phagocyte which is created by an infolding of the plasma membrane to engulf a foreign particle. The foreign particle is held inside the phagosome.

phenotype the observable traits of an organism.

phloem a tissue in plants that is used to transport dissolved sugars and other substances.

phosphodiester bond covalent bond found in DNA and RNA joining the 3' carbon atom of one nucleotide and the 5' carbon atom of another.

phospholipid specialised lipid molecule containing a phosphate group, two fatty acids, and a glycerol molecule. Consists of a hydrophilic head (phosphate group) and a hydrophobic tail (fatty acids). In water they form bilayers that make up cell membranes.

photolysis an enzyme-catalysed reaction of photosynthesis that uses light energy to split water.

photosynthesis the set of reactions in chloroplasts that use light energy, water and carbon dioxide to produce organic molecules.

phylogeny the study of the evolutionary relationships between organisms.

phytochrome a photoreceptor (light-sensitive protein) that enables plants to measure changes in day length.

pili hair-like structures found on prokaryotic cells and used in the exchange of genetic material between bacterial cells.

pituitary gland the major endocrine gland, found in the brain. It secretes hormones which are important in controlling growth and development and the functioning of the other endocrine glands.

placebo a 'dummy' pill or injection that resembles the real drug but contains no active ingredient.

plant assimilates the newly formed compounds that result from the incorporation of carbon, from carbon dioxide, into organic substances during photosynthesis.

plasma yellow fluid that transports blood cells and platelets around the body and contains a number of substances, including proteins.

plasmid small circular piece of DNA found in prokaryotic cells.

plasmodesmata microscopic channels which traverse the cell walls of plant cells, enabling transport and communication between them through cytoplasmic connections.

polar having a negative or positive charge.

pollen structures that contain male gametes in plants.

polymer a large molecule made up of small repeating sub-units or monomers. For example, polysaccharides are made up of monosaccharides joined together.

polymerase chain reaction (PCR) a technique used to increase the amount of DNA in a sample.

polypeptide chain more than four amino acids joined together by peptide bonds as a result of a series of condensation reactions.

positive feedback any process in which a change in a parameter causes a further increase in that change.

positron emission tomography (PET) neuroimaging procedure where radioactive glucose(FDG) is used, scanner detects positrons releasedby glucose to form an image.

potential difference the difference in electric potential between two points, such as across a membrane.

pressure gradient (heart) the difference in blood pressure across the vessel length or across the valve.

prevalence the proportion of individuals in a population having a disease or characteristic.

primary consumer an organism that gains its organic molecules and chemical energy from the consumption of producers.

primary immune response the response that the immune system displays when first exposed to an antigen – the initial production of antibodies.

primary productivity the energy captured by plants for photosynthesis.

primary structure the sequence of amino acids in a polypeptide chain that is determined by the sequence of nitrogen-containing bases in a gene.

primer a short single-stranded DNA sequence that is required in PCR because DNA polymerase cannot bind directly to single-stranded DNA.

producer an autotrophic organism that converts light energy to chemical energy.

progesterone a hormone that stimulates the development of blood vessels in the endometrium and inhibits FSH and LH.

prokaryotic cell a cell that does not have a true nucleus or any true (membrane-bound) organelles, for example, Blue-green algae, bacteria.

prophase the stage of mitosis in which chromosomes become visible and the spindle forms.

proto-oncogene a gene that helps regulate cell division.

pulmonary artery the artery carrying blood from the right ventricle of the heart to the lungs for oxygenation.

pulmonary vein a vein carrying oxygenated blood from the lungs to the left atrium of the heart.

pulmonary ventilation the total volume of gas per minute inspired or expired.

pulse as the heart pushes blood through the arteries, the arteries expand and recoil with the flow of the blood, causing a pulse.

purine nitrogen-containing base with a double ring structure i.e., adenine and guanine.

purkyne tissue (Purkinje tissue) specialised muscle tissue in the septum of the heart that conducts electrical stimulation from the AV node to the ventricles.

pyrimidine nitrogen-containing base with a single ring structure i.e., Thymine, cytosine and, uracil.

qualitative data observations or recordings that are not numerical e.g., presence/absence, colours.

quaternary structure the structure of a protein formed by two or more polypeptide chains and/or the presence of a prosthetic group(s).

quantitative data readings and recordings that are numerical.

radiotherapy treatment that destroys cancerous cells using ionising radiation (radiation that can dislodge electrons from atoms) e.g. X-rays, alpha rays and gamma rays.

random sampling obtaining, without bias, a respresentative subset of a population. When a sample is taken randomly, all members of the population have an equal chance of being chosen.

recessive an allele that is expressed only if no dominant allele is present.

reducing sugar a monosaccharide or disaccharide that has the ability to donate electrons to reduce copper sulfate from Cu^{2+} to Cu^+ and give a positive result in a Benedict's test.

refractory period the short period of time after firing during which it is more difficult to stimulate a neurone.

repolarisation the reestablishment of polarity, especially the return of cell membrane potential to resting potential after depolarisation.

residual volume the volume of air remaining in the lungs after a maximal expiratory effort.

resolution the ability to distinguish two distinct objects separately and to see detail.

resting potential the potential difference or voltage across a neurone cell membrane while the neurone is at rest.

restriction enzymes enzymes found in bacteria that are utilised in genetic engineering and research to cut DNA double helices at specific points.

retrovirus an RNA virus which also contains the enzyme reverse transcriptase e.g., HIV.

reverse transcriptase an enzyme originally derived from retroviruses. The enzyme catalyses the construction of a DNA strand using an RNA strand as a template.

Rf value ratio of the distance travelled by the centre of a spot to the distance travelled by the solvent front.

rhesus antigen the presence of the immunogenic D antigen on the surface of the erythrocytes (i.e., Rh positive).

ribosomes structures composed of RNA and proteins which may be free in cytoplasm or membrane bound and are the site of protein synthesis.

ring vaccination the vaccination of all susceptible individuals in a prescribed area around an outbreak of an infectious disease.

RQ respiratory coefficient the ratio of carbon dioxide given out compared to the oxygen consumed used to indicate the type of respiratory substrate being used.

rRNA folded polynucleotide chain found in both the small and large subunits of ribsosomes.

saltatory conduction the mechanism by which an impulse is transmitted along a myelinated nerve cell.

Schwann cells a cell that produces the myelin sheath around a nerve cell.

secondary consumer Carnivores (or omnivores) that eat primary consumers.

secondary (specific) immune response an integrated bodily response to an antigen, especially one mediated by lymphocytes and involving recognition of antigens by specific antibodies or previously sensitised lymphocytes.

secondary (specific) immune response an integrated bodily response to an antigen, especially one mediated by lymphocytes and involving recognition of antigens by specific antibodies or previously sensitised lymphocytes.

secondary structure the folding of a polypeptide chain into alpha helices and/or beta pleated sheets which are held in place by the presence of many hydrogen bonds.

selection pressure a factor that drives evolution in a particular direction.

selective reabsorption The absorption of particular molecules and ions back into the blood from fluid in the kidney nephron.

semi-conservative replication process by which DNA makes an exact copy of itself. Each DNA molecule will have one parent strand and one newly synthesised strand.

semilunar valves also known as pocket valves and found in veins and between the ventricles and the main arteries. They prevent the backflow of blood.

senescence The deterioration of the human body as it ages.

serum the serum is the component of blood that is neither a blood cell nor a clotting factor – it is the blood plasma not including the fibrinogens.

sinoatrial node (SA node) the patch of tissue that initiates the heartbeat by sending waves of excitation over the atria.

SNP (single nucleotide polymorphism) SNPs are DNA sequences that differ by a single nucleotide (e.g., GGATTC and GGATTG). SNPs within coding regions of DNA (i.e., genes) give rise to alleles.

solute a solid that dissolves in a liquid (the solvent).

solvent a liquid that dissolves solids (the solute).

somatic nervous system the section of the nervous system responsible for sensation and control of the skeletal muscles.

source (solutes) a part of a plant that releases sugars and other solutes to the phloem.

species a group of organisms whose members can interbreed to produce fertile offspring and possess similar genetics, physiology, appearance and behaviour.

sphygmomanometer an instrument used to measure blood pressure.

spindle microtubule fibres that attach to centromeres and separate sister chromatids during eukaryotic mitosis.

squamous epithelium a type of thin, flat cell found in layers or sheets covering surfaces such as skin and the linings of blood vessels.

stage micrometer a microscope slide with a finely divided scale marked on the surface.

stem cells undifferentiated cells that are capable of differentiating into a range of cell types.

sticky ends the staggered cut produced by some restriction enzymes in double-stranded DNA. Sticky ends comprise two short sequences of exposed, unpaired bases and enable a gene to be integrated into a plasmid.

stigma the part of the flower onto which pollen is transferred.

stroke volume the volume of blood pumped by the left ventricle of the heart in one contraction.

stroma the fluid-filled matrix of chloroplasts and site of the light-independent stage of photosynthesis.

succession a directional change in the community of organisms over time.

summation (nerves) the way that several small potential changes can combine to produce a larger change in potential difference across a neurone membrane.

supernatant the liquid component of a mixture left at the top of the centrifuge tube when suspended organelles and molecules have settled to the base of the tube after centrifugation.

surfactant a chemical that can reduce the surface tension of a film of water and is found in the alveoli

sustainability the ability to meet the energy and food requirements of a population now without compromising biodiversity or the ability to meet requirements in the future.

sympathetic system a division of the autonomic nervous system that is chiefly involved in producing an immediate and effective response (fight-or-flight response) during stress or emergency situations.

sympathetic system a division of the autonomic nervous system that is chiefly involved in producing an immediate and effective response (fight-or-flight response) during stress or emergency situations.

symplast pathway the transport route taken by water and dissolved substances through the cytoplasm of plant cells.

synapse the junction between two nerve cells, consisting of a minute gap across which impulses pass by diffusion of a neurotransmitter.

systole the contraction of the muscular walls of the atria or ventricles.

tachycardia an abnormally rapid heart rate, usually defined as greater than 100 beats per minute.

taxon (pl: taxa) one of the groups (e.g. domain, kingdom, phylum, class, order, family, genus or species) used in classification.

telophase the stage of mitosis in which two new nuclear envelopes form around the two sets of daughter chromosomes formed in nuclear division.

tertiary consumer a carnivore (or omnivore) that eats secondary consumers.

tertiary structure the further folding of a polypeptide chain into a fibrous or globular 3D structure – this is held in place by the presence of further hydrogen bonds, ionic bonds, disulfide bonds and/or hydrophobic interactions between R groups.

thrombin a protease enzyme that converts soluble fibrinogen into insoluble strands of fibrin.

thrombocyte correct name for the platelets involved in blood clotting.

thromboplastin a plasma protein that aids blood coagulation through catalysing the conversion of inactive prothrombin to active thrombin.

thylakoid the inner membrane of a chloroplast and site of the light-dependent stage of photosynthesis.

thymus an organ that is located in the upper chest in which T lymphocytes mature and differentiate.

tidal volume the volume of air inspired or expired in a single breath during regular breathing.

tissue a group of cells, with a common origin and similar structures, which performs a particular function.

tissue fluid the fluid that surrounds the cells of the body. It has a similar composition to blood plasma except it contains no large proteins and no cells. Its function is to supply the cells with nutrients and remove wastes products.

T lymphocyte a specialised leucocyte that is produced in the bone marrow and matures in the thymus gland. They coordinate the immune response and destroy infected cells.

toxin a poisonous substance produced within living cells or organisms.

trachea the windpipe leading from the back of the mouth to the bronchi.

transcription the formation of mRNA from a section of the template strand of the DNA that corresponds to a gene.

translation the production of a polypeptide gene using the sequence of codons on the mRNA.

translocation *(plants)* the movement of sucrose and other substances up and down a plant. *(Genetics)* rearrangement of DNA between non-homologous chromosomes.

transpiration the loss of water vapour from the aerial parts of a plant due to evaporation.

T regulatory cells (Tregs) T lymphocytes which suppress the function of other T cells to limit the immune response.

triplet code a three-nucleotide codon in a nucleic acid sequence codes for a specific single amino acid.

tRNA single RNA-polynucleotide chain that carries a specific amino acid to the ribosome during translation.

trophic level the level at which a particular organism feeds in a food chain.

tropomyosin fibrous protein twisted around the actin filament (made from two chains of actin molecules) in a myofibril – tropomyosin molecules can undergo a conformational change to expose parts of the actin molecules which act as binding sites for myosin during muscle contraction.

troponin protein attached to the actin filament at regular intervals in the myofibril – troponin can undergo a conformational change to expose parts of the actin molecules which act as binding sites for myosin during muscle contraction.

tumour a tumour is a lump or growth in a part of the body and is formed from unregulated cell division by abnormal cells. Benign tumours are not cancerous and are not usually life-threatening.

tumour suppressor genes a gene that protects a cell from becoming potentially cancerous by halting cell division. When this gene mutates to cause a loss or reduction in its function, the cell can become cancerous, usually in combination with other genetic changes.

turgid describes a cell that is full of water as a result of entry of water due to osmosis when the pressure of the cell wall prevents more water entering.

ultrafiltration filtration of molecules in the glomerulus of the kidney nephron.

ultrasound scan a procedure that uses high frequency sound waves to create an image of a fetus or some internal organs such as stomach, liver, heart, tendons, muscles, joints, or blood vessels.

vaccination the administration of antigenic material (a vaccine) to stimulate an individual's immune system to produce antibodies to a pathogen.

vacuolar pathway the pathway taken by water in plants as it passes from cell to cell via the cell cytoplasm and vacuole.

vacuole a fluid filled space found within the cytoplasm of some cells which is surrounded by a membrane (the tonoplast).

vascular bundles the transport tissue in a plant – usually found as a bundle containing both xylem and phloem.

vasoconstriction a decrease of blood flow through capillaries near the surface of the skin.

vasodilation an increase of blood flow through capillaries near the surface of the skin.

vector a structure that carries a gene into a recipient cell during genetic engineering.

vein a vessel that carries blood towards the heart.

vena cava either of two large veins that carry deoxygenated blood from the body, back to the heart.

ventricles the muscular lower chambers of the heart, which pump to the organs of the body.

vernalisation an extended period of exposure to low temperature that stimulates flowering in plants.

virus a metabolically inert, infectious agent that replicates only within the cells of living hosts: composed of an RNA or DNA core, a protein coat, and a surrounding envelope.

vital capacity the amount of air that can be forcibly expelled from the lungs following breathing in as deeply as possible.

VNTR variable Number Tandem Repeat – VNTRs are base sequences that show variation between individuals and can be used in genetic research and forensics.

water potential a measure of the ability of water molecules to move freely in solution. Measures the potential for a solution to lose water – water moves from a region of high water potential to one of lower water potential.

xylem a plant tissue containing xylem vessels (and other cells) that are used to transport water in a plant and provide support.

zygote a diploid cell formed from two haploid gametes.

Answers

16.1

1 a ATP is water soluble (1); easily transported within the cell (1); easily hydrolysed (to release energy) (1)

b Some energy is lost as heat (1)

2 Direct hydrolysis of glucose would release too much energy (1); also would release too much heat (1); this energy would be wasted as it would be higher than the energy demand (i.e., inefficient) (1); hydrolysis of ATP releases energy in small 'packets' to meet the energy demand (1). (max 3)

3 *Any two from* Idea that glucose is the substrate for the first enzyme in the pathway only (1); other substances required in the metabolic pathway may also be limiting (1); named example of a limiting substance e.g., ADP, NAD; some glucose (phosphate), will also be converted to glycogen or enter another metabolic pathway (1)

16.2

1 ATP used in other cellular reactions e.g., active transport (1); reduced FAD enters electron transport chain (1); reduced NAD enters electron transport chain (1); CO_2 removed from cell as a waste product (1); oxaloacetate combines with another acetyl coA to form another citrate molecule (1).

2 a Dehydrogenase enzymes catalyse removal of electrons (1); and protons from the substrate (1); and use these to reduce coenzymes such as FAD and NAD (1).

b Mitochondrial matrix (1); because the dehydrogenase enzymes and coenzymes are not found in the cytoplasm (only the matrix) (1)

3 a *Any two from*: product inhibition (1); and substrate availability (1); allosteric inhibition (1)

b Pyruvate (by dehydrogenase enzyme and using triose phosphate as the substrate) (1); Acetyl Co-A (by pyruvate dehydrogenase enzyme and using pyruvate as the substrate) (1)

16.3

1 Chemical substance (protons) moves across membrane (1); down a gradient of electrical potential (1); (similar to osmosis in that water moves down a water potential gradient)

2 a *hepatocytes* energy required to drive large and varied number of biochemical reactions occurring in cells (1)

b *striated muscle cells* energy needed for muscle contraction, especially where this is rapid (1)

c *spermatozoan tails* provide energy for sperm motility (1)

d *neurones* provide energy for the production <u>and</u> secretion (exocytosis) of neurotransmitter (1)

3 prokaryotes lack mitochondria (1); biochemical reaction sequence completed in cytosol with proton gradient for ATP production being across cell's surface membrane (rather than the inner membrane of the mitochondrion) (1)

16.4

1 Basis of brewing industry (1); and baking industry (1)

2 Lower yield of ATP from anaerobic respiration (1); (so less ATP available) for protein synthesis, DNA replication, active transport (or other ATP requiring activity) (1)

3 *location1* in cytoplasm (explanation 1 =) involved in part of anaerobic respiration (1); *location 2*: in mitochondria (explanation 2 =) involved in the formation of acetyl CoA (1)

16.5

Double chambered respirometer

Temperature ($^{\circ}$C)	Change in volume (cm^3)	Rate of oxygen consumption ($cm^3 \, O_2 \, min^{-1}$)	Mean rate of oxygen consumption ($cm^3 \, O_2 \, min^{-1}$)
5	1.70	0.56	0.67
5	2.00	0.67	
5	2.30	0.77	
10	3.00	1.00	1.02
10	3.20	1.07	
10	3.00	1.00	

1 Oxidation of lipids produces higher number of hydrogen atoms (1); these can be used to reduce NAD and FAD (1); which can then enter the electron transport chain (1); to produce a higher number of ATP (1)

2 a $C_6H_{12}O_6 + 6O_2 \rightarrow 6CO_2 + 6H_2O$ (1);

$RQ = \dfrac{6CO_2}{6O_2} = 1.0$ (1);

b $C_{55}H_{104}O_6 + 78O_2 \rightarrow 55CO_2 + 52H_2O$ (1);

$RQ = \dfrac{55CO_2}{78O_2} = 0.71$ (1)

3   a   Organisms respire more than one type of substrate at the same time (1)

   b   Organisms rarely respire protein (1); so value indicates the respiration of a mixture of lipid and carbohydrate (1)

   c   Value would decrease (1); as carbohydrate stores would be used up the body would begin to use more lipid stores (1)

17.1

1   $\frac{35}{5} \times 100 = 700$ (2)

2   Carbon dioxide dissolves in water to form carbonic acid (1); this dissociates for hydrogen ions and carbonate ions (1); increase in hydrogen ions lowers blood pH (1); can lead to respiratory acidosis (1); can lead to reduced enzyme activity (1)

3   a   *Advantages* easy to measure (1); no specialist equipment needed (1) *Disadvantages* can really only be used on one individual (1); as various individual factors other than cardiorespiratory fitness play a role in how quickly heart rate returns to a resting level, named example e.g., smoking, asthmatic (1)

   b   Lipids can only be broken down aerobically (1); as oxygen supplies to active muscle cells improve (1); more lipids are broken down to fuel exercise (reducing BMI) (1) max 2

17.2

1   a   During first seven weeks as training programme progresses VO_2 max increases (1); until week eight when VO_2 max reaches a plateau (1); data given to support with units (for both VO_2 max and time) (1)

   b $\frac{(3.5 - 3.26)}{3.26} \times 100 = 7.36 = 7.4\%$ (2)

2

Level 3 (7–9 marks) Details of apparatus and method to produce reliable data provided to include use of a range of intensities/durations and sufficient heart rate/recovery time readings taken over valid time period. Most variables identified, and method states how most variable controlled. *There is a well-developed line of reasoning which is clear and logically structured. Information presented is relevant and substantiated.*		Indicative scientific points could include: *Apparatus and method:* timer; apparatus for varying exercise e.g. treadmill; selection of appropriate number of individuals (minimum of 20); reference to initial heart rates/recovery times (at time 0); details of need for control group; method of determining heart rate/recovery rate
	6	

Level 2 (4–6 marks) Apparatus and method to provide reliable results are provided although some details may be missing. Outline given of how to use measure heart rate/recovery rate. Some variables identified and method states how some variables controlled. *There is a line of reasoning presented with some structure. Information presented is in the most-part relevant and supported by some evidence.*	*Variables:* independent variable = exercise, intensity/duration/type; dependent variable = heart rate or recovery time; appropriate units included; control variables e.g. age, type/duration of exercise (depending on programme selected), smokers/non-smokers, ethnicity (variations in muscle type)
Level 1 (1–3 marks) Apparatus and an outline method are suggested to provide some results but information, such as how to obtain measurements of heart rate/recovery rate, may be missing. Some variables omitted. *Information is basic and communicated in an unstructured way. The information is supported by limited evidence and the relationship to the evidence may not be clear.*	*Reliability:* repeats; reference to quantitative processing of data e.g., calculation of mean heart rate/recovery time; reference to use of standard deviation. *Risk Assessment:* health assessment prior to undergoing programme; awareness of maximum heart rate to work at (220–age)

3   Improvements in oxygen intake, oxygen transport, oxygen delivery, and oxygen storage (in muscle cells) (1); details of any of the above e.g., increases – vascularisation, red blood cell numbers, myoglobin stores (1); reduce requirement for anaerobic respiration (beginning of exercise)/delay onset of anaerobic respiration (end of exercise) (1).

17.3

1   As a result of training individual will have improved VO_2 max (1); e.g., increased stroke volume; increased myoglobin; so less deficit occurs (1)

2   Initial period of exercise is always anaerobic (1); as HR and BR take time to reach level required to meet oxygen demand (1); during this time active muscles will have an oxygen demand greater than their oxygen supply (1)

3 Exercise increases metabolism (and increases CO_2 production) (1); this causes decrease in blood pH (1); as a result of carbonic acid dissociating to release H^+ ions (1)

17.4

1 a Footballers do not have long enough between matches to fit in several days of carbohydrate loading (1)

   b Insoluble (1); compact (1); many terminal end points for enzyme activity (1); easily hydrolysed to release glucose (1)

2 If blood from a different blood group it may agglutinate (1); also reduces the risk of blood-borne diseases (1); named example e.g., hepatitis C, HIV. (1)

3 Steroid hormone enters its target cell (1); where it combines with a complementary receptor molecule in cell's cytoplasm (1); this results in increase in transcription of a particular gene (1); resulting in increased translation (i.e., protein synthesis) (1)

17.5

> ### Hiroshima haemoglobinopathy
>
> To the left – due to higher affinity for oxygen. Haemoglobin oxygen saturation remains higher at low partial pressures of oxygen.

1 a Presence of: alpha helices (1); beta pleated sheets (1); prosthetic group (1); folded 3-D shape (1)

   b Haemoglobin has four polypeptide chains (myoglobin has one) (1); haemoglobin has four prosthetic groups / four Fe^{2+} ions / four haem groups (myoglobin has one) (1)

   c All contain a prosthetic group (1); haemoglobin and myoglobin have haem group(s) and carbonic anhydrase has zinc atom (1)

2 Saturation of HbF minus Saturation of HbA as read from graph (1)

3 Haemoglobin becomes saturated in alveoli (1); plateau so small differences in pO_2 don't affect saturation (1); when red blood cells reach capillaries in muscles oxygen easily unloaded (1); steepest part of curve in this range so small drop in pO_2 giving big increases in oxygen being unloaded (1); in areas with a low pO_2; e.g., active skeletal muscles; (1) (max 4)

17.6

1 a Based on width of A band being 10 mm = 10 000 μm. 10 000 μm × 5000 = 5×10^7 μm (2)

   b Based on width of sarcomere being 15 mm/15 000 μm. 15 000 μm × 5000 = 75×10^6 μm (2)

2 key: red = actin, blue = myosin

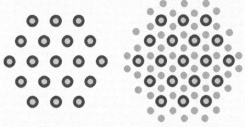

H zone   zone of overlap

3 Increased mitochondria to generate more ATP (for muscle contraction) (1); increased number of ribosomes to synthesis more protein (myosin and actin) (1); increased amount of endoplasmic reticulum/sarcoplasmic reticulum for increased transport of proteins within cell (1); increased number of nuclei as a result of cell surface membranes in embryonic muscle cells fusing together (1) (max 3)

17.7

> ### Studying muscle contraction using fluorescence
>
> 1 Error bars plotted above and below the mean by +/−1SD
> 2 Data becomes less reliable (as error bars become larger)
> 3 Actin–myosin interaction cannot proceed any faster

1 a (i) no changes (1); (ii) no changes (1); (iii) shortens (1); (iv) no change (1); v) narrows (1)

   b They move closer together (1)

2 a Myosin heads can only tilt in one direction (1); they can only pull in a direction that shortens the sarcomere (1); they can't pull actin in the opposite direction (1); they will remain in relative positions unless another muscle pulls them back again (1)

   b Muscles work in antagonistic pairs (1); its partner will contract to pull it back to its original position, description of named example of antagonistic pair e.g., bicep and tricep (1)

3   No ATP is produced (after death) (1); calcium ions are no longer pumped into the cisternae of the sarcoplasmic reticulum (1); so the mysoin heads bind to actin filaments (1); as no hydrolyis of ATP occurs; no separation of the actin and mysoin heads can occur (1) so muscles remain in contracted state.

18.1

> ### Examining the histology of reproductive systems
>
> 1   Clear diagram with testis labelled as a set of seminiferous tubules and epididymis shown (as a coiled tube) adjacent to the testis.
> 2   Epididymis is a coiled tube. During preparation of micrograph, epididymis tube was sliced (cross-sectioned) in many places.

1   a   Ovary (1)   b   Epididymis (1)   c   Uterus (1)
    d   Urethra (1)

2   a   acts as a barrier (1); for example, to pathogenic organisms (1)

    b   sperm will survive for longer (1); so more chance of oocyte being fertilised once it has been released (1)

3   *Alkaline fluid* (1); source = prostate gland (1); role = counteract acidity in vagina (1); *seminal fluid* (1); (containing proteins and fructose) source = seminal vesicles (1); role = nourishes sperm (1); *sperm* (1); source = testis (1); role = carries male genetic material to oocyte for fertilisation (1)

18.2

1   a   mitosis (1)   b   meiosis (1)   c   meiosis I (1)
    d   mitosis (1)

2   a   Streamlined structure (1); high density of mitochondria in the midpiece (1); tail for locomotion (1).

    b   acrosome (1); this contains hydrolytic enzymes to digest the zona pellucida of a secondary oocyte (1) (max 4)

3   *Similarities* both have multiplication, growth, and maturation phases (1); both have mitotic divisions followed by meiotic divisions (1); primary gametes are produced at end of growth phase (1); both result in haploid gametes being produced (1). *Differences* final gametes

have different structures (1); no polar bodies are produced in spermatogenesis (1); meiotic divisions are separated by greater time period in oogenesis (1)

18.3

> ### Do other species have menstrual cycles?
>
> 1   Behavioural changes; colour and anatomical changes (e.g., exaggerated sexual swellings in females of some African primate species such as mandrills, macaques, and chimpanzees); the production of particular smells (in the form of pheromones)
> 2   Theories of the evolution of concealed ovulation include – maintenance of a stronger pair bond (i.e., encouraging monogamy); opportunity for early human females to mate with other males while retaining parental care from their partner; reduction in short periods of male aggression during oestrus (when they would compete for sexually receptive females).

1   Oestrogen thickens lining of uterus (1); inhibits FSH (1); stimulates surge in LH secretion (1)

2   Only certain cells have receptors for a particular hormone (1); receptors are specific to one hormone (1) and must have the correct shape for the hormone to bind (1)

3   FSH stimulates release of oestrogen (1); as oestrogen levels rise, it inhibits further FSH secretion (1); LH stimulates progesterone secretion (1); progesterone inhibits further LH secretion (1)

18.4

1   Intra-cytoplasmic sperm injection (1)

2   Cortical granules (activated by sperm entry) release enzymes by exocytosis (1); enzymes digest membrane proteins used to bind sperm (1); sperm entry causes the zona pellucida to harden (1)

3   a   Monoclonal antibodies (MABS) are produced that are specific to hCG (hCG produced in early pregnancy) (1); MABS bind to hCG causing attached enzyme to produce a colour change (1).

    b   Female antibodies can destroy sperm in uterus (1); male antibodies can destroy sperm in semen (1); in both cases, the immune system has malfunctioned and produced inappropriate antibodies (1)

19.1

> ### Surgical menopause
>
> 1 Blood supply to uterus has been cut off, radiation treatment following hysterectomy for cancer may also cause ovaries to fail prematurely.
> 2 Testosterone helps regulate a woman's libido, energy, and mental state. Following a natural menopause testosterone will continue to be produced by ovaries in significant amounts (for ~12 years), therefore a woman that has ovaries removed will no longer produce testosterone and this may be responsible for poor libido, depression, and lack of energy following surgery. Testosterone is also important in conserving bone after menopause and testosterone supplementation may be more suitable for women unable to take oestrogen who have an increased risk of osteoporosis.
> 3 Oestrogen reduces chance of CHD, decrease in oestrogen means women have the same risk of developing CHD as men of the same age.

1 Males do not have sudden drop off in their sex hormones (1); changes in their physiology are not related to hormonal changes as seen in women (1)

2 lower concentrations of oestrogen and progesterone; higher concentrations of FSH; symptoms include – ovulation stops; menstruation stops; reproductive organs become smaller; vaginal dryness/irritation of outer reproductive structures; vaginal discharge; calcium loss from bone/reduction in bone mass; osteoporosis/description of osteoporosis; increase in % body fat; increase in blood cholesterol (max 3)

3 a Line graph plotted with three lines and key for each smoking category (1); lines should not extend beyond first and/or last plot (1); scale equidistant on both axis (1); plot area covers at least 50% of the available space (1); x-axis labelled "Age/years" and y-axis labelled "Percentage of women who smoke that are postmenopausal" (1); all plots accurate to +/–1 mm (1) (max 4)

  b Genetics (1); birthweight (1); childhood growth and early life nutrition (1) (max 2)

19.2

1 Continuous regimes involve hormones being taken every day, whereas cyclic regimes include gaps in which no hormones are taken (1); these gaps cause withdrawal bleeding (1)

2 Phytoestrogens appropriate when women pregnant or have conditions such as heart disease, high blood pressure, or a history of cancer (1); women may prefer phytoestrogens as an alternative to HRT, depending on the severity of symptoms (1)

3 Progestin reduces risk of uterine cancers when taken with oestrogen in combined HRT (1); it does this by mimicking effects of progesterone (1); evidence suggests taking progestin for at least 10 days in a 28 day cycle reduces risk of uterine cancer (1); however, progestin can result in side effects (e.g., fluid retention, headaches, mood swings, depression) (1); overall, the benefits of including progestin are generally considered to outweigh the costs, but this must be assessed on a case-by-case basis (1)

19.3

1 Condition involves a mass of cells that are non-cancerous (1)

2 Both conditions involve abnormal increase in number of cells (1); which may press on urethra and produce similar symptoms (1)

3 Volume of fluid ejaculated remains similar (i.e., seminal fluid still produced) (1); sperm are present (1); although there may be fewer living sperm in the fluid, reducing fertility (1).

20.1

> ### Using chromatography to investigate photosynthetic pigments
>
> A = 0.750 B = 0.375

> ### What determines the colour of photosynthetic pigments
>
> 1 Chlorophyll b is blue-green because it absorbs wavelengths for red light (620–700 nm) and absorbs few green wavelengths (490–575 nm), but it absorbs fewer blue/violet wavelengths (400–490 nm) than chlorophyll a. Carotenoids are yellow/orange/red because they absorb

blue/violet/green wavelengths but absorb no wavelengths beyond 550 nm (i.e., all yellow/orange/red light is reflected or transmitted)

2 Approximately 4.90×10^{-7} m

3 These wavelengths show the greatest absorption by photosynthetic pigments, which will maintain a high photosynthetic rate. However, no wavelengths in the red spectrum (600–700 nm) will be absorbed. Different wavelengths are likely to be important during particular stages of plant growth. For example, early vegetative growth relies principally on blue light, whereas flowering in plants requires red light.

1 a 0.83 (1) b 3.55 cm (1)

2 Double (inner and outer) membrane (1); intermembrane space (1); high density of ATP synthase and electron transport proteins embedded in inner membrane (1); membranes are surrounded by fluid (called the matrix in mitochondria and stroma in chloroplasts) (1); contain ribosomes and circular DNA (1) (max 2).

3 Chloroplast X contains grana and will produce ATP and reduced NADP in light-dependent reactions (1) chloroplast Y contains no grana (i.e., thylakoids are visible but not stacked into grana) (1) this suggests light-dependent reactions are less likely in Y (1); the presence of starch grains indicates that light-independent reactions of Calvin cycle are occurring in Y (1).

20.2

1 Oxygen (1)

2 Light energy to chemical energy (in photosystems) (1); kinetic energy of proton flow through ATP synthase is used to form ATP (chemical energy) (1)

3 Photosynthesis only occurs in the light (1); the rate of production is insufficient to supply plant with concentrations of ATP required (1); some plant cells lack chloroplasts and would not be able to generate ATP (1)

20.3

1 Rubisco is an enzyme (1); it catalyses the reaction between CO_2 and RuBP to produce GP (1).

2 ATP and reduced NADP both produced in the light-dependent reactions (1); both molecules are required for reactions in Calvin cycle (1)

3 102 kg (2)

20.4

What changes when GP is converted?

1 a ATP
  b reduced NADP

2 CH_2OH

3 NO_3^- would be required to provide N for the amine group.

1 *Fatty acids* uses (when combined with glycerol) include: plasma membrane formation; waterproofing (e.g., waxy cuticles); energy source (2); *Amino acids* uses: protein formation (e.g., enzyme production) (2)

2 a 2 (1); b 12 (1); c 16 (1)

3 *Respiration* produced from decarboxylation of pyruvate in link reaction (1); acetyl CoA is then converted to acetate, which enters Krebs cycle (1). *Photosynthesis* can be produced from GP during Calvin cycle (1); acetyl CoA then converted to fatty acids (1)

20.5

Methods for investigating the factors affecting photosynthesis

Low temperatures slow down enzyme activity before experiment begins, and 2% sucrose provides correct water potential to prevent chloroplasts disrupting.

1 a Photosynthetic rate decreases (1); lower temperature reduces kinetic energy of molecules involved in photosynthetic reactions (1)

  b Photosynthetic rate decreases (1) temperature becomes too high, denaturing proteins involved in photosynthetic reactions (1).

2 Initially increase in rate of photosynthesis is proportional to increase in light intensity (i.e., linear relationship) (1); at this point, light intensity is a limiting factor (1); graph plateaus at a high light intensity (1); at this point, CO_2 concentration is a limiting factor (1)

3 Distance of 0.5 m: 0.04 cm^3 min^{-1} (1); distance of 1.0 m: 0.02 cm^3 min^{-1}

20.6

> ### Culturing Rhizobium
>
> Stage 2: sterilisation of nodule prior to it being opened reduces possibility of culture being infected by microorganisms other than *Rhizobium*. Stage 5: incubation at 30°C provides optimum temperature for bacterial enzymes and ensures bacterial culture has high growth and reproductive rates.

> ### Luminescent microorganisms as biosensors
>
> Many biosensor roles can be researched. For example: screening for chemicals (e.g., tetracyclines) in chicken muscle before meat is sold; environmental monitoring of arsenic using luminescent *E. coli*; monitoring heavy metal concentrations at water treatment plants.

1 Nitrogen fixation (1); ammonification (1); nitrite (1); denitrification (1)

2 Temperature of 25–30 °C maximises bacterial metabolism and growth (1); sterile conditions ensure no other microorganisms are growing on the petri dishes prior to the Rhizobium culture (1); mannitol provides energy source (1); yeast extract a source of nitrogen (1); dipotassium phosphate and sodium chloride pH and osmotic buffering (1)

3 Legumes and nitrogen-fixing bacteria have a mutualistic relationship (1); leghaemoglobin is produced to transport oxygen to the bacteria (1); leghaemoglobin also maintains low oxygen concentrations, enabling nitrogenase to function (1); this enables them to survive in the roots and fix nitrogen for the plant (1).

20.7

1 Carp (1); wild fish stocks not depleted to same extent as they would be to feed salmon in order to provide food (1)

2 Size of species at particular trophic levels will vary (e.g., trees are large producers, each with large amount of stored energy/biomass, whereas grass plants are smaller producers, each containing less stored energy) (1); biomass is proportional to amount of stored chemical energy at each trophic level (1)

3 The mean trophic level has decreased (1); as fishing industry tends to target larger, predatory fish at high trophic levels, and their numbers have declined as a result (1); decline might be more pronounced in freshwater ecosystems as populations smaller and overfishing has been happening for longer than in marine ecosystems (1)

20.8

1 Offers incentives for farmers to use methods that promote conservation e.g., reduced use of chemicals such as fertilisers (1); growing crops that attract bird and insect species (1); maintaining hedgerows and dry stone walls (1); creating buffer strips (1) (max 3)

2 Intensive farming increases yields (1); but uses more labour (1); fertilisers (1); herbicides (1); and pesticides (1); and exhibits lower animal welfare standards than extensive farming (1) (max 3)

3 a $0.1 \, \text{kg} \, \text{m}^{-2} \, \text{yr}^{-1}$ (2)

  b Genetic modification of crop could increase resistance to disease (1); pests (1); herbicides (1); or drought (1). This enables increased growth and higher yields (1) (max 3)

20.9

1 *Plagioclimax* – community resulting from deflected succession, where human influence has prevented succession from producing climax community (1); examples include managed forests, grazing grassland, arable crop fields (1).

2 *Succession* – natural development of an ecosystem to form a climax community; deflected succession is when human activity interrupts succession, producing a plagioclimax (2); succession tends to produce greater biodiversity because deflected succession reduces the number of ecological niches available (1).

3 Number of samples taken should be maximised, but will depend on number of surveyors (1); and the time and resources available (1); stratified random sampling should be used (1); number of samples in each of the three areas should be proportional to their size (e.g., 100 samples taken in total, 70 on grassland, 20 in shrubby area, 10 in orchard) (1); random sampling should be used within each area (e.g., selecting coordinates using a random number generator) (1); quadrat can be used to estimate percentage cover of each species (1)

Answers

21.1

1 a Removal of plant species can affect soil structure and content (1); increased soil salinity due to human activity (1)

  b Reduced biodiversity due to eutrophication (1); toxic chemicals can be washed into aquatic ecosystems and kill organisms directly (1)

2 An increase, with a relatively constant rate of change for much of the two stages (1); increase occurs because death rate declines at an earlier stage than birth rate (1); this means difference in the two rates widens during these stages (1) (max 2)

3 Country A has the faster population growth (1); a greater proportion of its population can be found in the younger age categories (1)

21.2

1 *Max 1 mark per aspect*

  a Recycling crop waste; crop rotations; farming without chemical fertilisers; fish farming species low in the food web.

  b Contamination of food with chemicals (such as dioxins); disease within livestock (e.g., foot-and-mouth).

  c Deliberate mislabelling of foods

2 *Max 3 from the following* aesthetic pleasure; greater stability of ecosystems/food webs; increases chances of discovering useful chemicals in organisms.

3 Habitat A = 0.500 (1); habitat B = 0.759 (1); habitat A has less biodiversity and will be more sensitive to environmental change (1)

22.1

Investigating flowering – the long and short of it

1 Red light interrupts periods of darkness and prevents flowering. Far red light promotes flowering, it reduces length of darkness required for flowering and reverses effect of red light

2 A short-day plant. Flowering occurs when period of darkness is lengthened (from 6.5 to 8.5 hours).

---

1 Bright/colourful flowers (1); scents (1); flower shape and size (1); nectar (1) (max 3)

2 a increased concentration (1)

  b decreased probability (1)

3 Plants would have insufficient time to reproduce before frost (1); frost likely to damage ragweed seeds (1)

22.2

1 Pollen lands on stigma (1); pollen tube grows to ovary (1); male gamete fertilises ovule (1)

2 Select and set up experimental conditions suitable for species being studied (1); choose appropriate range of temperatures to test (e.g., 10, 20, 30, 40°C) and set up seeds at each temperature, with same number in each group (1); maintain other variables at constant values across all groups (e.g., same solution content and concentrations, water volume, and light regimes) (1); select appropriate dependent variable(s) to measure at regular intervals (1)

3 Gibberellins initiate starch breakdown in endosperm (1); gibberellins induce synthesis of amylase (1); endosperm represents nutrient source for embryo as it develops (1); most of an endosperm is starch, thereby providing a carbohydrate source for respiration in embryonic cells (1)

23.1

1 a Homozygous dominant (2)

  b Homozygous recessive (2)

  c Recessive allele/gene variant (2)

2 All mutated gene variants fail to produce a functional regulator protein (1); this means they all result in same phenotype (1).

3 a 0% (1) b 50% (1) c 12.5% (1)

  d 50% (1)

23.2

The FTO 'hunger' gene – an example of the subtle effects of gene mutation?

1 Point mutation/substitution. Gene variants all produce functional proteins, albeit differing in their effect. Insertion and deletion mutations tend to have more severe effect on phenotypes

2 It might regulate production of ghrelin 'hunger hormone' (i.e., influences gene expression).

---

1 a 2 (1) b 1 (1) c 4 (1)

2 a 75% (1) b 25% (1)

3 3 point mutations (A to T (10th base) (1); A to T (12th base) (1); C to T (22nd base) (1); deletion of 8 nucleotides (GATTATGG) (1)

23.3

1 In most cases, parents will only share 50% of their HLA alleles with a child (1); there is a possibility that HLA alleles two siblings inherit will have a greater than 50% similarity (1); for example, child 1 and child 4 share all of their HLA alleles (1)

2 $C^R C^R$ and $C^R C^W$ (2)

3 Cannot be certain either way (1); man's genotype could be $I^A I^A$ (which would mean he cannot be the father) or $I^A I^O$ (which means he could be the father) (2)

23.4

> ### The inheritance of mitochondrial DNA
>
> **1** Many mitochondria present in mother's egg. Although mitochondria are present in tail of a sperm cell, these are lost or destroyed in fertilised egg and are therefore absent from developing embryo
>
> **2** mtDNA has a fast mutation rate, which means it shows more genetic differences between individuals than nuclear DNA. This enables genetic differences between individuals to be observed more easily

1 The X chromosome is larger (1); and contains more genes than the Y chromosome (1)

2 (3)

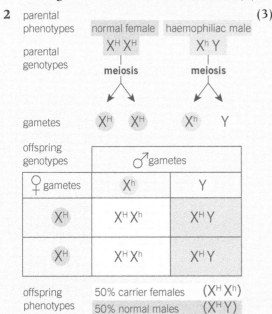

offspring phenotypes 50% carrier females ($X^H X^h$) 50% normal males ($X^H Y$)

50% probability (1)

3 a 50% (1) **b** A or AB (2)

23.5

> ### A *Drosophila* dihybrid cross
>
> You assume that the two genes are not linked (i.e., they are not on the same chromosome) and that the two genes do not interact with each other in any way (i.e., one gene does not affect the expression of the other).

1 Clear phenotypic traits that are easy to identify – ensures accurate observations and makes process more efficient (1); prior knowledge of species' genetics – enables predictions to be made and allows certain gene variants to be focussed on if necessary (1); the production of many offspring because this increases sample sizes (1); short generation time and straightforward maintenance requirements because this reduces financial and time costs (1)

2 Round, yellow = $\frac{12}{16}$ (75%) (1); round, green = $\frac{4}{16}$ (25%) (1)

3 Chi-squared value = 1.00 (1); this indicates a greater than 50% probability that differences between observed and expected results are due to chance (1); the scientist's predictions are supported (1); the genotypes are RrGG and rrGG (or rrGg) (1)

> ### Calculating chi-squared
>
> Value is only slightly lower than the 0.05 significance level (7.815). This means that there is a 5–10% probability that the differences between the observed and expected results can be explained by chance. Conducting the genetic cross again, possibly with a larger sample size, would be advisable. This would help to confirm whether or not the original hypothesis for the pattern of inheritance can be accepted or rejected.

23.6

1 Translocation involves part of a chromosome being transferred to another chromosome (1); non-disjunction involves two chromosomes failing to separate during meiosis and therefore ending up in same haploid cell (1)

2 Pedigree analysis can be used to check for family history of disease (1); genetic testing can assess whether the mother is a carrier of the haemophilia allele (X^h) (1) were the father to have the X^h allele, he would have haemophilia and testing would be unnecessary.

3 a Yes, albinism is caused by a recessive allele (1); this can be deduced because neither

Answers

parent in generation 1 has albinism, yet two of their children have the disorder (1)

  b No, albinism is not sex-linked (1); this can be deduced because one daughter and one son have albinism (1)

24.1

1 Decrease in population size (1); some alleles are lost from the population (1)

2 The founder effect (1); EVC allele present at relatively high frequency in new colony (1); inbreeding increased its frequency (1)

3 Higher frequency in Africa (1); allele is not beneficial in areas without malaria (e.g., Europe) but is beneficial in malaria areas (e.g., Africa) (1); natural selection maintains a high frequency of the H^S allele in Africa (1)

24.2

1 Animals in zoos have small populations (1); and mating is unlikely to be random (1)

2 0.008 ($q^2 = 0.0000588$ (1); $q = 0.007669$ (1))

3 13.1% are carriers ($q^2 = 0.005$; $q = 0.0707$ (1); $p = 0.92929$ (1); $2pq = 0.13142$) (1); 0.8% of the general population are carriers, which is 12.3% less (1).

24.3

1 a geographical isolation (1)

   b mechanical/anatomical isolation (1)

2 Geographical isolation often does not exist (i.e., ranges of different species overlap) (1); facial differences aid species identification and reduce breeding between species (1).

3 *No* the genomes/DNA sequences of the two species are sufficiently different to consider them separate species (1); *Yes* geographical isolation has caused their phenotypes to diverge, but not yet to point where they are unable to breed together (1)

24.4

Of mice and memories?

1 DNA methylation tends to reduce gene expression, a lack of methylation allows more transcription and therefore more odour receptor proteins to be translated

2 Two aspects that are unexplained are how the epigenetic changes were present in sperm cells of mice and the how the links between smell perception and fear responses were passed to subsequent generations

1 DNA mutations produce change in nucleotide sequence (1); epigenetic changes do not alter nucleotide sequence but affect gene expression (1)

2 Environmental conditions (i.e., lack of nutrition) during Dutch hunger winter caused epigenetic changes (1); these changes affected genes linked to heart disease and diabetes (1); the epigenetic changes were inherited (1)

3 Identical twins have 100% of the same DNA and fraternal twins have approximately 50% of the same DNA (1); there is a greater genetic influence on height than risk of strokes (1); evidence for this is that more than 90% of identical twins have same height, whereas approximately 15% of identical twins share the risk of having a stroke (1)

25.1

The use of log scales to illustrate DNA amplification

1 048 576 is the theoretical maximum number of fragments. A good estimate from graph would therefore be 1 000 000 (10^6).

1 Human DNA polymerase would denature at the high temperatures used in PCR (1)

2 a $10^{1.50}$ (1) b $10^{3.61}$ (1) c $10^{5.12}$ (1)

3 *Similarities* each new DNA molecule consists of one old (template) strand and one new strand (1); free complementary nucleotides are joined to a template strand (1) *Differences* only short fragments are replicated in PCR, whereas entire chromosomes are replicated naturally (1); PCR requires primers to be used (1); the DNA helicase enzyme separates strands in nature, whereas temperature cycling controls the process in PCR (1) (max 4)

25.2

The steps in gel electrophoresis

The ions carry charge through electrophoresis apparatus. Constant pH preserves structure of DNA; changes in pH can alter charges on molecules.

1 Suspect 1 (1); VNTR bands match those of specimen (1).

2 In both procedures, molecules are separated by size (1); some molecules are slowed by the stationary phase more than others (1)

3 Founder effects during human population migrations produced gene pool differences between populations of humans (1); haplotypes are genes that are near each other, are inherited together and can be analysed as a set (1); haplogroups are sets of haplotypes (1); different ethnic groups display different haplogroups (1) (max 3)

25.3

1 Restriction enzymes cut at specific recognition sites (1); the gene and the plasmid must have complementary sticky ends to be able to join together (1)

2 Beta cells in pancreas produce insulin (1); insulin gene will be transcribed in these cells, producing mRNA for insulin (1); insulin gene will be switched off in other cells (1)

3 Insulin from transgenic bacteria is from a human gene and will produce no side effects or immune response (1); animals do not need to be killed to obtain the insulin (1); there is a reduced risk of transferring disease (1)

25.4

1 Gene is isolated and inserted into a *ti* plasmid (1); plasmid inserts DNA into plant cell's chromosomes (1); cells are screened for new gene using reporter genes (1); a transgenic plant is grown from a single genetically modified cell (1)

2 Herbicide will kill only weeds that are competing with crop for resources (1); crop yield will increase and food prices will remain low (1)

3 Gene being studied is inactivated (1); gene is inactivated in embryonic stem cells of mice by inserting an additional nucleotide sequence (1); knockout stem cells are fused with embryos to create mice that lack a functional version of studied gene (1); phenotypes of knockout mice and normal mice can then be compared (in this case, in terms of the incidence of cancer) (1)

25.5

RNA interference in model organisms – genes down but not out?

1 The knockout procedure removes a gene, the use of RNA interference only reduces a gene's activity

2 Mammalian immune systems recognise foreign DNA and RNA and produces an immune response. Double-stranded RNA is larger and more likely to be detected than siRNA. Less complex organisms do not have the same immune responses

1 Introns removed from pre-mRNA (1); remaining exons spliced together to form mature mRNA (1)

2 Alternative splicing enables many different proteins to be translated from a single gene (1); this increases genetic diversity, which allows greater complexity in organisms (1)

3 Other, non-target mRNA molecules could be silenced if nucleotide sequence on target mRNA is similar to that of a non-target mRNA molecule (1); nucleotide sequence of siRNA is entirely complementary to target mRNA, whereas miRNA is only a partial match (1); means miRNA is much more likely to bind to and silence non-target mRNA (1)

25.6

1 Somatic cell therapy is used; this treats only affected somatic tissue and the new alleles are not passed on to future generations (1); treated cells are replaced, meaning additional treatment is required (1)

2 Delivering a single gene via a vector is problematic (1); delivering more than one gene would require a vector large enough to carry even more DNA and the probability of all the new genes being integrated and functioning would be very low (1)

3 Use of viruses as vectors taps into and adapts their natural mode of infection (i.e., inserting their DNA into specific cells) (1); liposomes are hydrophobic and are able to move through cell membranes (1)

26.1

1 *Structure* (sensory) cell body central OR long dendron (motor) (1). *Function* (sensory) impulses from receptors to CNS, (motor) impulses from CNS to effectors (1)

2 a Parasympathetic branch of autonomic nervous system (1)

   b Prevent excessive secretions (of saliva, mucus) (1); and prevents bronchial constriction (1) (and stops heart rate slowing)

3 Identifies primates as chimpanzees, squirrel monkeys, and humans (1); comment on line showing correlation (1); comment that humans are above line so brain size is larger than predicted by body mass (2)

26.2

> ### Practical investigation into reflexes
>
> Experimental results have 18 degrees of freedom giving a critical value for t of 2.1. There is a significant difference in reaction times between the right and left hand at the 95% confidence level so subjects were significantly faster using their right hands.

1 Pathway taken from receptor – the sensory neurone, to association neurone onto motor neurone and then finally to the effector (1); reflex arc important as has a protective function (1); as response made so quickly that damage to body or tissues may be avoided (1).

2 Myelinated nerve conduction is much quicker as impulse jumps from one Node of Ranvier to the next (1); there is a long circuit (1); conduction known as saltatory conduction (1); in MS myelin sheath is destroyed especially in motor neurones so nerve conduction is very slow leading to loss of motor coordination (1) (max 3)

3 Light shining in right eye produces no constriction in the left (or right). Light in left eye produces constriction in both eyes (1); untreated glaucoma causes damage to optic nerve (1)

26.3

1 *MRI* gives detailed image (1); useful for detecting tumours and demyelination of neurones in CNS but expensive and magnetic nature can be disruptive to body tissues (1); *fMRI* detects differences in blood flow to different areas of brain due to haemoglobin iron having magnetic properties that vary as they become oxygenated or deoxygenated so useful in brain surgery (1); *CT scans* give computer image of sections through an area of the body (1); CT perfusion increases differentiation so better image but neither are as detailed as MRI (1); *PET scan* detects injected radioactive tracers to determine healthy and non-healthy or cancerous tissue (1); not usually used in diagnosis but very good for long term care (1); *EEG* detects electrical impulses between brain cells and so indicates trauma (1); brain inflammation or other unusual brain activity so useful for TBI (1) (max 6).

2 Stroke may damage areas of brain controlling release of hormones (hypothalamus or pituitary) (1); so hormones not released or released at an inappropriate time or in inappropriate levels (1); TBI may cause trauma to hypothalamus or to pituitary so will disrupt release of hormones

such as ADH that controls water balance (1); or sex hormones FSH and LH that help to control reproductive cycle (1).

3 Anadamide broken down rapidly as is a naturally occurring chemical produced by body to stimulate temporary release of dopamine, constant release of dopamine not necessary or desirable for body as so it is broken down quickly to allow another stimulation to occur when needed (1); THC not produced by body and not naturally present in body so no mechanisms / enzyme pathways to break it down (1).

27.1

> ### Focusing the light onto the retina
>
> 1 The light rays focus in front of retina
> 2 Laser removes some of the cornea, this adjusts the curvature of the cornea, which alters the refraction of the light rays, in short-sightedness less refraction is needed/in long-sightedness more refraction is needed
> 3 Using corrective lenses such as glasses or contact lenses to compensate for structural defects in the eye, implanting an artificial lens into the eye to compensate for the longer eye length in people who are severely short-sighted

1 To supply oxygen (and glucose) (1); to cells of retina (which is attached to it) (1)

2 Lighter coloured irises are more common in white Caucasions (1); (some evidence suggests) that increased sunlight exposure contributes to development of ciliary body melanoma (1)

3 (High blood sugar in diabetes) causes lens of eye to swell (1); lens becomes distorted and can't refract the light rays correctly (1)

27.2

1 a To absorb any light that is not absorbed by rod and cone cells (1)

b Mitochondria produce ATP (1); which is required to resynthesise rhodopsin (1)

c Rod cells can't distinguish between different wavelengths of light (1); hence only produce images in black and white (1)

2 a In bright light all the retinal was in the *trans* isomer form (1); it takes time for the *cis* isomer to form (1); and respond to lower light intensity (1); which is too low to stimulate cone cells (1); ref to 20 min to produce sufficient rhodopsin to see fully in the dark (1) (max 4)

b 2 mm = 2000 μm

$$\frac{2000}{2} = \times 1000 \text{ magnification}$$

3 Vision will be less influenced by changes in light intensity (1); red light stimulates rhodopsin (1); so less bleaching occurs (1); so less time needed for eyes to become adapted to the low light intensity (1)

27.3

1 a Fast (1), simple (1), and effective at identifying broad categories of colour-blindness (1).

   b Patient does not know their numbers e.g., age, brain damage, mental retardation (1)

   c Use shapes, objects etc rather than numbers (1)

2 *Any two from* therapeutic follow-up of glaucoma; diabetic retinopathies; ARMD

3 Autonomic nervous system blood vessels in choroid supply oxygen and glucose to retinal cells (1); oxygen and glucose essential for aerobic respiration in rod (and cone) cells (1), No ATP means rhodopsin cannot be regenerated (remains 'bleached') (1) (max 2)

28.1

Treating AD

1 Hydrolysis (which is a catabolic reaction)

2 Inhibitor could be a competitive inhibitor, hence has a shape similar to acetylcholine which binds to active site of acetylcholinesterase preventing acetylcholine from binding, *OR* inhibitor could be non-competitive inhibitor which binds to binding site on acetylcholinesterase separate from active site, causes an allosteric effect, which alters shape of active site, preventing acetylcholine from binding (to active site), therefore no breakdown of the substrate

3 Stem cells could be supplied with necessary growth factors to stimulate production of neurones OR to replace damaged or destroyed neurones.

1 a *Any 3 from* eating; washing; dressing; going to toilet;

   b *Any 3 from* providing residential care; holding their own job whilst caring for a member of the family; distress of 'losing' a loved one; dealing with patient's legal matters

2 a Presence of α-helices (1); presence of β-pleated sheets (1); further folding and/or twisting of helices (1) (max 2)

b Acetyl cholinesterase has more than one polypeptide chain (1); and a cofactor (1)

3 *At 50 years* = 100 neurones, *at 60 years* =

$$100 - \left(\frac{5}{100} \times 100\right) = 95 \text{ neurones, } at \ 70 \ years$$

$$= 95 - \left(\frac{5}{100} \times 95\right) = 90.25 \text{ neurones (accept}$$

90 neurones), *at 80 years* = $90.25 - \left(\frac{5}{100} \times 90.25\right)$

= 81.23 neurones (accept 81 neurones),

At 90 years = $81.23 - \left(\frac{20}{100} \times 81.23\right) = 64.98$

neurones (accept 65 neurones). Award one mark

for $\frac{81.23}{81}$ neurones (at the age of 80 years).

Award one mark for $\frac{64.98}{65}$ neurones (at the age

of 90 years). Award one mark for working

28.2

1 a Lens becomes less elastic (1); so more difficult to focus on objects at different distances from eye (1)

   b Plastic lens will be set up for a certain level of vision (1); person may need to wear glasses in order to see objects that are either far away or close to them (1)

2 a Area of macula comprises only about 2.1% of retina (1); and remaining 97.9% (peripheral field) remains unaffected by disease (1)

   b If retina fully detaches and can't be reattached through surgery (1)

3 a Lens proteins become denatured (1)

   b *Any 2 from* (as proteins denature) they lose their tertiary structure (1); polypeptide chain uncoils (1); causing polypeptide chains to become entangled (1)

29.1

1 a Desired value around which negative feedback operates (1)

   b Range of values across which a physiological factor varies and negative feedback operates (1)

2 Oxytocin increases strength and frequency of contractions (1); increase in contractions causes more oxytocin to be release (1); contractions are therefore intensified further and gradually increased (1); gradual increase in contractions enables a baby to be born using minimum intensity of contraction (1)

3 Homeostasis maintains conditions within a narrow range around an optimum value (set point) (1); negative feedback reverses any deviation away from set point (1); positive feedback increases any change made to a physiological factor (1)

29.2

> ### Measuring core body temperature
>
> The strip-type thermometers measure skin temperature (i.e., peripheral temperature), which is likely to be very different to core temperature.

1 Arterioles (and shunt vessels) (1); sweat glands (1); hair erector muscles (1); fat tissue (1)

2 Thermoreceptors detect temperature rise (1); heat loss centre of hypothalamus transmits nervous impulses to effectors (1); arterioles near skin surface vasodilate (1); sweat is produced from sweat glands (1); heat lost from skin surface (1); hairs lie flat on skin reducing insulation (1)

3 More food provides additional substrates for respiration (1); more respiration increases metabolic rate (1); more heat is generated in response to colder winter temperatures (1); some species, especially hibernating species, increase their fat reserves (1) (max 3)

29.3

1 Chemoreceptors detect changes in blood pH that reflect carbon dioxide levels (1); pressure receptors detect changes in blood pressure (1); receptors pass information to medulla oblongata, which initiates necessary response (1)

2 Increased physical activity raises amount of energy required by muscles (1); muscles require more oxygen and glucose for respiration (1); cardiac output must be increased to supply muscles with these molecules at necessary rate (1)

3 Blood pressure would remain high (1); parasympathetic nervous system no longer able to transmit to the heart (1); medulla oblongata cannot stimulate SAN to lower heart rate (1)

30.1

1 Endocrine glands secrete hormones into blood (1); exocrine glands secrete other substances such as enzymes (1); pancreas operates as both types of gland (1); it secretes digestive enzymes and the hormones insulin and glucagon (1)

2 9×10^{-4} (or 0.0009) (1)

3 Liver cells contain stores of glycogen (1); glucagon is secreted by the pancreas in response to low blood glucose concentration (1); when glucagon binds to receptors on liver cells, glycogen within these cells is broken down into glucose molecules (1)

30.2

> ### Maturity-onset diabetes of the young
>
> 1 Condition inherited via a single gene, whereas type 1 and type 2 diabetes appear to have more complicated genetic influences (e.g., likely that many different genetic variants increase person's susceptibility for developing type 2 diabetes)
>
> 2 75–100% (75% if both parents are heterozygous and 100% if either parent is homozygous dominant for the MODY allele)
>
> 3 Homozygous genotype (i.e., having two copies of defective MODY gene variant) produces symptoms that are more severe

1 Type 1 is insulin-dependent because people with this form produce insufficient insulin (1); type 2 is insulin-independent because people with this form often produce sufficient insulin but the target cells are insensitive to the hormone (1)

2 Blood glucose increases to a greater extent in the diabetic (1); blood glucose takes longer to return to the original level in the diabetic (1); the diabetic produces no additional insulin because their pancreatic beta cells have been destroyed (1)

3 Lifestyle is not thought to influence the risk of developing type 1 diabetes to any significant extent (1); type 1 has a strong genetic component (several genes have been identified) (1); certain environmental factors, notably viral infections, are thought to trigger the onset of type 1; type 2 has a genetic component (1); however, lifestyle (e.g. diet) plays a much greater role in the development of type 2 (1).

30.3

1 Insulin is a protein (1); if ingested it would be broken down by digestive enzymes (1).

2 Glucose can bind to haemoglobin (1); the binding is permanent (for the lifespan of the cell) (1); the rate at which glycosylated haemoglobin forms is proportional to blood glucose concentration (1); a patient with a high concentration of glycosylated haemoglobin would have had a high average blood glucose concentration during the preceding weeks (1); a relatively high level of glycosylated haemoglobin suggests the treatment is ineffective (1).

3 Type 1 diabetes is not affected by lifestyle to a great extent (1); prevalence of type 1 diabetes is unlikely to change much (1); prevalence of type 2 diabetes is increasing (1); due to an aging

population and an increase in obesity levels (1); the increasing prevalence will cost health care services more money to treat and manage in the future (1) (max 3).

31.1

> ### Investigating the composition of body fluids
>
> Calibrate colorimeter using cuvette of distilled water. Use colour filter to improve accuracy (e.g., red filter when testing concentration of blue Benedict's solution that remains in each urine sample). Construct calibration curve by carrying out Benedict's test on solutions with known glucose concentrations and obtaining colorimeter readings for each solution (once precipitate has been removed). Carry out Benedict's test on urine sample (using same concentration and volume of Benedict's reagent that was used to construct calibration curve). Remove precipitate following the test. Measure transmission or absorption of sample in colorimeter. Find concentration of glucose in urine by comparing colorimeter reading to values in calibration curve.

1   Renal vein will have a lower concentration of urea and other toxins (1); water potential and certain ion concentrations may be different in the two blood vessels (1).

2   Sodium ions actively transported from PCT cells into tissue fluid (1); glucose co-transported with sodium ions from PCT lumen into cells in PCT wall (1); this is called facilitated diffusion (1); glucose moves into tissue fluid by passive diffusion (1); glucose diffuses from tissue fluid into capillaries (1).

3   To transport the few proteins that have been filtered into nephron (1); endocytosis transports proteins from PCT lumen into cells in its wall (1); exocytosis transports proteins from cells in the PCT wall into tissue fluid (1).

31.2

> ### Diuretics – treating, cheating, and energy-depleting?
>
> 1   Fewer sodium and chloride ions are co-transported from the distal convoluted tubule into the tissue fluid, the DCT water potential remains lower than it otherwise would be, less water diffuses out of the nephron by osmosis.

> 2   Alcohol inhibits ADH, which causes a greater volume of urine to be produced; drinking water replaces some of the lost fluid, this helps to lower solute concentration of blood.

1   Receptor = osmotic cells in hypothalamus (1); controller = hypothalamus/pituitary gland (1); effector = kidney (1).

2   Constricts blood vessels (1); causes ADH production to be increased, which results in more water being reabsorbed from kidneys into blood (1); activating thirst reflex in hypothalamus, which causes an individual to replace water that has been lost from the blood (1).

3   a   Positive correlation (1); as solute concentration increases, blood water potential decreases (1); causes more renin to be secreted, which activates the thirst reflex.

    b   ADH concentration would increase as solute concentration increases (1); ADH would cause more water to be reabsorbed from collecting duct to lower solute concentration and raise water potential in blood (1).

31.3

1   Reduced urine volume (1); cloudy or bloody urine (1); oedema (1); blood cells or proteins in the urine (1); high blood plasma concentration of creatinine (1); toxin accumulation in tissues (1); increased renin production (1) (max 4).

2   Stem cells (e.g., iPSCs from the patient) could be grown into kidney tissue and transplanted into patient (1); rejection will not occur because cells are taken from patient originally and are therefore genetically identical to his or her other cells (1).

3   Dialysis does not provide a cure, only an improvement in condition for patient (1); successful kidney transplantation provides a cure (1); however, transplant surgery carries a high risk of organ rejection (1); future transplant surgery may use therapeutic cloning to reduce the possibility of rejection (1); haemodialysis usually requires regular trips to a health clinic and is time-consuming (1); peritoneal dialysis can be carried out by patient while they work, but must be done every day (1).

Index

Appendix

▼ **Table 1** *Table of values of chi-squared*

df	*p* values								df
	0.99	0.95	0.90	0.50	0.10	0.05	0.01	0.001	
1	0.0016	0.0039	0.016	0.46	2.71	3.84	6.63	10.83	1
2	0.02	0.10	0.21	1.39	4.60	5.99	9.21	13.82	2
3	0.12	0.35	0.58	2.37	6.25	7.81	11.34	16.27	3
4	0.30	0.71	1.06	3.36	7.78	9.49	13.28	18.46	4
5	0.55	1.14	1.61	4.35	9.24	11.07	15.09	20.52	5
6	0.87	1.64	2.20	5.35	10.64	12.59	16.81	22.46	6
7	1.24	2.17	2.83	6.35	12.02	14.07	18.48	24.32	7
8	1.65	2.73	3.49	7.34	13.36	15.51	20.09	26.12	8
9	2.09	3.32	4.17	8.34	14.68	16.92	21.67	27.88	9
10	2.56	3.94	4.86	9.34	15.99	18.31	23.21	29.59	10
11	3.05	4.58	5.58	10.34	17.28	19.68	24.72	31.26	11
12	3.57	5.23	6.30	11.34	18.55	21.03	26.22	32.91	12
13	4.11	5.89	7.04	12.34	19.81	22.36	27.69	34.53	13
14	4.66	6.57	7.79	13.34	21.06	23.68	29.14	36.12	14
15	5.23	7.26	8.55	14.34	22.31	25.00	30.58	37.70	15
16	5.81	7.96	9.31	15.34	23.54	26.30	32.00	39.29	16
17	6.41	8.67	10.08	16.34	24.77	27.59	33.41	40.75	17
18	7.02	9.39	10.86	17.34	25.99	28.87	34.80	42.31	18
19	7.63	10.12	11.65	18.34	27.20	30.14	36.19	43.82	19
20	8.26	10.85	12.44	19.34	28.41	31.41	37.57	45.32	20
21	8.90	11.59	13.24	20.34	29.62	32.67	38.93	46.80	21
22	9.54	12.34	14.04	21.34	30.81	33.92	40.29	48.27	22
23	10.20	13.09	14.85	22.34	32.01	35.17	41.64	49.73	23
24	10.86	13.85	15.66	23.34	33.20	36.42	42.98	51.18	24
25	11.52	14.61	16.47	24.34	34.38	37.65	44.31	52.62	25
26	12.20	15.38	17.29	25.34	35.56	38.88	45.64	54.05	26
27	12.88	16.15	18.11	26.34	36.74	40.11	46.96	55.48	27
28	13.56	16.93	18.94	27.34	37.92	41.34	48.28	56.89	28
29	14.26	17.71	19.77	28.34	39.09	42.56	49.59	58.30	29
30	14.95	18.49	20.60	29.34	40.26	43.77	50.89	59.70	30
40	22.16	26.51	29.05	39.34	51.81	55.76	63.69	73.40	40
60	37.48	43.19	46.46	59.33	74.40	79.08	88.38	99.61	60
80	53.54	60.39	64.28	79.33	96.58	101.88	112.33	124.84	80
100	70.06	77.93	82.36	99.33	118.50	124.34	135.81	149.45	100

▼ Table 2 *Critical values for Spearman's rank correlation coefficient, r_s*

	$p = 0.1$	$p = 0.05$	$p = 0.02$	$p = 0.01$			$p = 0.1$	$p = 0.05$	$p = 0.02$	$p = 0.01$
	5%	$2\frac{1}{2}$%	1%	$\frac{1}{2}$%	**1-Tail Test**		5%	$2\frac{1}{2}$%	1%	$\frac{1}{2}$%
	10%	5%	2%	1%	**2-Tail Test**		10%	5%	2%	1%
n						n				
1	–	–	–	–		31	0.3012	0.3560	0.4185	0.4593
2	–	–	–	–		32	0.2962	0.3504	0.4117	0.4523
3	–	–	–	–		33	0.2914	0.3449	0.4054	0.4455
4	1.0000	–	–	–		34	0.2871	0.3396	0.3995	0.4390
5	0.9000	1.0000	1.0000	–		35	0.2829	0.3347	0.3936	0.4328
6	0.8286	0.8857	0.9429	1.0000		36	0.2788	0.3300	0.3882	0.4268
7	0.7143	0.7857	0.8929	0.9286		37	0.2748	0.3253	0.3829	0.4211
8	0.6429	0.7381	0.8333	0.8810		38	0.2710	0.3209	0.3778	0.4155
9	0.6000	0.7000	0.7833	0.8333		39	0.2674	0.3168	0.3729	0.4103
10	0.5636	0.6485	0.7455	0.7939		40	0.2640	0.3128	0.3681	0.4051
11	0.5364	0.6182	0.7091	0.7545		41	0.2606	0.3087	0.3636	0.4002
12	0.5035	0.5874	0.6783	0.7273		42	0.2574	0.3051	0.3594	0.3955
13	0.4835	0.5604	0.6484	0.7033		43	0.2543	0.3014	0.3550	0.3908
14	0.4637	0.5385	0.6264	0.6791		44	0.2513	0.2978	0.3511	0.3865
15	0.4464	0.5214	0.6036	0.6536		45	0.2484	0.2945	0.3470	0.3822
16	0.4294	0.5029	0.5824	0.6353		46	0.2456	0.2913	0.3433	0.3781
17	0.4142	0.4877	0.5662	0.6176		47	0.2429	0.2880	0.3396	0.3741
18	0.4014	0.4716	0.5501	0.5996		48	0.2403	0.2850	0.3361	0.3702
19	0.3912	0.4596	0.5351	0.5842		49	0.2378	0.2820	0.3326	0.3664
20	0.3805	0.4466	0.5218	0.5699		50	0.2353	0.2791	0.3293	0.3628
21	0.3701	0.4364	0.5091	0.5558		51	0.2329	0.2764	0.3260	0.3592
22	0.3608	0.4252	0.4975	0.5438		52	0.2307	0.2736	0.3228	0.3558
23	0.3528	0.4160	0.4862	0.5316		53	0.2284	0.2710	0.3198	0.3524
24	0.3443	0.4070	0.4757	0.5209		54	0.2262	0.2685	0.3168	0.3492
25	0.3369	0.3977	0.4662	0.5108		55	0.2242	0.2659	0.3139	0.3460
26	0.3306	0.3901	0.4571	0.5009		56	0.2221	0.2636	0.3111	0.3429
27	0.3242	0.3828	0.4487	0.4915		57	0.2201	0.2612	0.3083	0.3400
28	0.3180	0.3755	0.4401	0.4828		58	0.2181	0.2589	0.3057	0.3370
29	0.3118	0.3685	0.4325	0.4749		59	0.2162	0.2567	0.3030	0.3342
30	0.3063	0.3624	0.4251	0.4670		60	0.2144	0.2545	0.3005	0.3314

Important formula

Spearman's ranked correlation coefficient formula

$$r = 1 - \frac{6 \sum d^2}{n(n^2 - 1)}$$

Chi-squared formula

$$\chi^2 = \sum \frac{(O - E)^2}{E}$$

Paired t-test $= t = \dfrac{|\bar{x}_A - \bar{x}_B|}{\sqrt{\dfrac{s_A^2}{n_A} + \dfrac{s_B^2}{n_B}}}$

Unpaired t-test $= t = \dfrac{\bar{d}\sqrt{n}}{s_d}$

Acknowledgements

Header Photos: Ch16: Martin Shields/Science Photo Library; **Ch17**: Steve Gschmeissner/Science Photo Library; **Ch18**: Dr. Keith Wheeler/Science Photo Library; **Ch19**: Steve Gschmeissner/Science Photo Library; **Ch20**: Dr. Kari Lounatmaa/Science Photo Library; **Ch21**: Eye of Science/Science Photo Library; **Ch22**: Neil Hardwick/Shutterstock; **Ch23**: Studiotouch/Shutterstock; **Ch24**: Sovereign, ISM /Science Photo Library; **Ch25**: David Parker/Science Photo Library; **Ch26**: Microscape/Science Photo Library; **Ch27**: Eye of Science/Science Photo Library; **Ch28**: Laguna Design/Science Photo Library; **Ch29**: Outdoorsman/Shutterstock; **Ch30**: Astrid & Hanns-Frieder Michler/Science Photo Library; **Ch31**: ISM/Science Photo Library;

COVER: robert_s / Shutterstock; **p13**: Martin Shields/Science Photo Library; **p21**: Robert Kneschke/Shutterstock; **p22**: Philippe Psaila/Science Photo Library; **p26**: Robyn Mackenzie/Shutterstock; **p27**: Christophe Vander Eecken/Reporters/Science Photo Library; **p28**: Christophe Vander Eecken/Reporters/Science Photo Library; **p29**: Leonid Andronov/Shutterstock; **p36**: Steve Gschmeissner/Science Photo Library; **p40**: Steve Gschmeissner/Science Photo Library; **p44**(BL): Steve Gschmeissner/Science Photo Library; **p44**(BR): Prof. R. Wegmann/Science Photo Library; **p44**(T): Dr. Keith Wheeler/Science Photo Library; **p45**: Professor P.M. Motta, G. Macchiarelli, S.A Nottola/Science Photo Library; p54: Steve Gschmeissner/Science Photo Library; **p55**: Zephyr/Science Photo Library; **p58**: Steve Gschmeissner/Science Photo Library; **p59**: CNRI/Science Photo Library; **p62**(B): Image Point Fr/Shutterstock; **p62**(T): Sheila Terry/Science Photo Library; **p67**(B): Mr. Gordon Muir/Consultant Urological Surgeon/ Renaissance Healthcare Ltd/Science Photo Library; **p65**(T): John Bavosi/Science Photo Library; **p68**: Dr. Kari Lounatmaa/Science Photo Library; **p70**: Dr. Barrie Juniper, Department of Plant Sciences, Oxford University; **p71**: Leonid Andronov/Shutterstock; **p81**: Image Reproduced With Permission From SSERC (www.sserc.org.uk); **p82**: Dr. Jeremy Burgess/Science Photo Library; **p83**: Power and Syred/Science Photo Library; **p86**: Gusto Images/Science Photo Library; **p87**(B): Dr. Kari Lounatmaa/Science Photo Library; **p87**(T): Designua/Shutterstock; **p88**: Alvis Upitis/Agstockusa/Science Photo Library; **p89**: Martin Bond/Science Photo Library; **p90**: Simon Booth/Science Photo Library; **p92**: David Woodfall Images/Science Photo Library; **p93**: Martyn F. Chillmaid/Science Photo Library; **p97**: Geogphotos/Alamy; **p98**: Massimo Brega/Eurelios/Science Photo Library; **p100**: Viacheslav Nikolaenko/Shutterstock; **p101**: Eye of Science/Science Photo Library; **p105**(B): Neil Hardwick/Shutterstock; **p105**(T): DR JEREMY BURGESS/SCIENCE PHOTO LIBRARY; **p113**: Mauro Fermariello/Science Photo Library; **p116**: Scientifica/Visuals Unlimited, Inc./Science Photo Library; **p117**: Eye of Science/Science Photo Library; **p118**: Poonsap/Shutterstock; **p121**: Biophoto Associates/Science Photo Library; **p124**: Studiotouch/Shutterstock; **p128**: Biophoto Associates/Science Photo Library; **p135**: Sovereign, ISM /Science Photo Library; **p140**(B): Frans Lanting, Mint Images/Science Photo Library; **p140**(CB): Greg Courville/Shutterstock; **p140**(CT): Anton_Ivanov/Shutterstock; **p140**(T): Tony Camacho/Science Photo Library; **p142**: Copyright (2005) National Academy of Sciences, U.S.A.; **p146**: Vit Kovalcik/Shutterstock; **p149**: David Parker/Science Photo Library; **p150**: Peter A. Underhill/European Journal of Human Genetics (2015) & Macmillan Publishers Ltd.; **p155**: Oak Ridge National Laboratory/US Department of Energy/Science Photo Library; **p157**: Laguna Design/Science Photo Library; **p159**: NASA/Science Source/Science Photo Library; **p163**: Microscape/Science Photo Library; **p174**: Zephyr/Science Photo Library; **p175**: Zephyr/Science Photo Library; **p184**: Eye of Science/Science Photo Library; **p188**(B): Annabella Bluesky/Science Photo Library; **p188**(T): Adam Gault/Science Photo Library; **p189**(B): Public Health England/

Science Photo Library; **p189**(T): Annabella Bluesky/Science Photo Library; **p193**: Tony Craddock/Science Photo Library; **p194**(B): Science Photo Library; **p194**(T): Alfred Pasieka/Science Photo Library; **p195**: Laguna Design/Science Photo Library; **p196**: Diane Macdonald/Getty Images; **p197**(B): Arztsamui/Shutterstock; **p197**(C): Spencer Sutton/Science Photo Library; **p197**(T): Dr. P. Marazzi/Science Photo Library; **p203**: Ruth Jenkinson/MIDIRS/Science Photo Library; **p204**(B): Outdoorsman/Shutterstock; **p204**(T): Mitchii/Shutterstock; **p206**: Ian Boddy/Science Photo Library; **p212**: Astrid & Hanns-Frieder Michler/Science Photo Library; **p219**(B): Dmitry Lobanov/Shutterstock; **p219**(T): Greenland/Shutterstock; **p222**: Astrid & Hanns-Frieder Michler/Science Photo Library; **p226**: ISM/Science Photo Library; **p232**: Gopixa/Shutterstock; **p234**: Manfred Kage/Science Photo Library; **p245**: Ed Reschke/Getty Images; **p246**(B): Visuals Unlimited, Inc./Dr. Fred Hossler/Getty Images; **p246**(T): Ed Reschke/Getty Images; **p252**: Bob Gibbons/Science Photo Library; **p253**(B): J.C. Revy, ISM/Science Photo Library; **p253**(T): Dr. Jeremy Burgess/Science Photo Library;

Artwork by Q2A Media Private Inc.